Sociology / Technology

Sociology / Technology

Foundations of Postacademic Social Science

Jay Weinstein

Transaction Books
New Brunswick (U.S.A.) and London (U.K.)

This book is dedicated to my teachers, my students, and especially my colleagues in the technical fields, whose instinct of workmanship will, I hope, be stimulated by some of these ideas.

LIBRARY
The University of Texas
At San Antonio

Copyright © 1982 by Transaction, Inc.
New Brunswick, New Jersey 08903

All rights reserved under International and Pan-American Copyright Conventions. No part of this book may be reproduced or transmitted in any form or by any means, electronic or mechanical, including photocopy, recording, or any information storage and retrieval system, without prior permission in writing from the publisher. All inquiries should be addressed to Transaction Books, Rutgers — The State University, New Brunswick, New Jersey 08903.

Library of Congress Catalog Number: 80-24637
ISBN: 0-87855-404-1
Printed in the United States of America

Library of Congress Cataloging in Publication Data

Weinstein, Jay A   1942-
  Sociology/Technology.

  Includes bibliographical references and index.
  1. Technology and civilization. 2. Social change.
3. Technological innovations. I. Title.
HM221.W43   303.4'83   80-24637
ISBN 0-87855-404-1

Contents

Preface ix

PART I   SOCIOLOGY AND TECHNOLOGY: AN OVERVIEW
Introduction 3
1. Technology and the Three Births of Sociology 6
2. Sociology and Technology in the Wake of Fission: Max Weber and
   European Social Thought 24
3. Early Academic Approaches in the United States: The Sociology of
   Technology of Veblen and Ogburn. 41
4. Karl Mannheim and the Postacademic Program 67
5. C. Wright Mills and Pragmatist Sociology 82
6. The Frankfurt School: A Critical Theory of Technology 100

PART II   SOCIETY AND THE STUDY OF TECHNOLOGY
Introduction 119
7. Recent Landmarks in the Transition to Postacademic Social Science   121
8. Feeling Helpless: The Idea of Autonomous Technology in
   Social Science 141
9. Technology, Social Science, and Quality of Life 153
10. Sociology: Pure, Applied, and Interdisciplinary 169

PART III   SCIENCE AND TECHNOLOGY FOR DEVELOPMENT
Introduction 195
11. Development and Postacademic Social Science 198
12. The Soft State and Development Administration: Toward an
    Appropriate Planning Technology for the Third World 210
13. Appropriateness Versus Appropriation: Alternative Approaches to
    Technology Transfer 231
14. A Call for a Code of Ethics for the Transfer of
    Population Control Technology 242

PART IV SOCIOLOGY/TECHNOLOGY: LOOKING AHEAD

Introduction 257
15. Perspectives on Sociological Ideology 260
16. Ideological Principles of Sociology 270
17. Subjectivism Versus Natural Science 281
18. Technology, Society, and Social Science 294

References 300
Index 335

Acknowledgments

Earlier versions of the following chapters were coauthored: chapter 12, with John M. Weinstein, Kennesaw College; chapter 13, with K. Vijayan Pillai, The University of Iowa; and chapter 14, written with assistance of Daryl Chubin, Stanley Carpenter, Fred Rossini, and John Hines, School of Social Sciences, Georgia Institute of Technology.

I am grateful to these colleagues and to Louis Schneider, Irving Louis Horowitz, Jon Johnston, and Bill Martin for their comments on drafts of various chapters and sections. Thanks to Té Burt for her patience and persistence in typing endless drafts and revisions; to Jane Wilson and Genie Vidal for their assistance; and to Marilyn Weinstein for her constant help throughout the project.

The author gratefully acknowledges permission to reprint the following material:

"Science, Technology, and the Quality of life," by Jay Weinstein. *Social Development Issues* 2(Fall 1978):23-37).

"Appropriate Technology versus Appropriating Technology," by Jay Weinstein and K. Vijayan Pillai. *Social Development Issues* 3(Fall 1979):37-53.

"Science, Technology, and Society: A Cross-Disciplinary Perspective" (Review), by Jay Weinstein. *Social Forces* 57(December 1978):729-31.

"The Dialectic of Ideology and Technology: The Origins, Grammar, and Future of Ideology," by Alvin Gouldner. *Knowledge* 1(December 1979):325-28, by permission of Sage Publications, Inc.

"Feeling Helpless: The Idea of Autonomous Technology in Social Science." *Theory and Society* (forthcoming).

Tables 1 and 2 (chapter 9) are reprinted with the permission of Ralph H.Todd, Todd Research Associates, San Diego, California.

Partial support for research and writing were provided by The United States Educational Foundation in India, The University of Iowa Old Gold Grants Program, The Georgia Tech Foundation, and The Exxon Educational Foundation. I express my appreciation to these organizations. Responsibility for all errors, omissions, and oversights is the author's.

Preface

This book is about the role of technology in modern and modernizing societies. It is addressed to those who understand technology as a powerful, complex, and morally ambiguous force, and who are not satisfied with neat and simple answers to questions such as "What has technology done for (or to) people?" It views the relationship between technology and society as neither fixed nor unmediated; rather, this relationship is conceived as a highly variable part of a far more extensive system that always includes cultural and environmental factors, human will, and human error. Ours is very much a technological age and modern society is, in many ways, technological society. What does this mean in human terms? Does it mean that machines are taking control of our lives and minds, or that we are now more free than ever before to control nature for our own benefit, or that both nature and humanity are on the brink of being destroyed by technology? In the following chapters, these and related questions are raised, discussed, and, if not always answered, they are at least countered with alternative questions.

For over two hundred years, technology has been a major cause of social change; it has affected social life at all levels: in international relations, national institutions, values, norms, and in our most intimate personal habits and thoughts. While it is true that people have everywhere and always possessed, maintained, and been affected by a stock of tools and techniques, in Europe, North America, Japan, Oceania, and in the modern sectors of the Third World today, this technological stock, its improvement, application, and regulation are, in an unprecedented way, key public concerns. Modern technology and the technological innovation process require ever-increasing investments of money and manpower. They are frequent subjects of debate and legislation in local, national, and international forums, and they affect our jobs, housing, recreation, education, and other activities and roles.

Despite the fact that technology has played such a central role in modern society, until very recently its social sources and effects have not been taken into account in the innovation process. Knowledge of the social aspects of technological innovation has not been considered a necessary or even desirable part of the education of engineers and other technical professionals. Technological design, diffusion, evaluation, and other activities have not been pursued in explicit awareness of social antecedents and consequences, nor have social factors been treated as variables that can or should influence the course, scope, and timing of innovation.

During the past thirty years, a combination of circumstances has altered this situation dramatically. Innovations in communications, electronic, chemical, transport, and other technologies, growing public concern about technology's unwanted side-effects, the increasing transfer of modern technologies to underdeveloped countries and regions, and changes within the social and policy sciences have contributed to a new appreciation of the relationship between technology and society. With this appreciation has come a growing interest and commitment among educators and professionals to incorporate principles of social science into technological training and practice.

Today, these changes are accelerating and their effects are multiplying. The attempt to combine social knowledge with technical knowledge and skills is found not only in classrooms and laboratories, but also in newspapers, television, community forums, and everyday life. This book discusses these recent changes at several points. The current movement to incorporate knowledge of social factors into technical education, research, design, and innovation activities — where they have not been considered previously — can have significant consequences for society, technology, and social science. Because technology's impact is so effective and far-reaching, these changes could lead to one of the most important intellectual developments of the last decades of this century.

The title of this book emphasizes the close connection between technological factors and the disciplines that study social relations such as sociology. The subtitle, *Foundations of Postacademic Social Science,* isolates a major consequence for the social disciplines of current efforts to apply knowledge of this connection to practical contexts. The term *postacademic* is meant to indicate that a major shift is occurring in sociology and related fields as they assume an integral role in technical education and professional practice: from traditional academic settings, sources of support, and subject matter to a new, interdisciplinary, applied orientation. While a portion of the work of social scientists has always been performed outside of universities, in nonacademic activities, and under the auspices of government or private organizations, such work has until recently been the exception to a predominantly academic norm. In referring to this new type of social science as post- (as opposed to non-) academic, two of its characteristics are stressed. First, it is in no way inferior to or merely residual in relation to the academic orientation; rather it is a successor to it. Second, its practice is no longer exceptional;

rather the norm itself has changed and it is becoming the dominant (though not exclusive) orientation.

A final comment is required on the word *technology* as it is used in this book. Coined in Germany about two hundred years ago, it became a regular part of English and (in cognate form) many other languages. Although it can refer to several different objects or processes, it is defined more narrowly for the purposes of this discussion. Here *technology* refers primarily to a system of knowledge intended to have practical bearing — know-how. It is knowledge designed to extend the capabilities of humans in dealing with their environment, to overcome constitutional and environmental limitations. Thus, while people cannot naturally lift great weights, traverse large bodies of water, fly faster than the speed of sound, speak instantaneously with others halfway around the world, or become temporarily infertile at will, technology does permit us to perform these and other (more and less remarkable) feats.

Technology can be compared and contrasted with other types of knowledge in several ways. For example, unlike science, it focuses on the solution of specific practical problems rather than on the accumulation of knowledge and the resolution of theoretical or experimental issues. Unlike religious belief systems, technologies are not concerned with moral conduct, spiritual transcendence, or similar ends. Unlike magic, technology is based upon a materialistic — as opposed to spiritualistic — view of nature, and it is largely impersonal in its operation and purposes.

But the boundaries between science and technology or even magic or religion and technology are not rigid. As separate but interacting parts of culture, these types of knowledge share the patterns and styles that distinguish one people from another, that allow us to speak of Western or Eastern, modern or ancient, science, religion, magic, and technology; as alternate perspectives on a single reality, they constantly borrow from and influence one another.

In using *technology* to refer specifically to a type of knowledge, it can be kept distinct from several closely related things to which the term has also been applied: machinery, inventions, techniques, and engineering. Machines are the most immediate, concrete, and visible products of technology. They incorporate know-how that at one time existed as a theory, concept, design, plan, or prototype (or all of these). When we think of the impact of technology on society or environment, often the first thing that comes to mind is the effects of machines or of a particular machine, such as the cotton gin, the automobile, or the digital computer. Yet machinery per se is only a part of a much broader process of technological innovation and diffusion that begins with the development or application of knowledge to solve a practical problem and includes all the activities and resources needed to conceptualize, build, test, distribute, operate, and modify the machine. For example, the impact of automotive technology is not exclusively confined to the ways in which cars have changed our travel habits, but it extends to the effects of automobile production on our economic systems, the influence of automobile design and testing on industrial organization generally, the role of the automobile as a consumer good and a symbol of freedom, and more.

An invention — whether a machine, part of a machine, or a process — is not a technology; it is an artifact in which a new technology is applied or an old technology is applied in a new way. Inventions are material solutions to technological problems. When confronted with a need which cannot be satisfied in known ways, the inventor extends or otherwise revises known methods so that the need can be satisfied. Invention both depends upon and contributes to the stock of technology. Automotive technology has given rise to (and has been enriched by) scores of inventions: from something as basic as the modern internal combustion engine, to refinements such as automatic transmission, to social inventions such as certain assembly-line production methods.

A technique is a process, a formalized or regular way of performing a task. As a system of knowledge, a technology is composed of techniques as major parts; automotive technology consists, in part, of the set of related techniques designed to produce automobiles. A large share of the impact of technology on social life comes not only from the machines which it makes possible, but also from the techniques required in their production. Today technology is a very technical matter that requires specialized mathematical and formal languages, expertise and experts trained in technical disciplines, and elaborate support systems in which many operations are so precise that special machines or processes replace human judgment in monitoring other machines and processes.

In becoming a technological society we have also become a society of technicians: few modern occupations can be pursued without some formal technical training — technical specialists have increasingly gained public support, prestige, and power; even to pursue one's daily rounds in a modern city — to travel, work, shop — requires a high degree, by historical standards, of technical sophistication on the part of all.

Engineering refers to the activities associated with technological innovation. It can include the invention, design, construction, maintenance, evaluation, or modification of machines and processes. It is through engineering that technologies or specific techniques are developed and used. The growing importance of technology in modern society has brought considerable growth in engineering and related fields as professions. The highly technical character of modern technology has, in turn, made engineering heavily dependent on measurement, systems of units, standardization, and on rigorous training for practitioners in the application of these techniques. As the authors of a recent collection on the social dimensions of technology define it, technology is

> the means by which man undertakes to change or influence his environment. In this sense, it is close to the meaning of the French word *technique,* which, as Jacques Ellul explains it "does not mean machines . . . . It is the totality of methods rationally arrived at and having efficiency in every field of human activity." Thus, the concept transcends the engineering arts to include forms of organization of human endeavors, methods of rational analysis, and the structure of systems for achieving determined objectives (Evans and Adler 1979:24).

It is difficult to discuss technology without also referring to machines, inventions, techniques, and engineering. Some of the authors quoted or cited in this book have failed to carefully distinguish between them. In treating these separately and focusing on technology as a type of knowledge system, it is easier to see and understand the intimate character of the relationship between technology and society. As a cause, technology has been a factor in the highest achievements, the basest degradations, as well as the most trivial activities of modern man; but like any other type of knowledge, it has not been nor can it ever be a sole cause — operating without agents or in the absence of cultural, social, political, economic, environmental, and psychological conditions. As an effect, technology is subject to the same influences as other knowledge systems. The current state of knowledge, the inherited wisdom, and the future course of innovation in any technical field are the products not only of internal developments but also of external factors such as availability of funds and other resources, career pressures, values and predispositions of technicians and inventors, the state of the engineering professions, and the ideological climate in the laboratory and in society at large.

A broad model of the relationship between technological innovation and social change underlies the discussion in the following chapters. This model is both multidimensional and multidirectional. It views technology as one knowledge system among many (though a very important one) which has affected our way of thinking about and our interaction with other people, nature, and God. At the same time, changes in beliefs, behavior, and institutions — which are partly the direct and indirect effect of technology — can and often do affect the course of technological innovation. For example, automotive technology has, with other factors, contributed to the creation of automobile-dominated transportation systems. This has led to the growth of numerous subsidiary industries to service the needs of producers, buyers, and users of cars, and to the trend toward an automobile-dominated society. These changes, in turn, have increased the general demand for automobiles and, as conditions in these subsidiary industries change (e.g., oil supplies), they have increased demand for certain changes in automotive design, operation, costs, and distribution.

This model reveals a very close, complex, and sometimes conflict-laden connection between society and technology. Technology is an amoral — and substantially apolitical — type of knowledge. Since the early industrial era, it has been based increasingly on scientific principles and canons of technical precision and has been less concerned with normative, ethical, or religious matters. As a result, changes in the technological stock and changes in social institutions such as family, church, and state, in which normative, ethical, and religious matters are central, have often occurred at different rates or in divergent directions. Culture lag has arisen whereby people find themselves unprepared to accept a new technology or when technological innovation is felt to lag behind social needs. This has often given rise to psychological, interpersonal, and even international conflicts.

The relationship between technology and society has been made more complex, and the seriousness of culture lag and other potential sources of conflict has increased, with the growing institutionalization of technology. Like other knowledge systems such as science and religion, modern technology has achieved a high degree of standardization and universality. For it to be properly maintained and developed, an elaborate organizational network, formal training and recruitment, professional recognition for specialists, and other social supports are now required. These supports have given technology a status and power comparable to that of other institutions. Its effect upon our values and social relations are all the more significant because technological priorities are pursued within an institutional framework and with the legitimacy and resources which such a framework commands.

Institutionalization has not only made technology a more potent factor in social change, it has also subjected it to additional social constraints. Organization, formal procedure, and professional standards have acted to routinize and bureaucratize technology as they have done to religion and science. The innovation process has increasingly been influenced by formal bureaucratic requirements and regulations, with the result that a growing portion of the activity of engineers and related specialists is directed to the maintenance of organizations, formal accounting, and similar ends.

As a system of knowledge and especially as institutionalized knowledge, technology is not merely machines, techniques, artifacts, or inventions. It is as much a part of our culture and social order as religion, kinship systems, political structures, and urbanization. As a distinctively modern institution, technology is often at the forefront of social change. We depend upon it and we fear it; we take pride in it, praise it, and blame it. We have still not learned to live in harmony with it; in some fundamental ways, we have only begun to understand it. This book is intended to provide a clearer sense of what it means to view technology as a social phenomenon, to illustrate how social science responds to claims about the moral, economic, and political consequences of technological innovation, and most of all, to demonstrate the promise of greater harmony that lies in the conscientious application of social science principles to technological research and development.

PART I

Sociology and Technology: An Overview

Introduction

Part I discusses key trends in the history of the social sciences, particularly sociology. Its main focus is on the ways in which social theorists have viewed technology and technological innovation. The period of interest extends from mid-eighteenth to the mid-twentieth century, a two-hundred-year period during which modern technology and sociology were formulated, institutionalized, and diffused to all parts of the world. This period has been divided into three distinct eras corresponding to stages in the evolution of the role that society has assigned to science and scientists (Ben-David 1971). The first, beginning prior to 1750 and lasting until about 1850, is the era of amateur science. It is to this era, characterized by a system in which the predominant modes of support for science were patronage and private sponsorship of various types, that the origins of modern technology and social science can be traced. The second is the era of academic science, which extended from the 1850s to the World War II period. During these years, universities, colleges, and schools became the principal sites for and sponsors of scientific work; the academy replaced preacademic sources which, because of the growth of science and changes in economic and political conditions, were no longer adequate. This was also the era in which technology, as a group of engineering and related professions, and social science, as a set of disciplines such as economics, political science, sociology, and anthropology, were institutionalized. The current era, which began in the 1940s, is that of organizational or postacademic science. Since World War II, an ever-increasing proportion of scientific activity has taken place outside of normal academic settings and/or with the support of government agencies, public and private foundations, corporations, and similar organizations. As was the case in the mid-nineteenth century, growth in science and other factors have rendered traditional methods of sponsorship insufficient; universities no longer have the resources to sustain science as they once did.

During the academic era, both institutionalized technology and institutionalized social science achieved an accepted place in modern culture and society. Yet they have both been subjected to increasing scrutiny and criticism by the public, political leaders, and the press; and — despite their differences — on similar grounds: their lack of responsiveness to current social needs, their questionable commitment to human welfare, and the large resources required for their maintenance and growth. During the current era sociology and technology have achieved their greatest maturity but have also begun to experience their most serious public challenges.

In each of these three eras, a specific way of thinking about technology has dominated social theory and research in Europe and North America. Originally, no serious distinction was made between the scientific study of society and the pursuit of technological activity. The same individuals were innovating in both fields, and the approaches were explicitly viewed as complementary ways of solving the problems of development and modernization. In the transition from the amateur era a series of factors, including pressures toward specialization within the universities and general social unrest, brought about a separation between technology and social science. With institutionalization, distinct sets of standards and practices, types of education, professional criteria, and world views were established for engineering and related professions on one hand and sociology and related social sciences on the other. As a result, the two types of activity have evolved in relative independence of each other: technological research and development has proceeded largely in the absence of a formal social science component and the social sciences have developed with no formal ties to technological design, innovation, and evaluation. With the coming of the postacademic era, interest has been expressed by the engineering professions, by social scientists, and by people in all sectors of society in the creation of closer ties between technological activity and social research. Aided by the shift to organizational sources of support for much work formerly pursued under academic auspices, and largely because of the relative maturity of institutionalized technology and social science, these fields have begun to respond to their respective public challenges by turning to each other. Both are currently seeking to alter the separate courses on which they embarked more than one hundred years ago, and eliminate at least some of the distinctions that have existed between them.

The following chapters examine the main features of the amateur, academic, and postacademic social scientific approaches to technology. They discuss some of the causes and consequences of this cyclical movement in the relationship between sociology and technology, from no distinction to institutionalized separation, to integration. Among the causes highlighted is the predominance of a particular source of support for science; emphasis is placed on how amateur, university, and organizational settings have influenced the manner in which social scientists have viewed technology and engineers have regarded social science.

Among the principal consequences cited are the effects that particular social scientific perspectives have had on the conduct of technological research and development and on how R&D activity, in turn, has affected political relations, working conditions, and social development. Remarks are offered on how the one-hundred-year separation between social and technological research has affected the quality of life in modern society.

1.

Technology and the Three Births of Sociology

This chapter examines the origins of the modern social scientific perspective. This perspective and modern technology arose at the same times, in the same places, and in response to a common set of social and cultural changes. Although their prototypes, preindustrial craft and invention and preenlightenment social philosophy, are ancient and universal, in their present forms technology and social science are the specific products of Europe's industrial revolution. The attempt to understand and transform man and nature in an organized, sustained manner and on the basis of scientific theory and experiment began as the chief development project of English, French, German, and neighboring societies during the 1750-1850 era, the era of amateur science. Despite the fact that since that time technology and social science have been viewed as very different and even conflicting systems of knowledge, they retain the stamp of their common roots to this day.

We shall identify aspects of early modern social science and technology that have a bearing on current efforts to bring the two closer together. These include, first, the belief held by the founders of social science such as Adam Smith, Henri Saint-Simon, and Karl Marx, in the power of reason and science to bring about improvements in man's material well-being and social conditions. While this belief was often challenged and its optimism was frequently matched with skepticism, it remained basic to European thought and social reform until the Marxist "watershed" of the mid-nineteenth century.[1] A second aspect of early social science and technology to be discussed is their essential complementarity. From the standpoint of their amateur roles, in which no distinctions were made between strictly economic, physical, psychological, mathematical, sociological, or other disciplinary factors, the founders were able to appreciate the moral impact of the activities and products of technology and the influence of social structure and dynamics on it as interdependent parts of a single system.

Modern social scientific perspectives were formulated or reformulated on three separate occasions: once in England and Scotland during the middle to late eighteenth century, once in France in the late eighteenth and early nineteenth centuries, and once in Germany and elsewhere in Europe in the mid- to late nineteenth century. Although these three "births" have definite continuities, the later births were more than mere revisions or refinements of the first. Participants in each set out to create anew organized bodies of knowledge and technique in response to newly emerging social conditions and to unprecedented moral problems these changes had produced.

Each of the three births provided distinctive insights into the relationship between modern society and its technologies, and each was based upon a distinct set of assumptions concerning the nature of man, science, and history. These insights and assumptions continue to inform our approaches in social science and technology today. Perhaps the most important contribution of the era of amateur science lies beyond these differences and in the general program, discussed here: the commonly shared belief in the interdependence of society and technological activity, the undifferentiated nature of the social sciences, and a cautiously optimistic assessment of technology's social impact.

First Birth: The Development Pivot

More than any other single factor, what sets modern social science and technology apart from their predecessors is that they were formulated at a time and place in which social and economic development was occurring and was becoming a dominant cultural goal. Development brought to the fore a whole range of opportunities and problems never before encountered. Prior to the mid-eighteenth century, and outside of the major urban areas of northern Europe at that time, man's relationships to other men, nature, and God were viewed fatalistically. The belief that people could improve or otherwise change their life conditions through conscious individual and collective effort, or that such an effort could be made more effective if based on scientific knowledge of these life conditions, was not generally held; in many cases, such a belief was felt to be heretical. Yet lacking such a commitment it is difficult to conceive of, or support on any large scale, activity designed to understand and control nature and society in a precise and predictable fashion. It is in this sense that the ideals and realities of development which came to dominate

European society after 1750 or so are logical and historical preconditions for the first birth of social science and technology.

Society and Philosophy in Industrializing England

Over a two-hundred-fifty-year period that began in what Cipolla (1965) calls the Vasco Da Gama era, Portugal, Spain, Holland, and eventually England participated in the creation of a worldwide economic system that fostered the accumulation of vast amounts of economic and social capital in the cities of Europe (Wallerstein 1974, 1979). By the middle of the eighteenth century, Britain's dominance as a world naval power, the great material resources of its colonial empire, and a series of internal political changes combined to provide it with the means for history's first takeoff toward industrialization (Rostow 1960). In the major population centers of northern England and Scotland, there began the first full-scale movement toward systematic replacement of human by machine labor and the en masse transformation of raw materials into commodities.

The complex set of events surrounding England's industrial revolution changed people's assessment of their own powers to control the course of history; the old, fatalistic world views were challenged by the new industrial reality. Clergymen such as Richard Baxter, philosophers and inventors such as Condorcet in France and Benjamin Franklin in the United States, and the emerging middle classes all saw in the end of the feudal order and the rise of industrial society the need for a new, self-reliant ethic for humanity in its relations with God and nature (Weber 1958a). Industrialization provided "a new and robust dimension to the concept of development."

> [T]he perception of change quickly followed the fact of change. There was an intense interest in new forms of social production, ownership, and consumption to fit the changing circumstances of the industrial world. Standards of "normal" life span and infant mortality were no longer accepted; hygenic measures to guarantee maximum growth and life were demanded. Salvation became naturalized and "man's lot" was no longer synonymous with "man's fate" (Horowitz 1966:50).

This new conception of development allowed the amateur intellectuals of the Enlightenment to view man as an active and effective agent of change, and encouraged people to seek to understand and improve upon the remarkable transformations taking place around them. For many, this search led to inventions of new machines and processes. For others, it led to inventions of social systems, new constitutions, and the formulation of theories whereby social systems and constitutions could be studied and changed. For most, this search led to an intense interest in both machines and social innovation. Unconcerned with distinguishing themselves as scientists, engineers, or social philosophers, those who participated in the changes of the early industrial era treated all such matters as aspects of development, as complementary means of self-improvement through the application of rational knowledge (Ben-David 1971: ch.5; Kranzberg 1967, 1968; Rose and Rose 1969: 16ff). The Enlightenment belief in science and technology for development

> served as the basis for [many] philosophical and technological enterprises . . . . Gifted and creative individuals now had a much greater range of opportunities to exercise their creativity than ever before. It was possible to discuss matters of politics, economy, and philosophy without fear of violent conflict, and there was fair scope for actually influencing policy . . . .Experimentation with steam engines, textile machinery, and other technological schemes began and were spurred on . . .(Ben-David 1971:79; see also Kranzberg and Gies 1975,chs.10-11).

What most impressed observers during the earliest days of the industrial revolution were the powers that the new technologies and social changes had given the common person (Du Mont 1977). But the concerns of the moral philosophers of the period show that these new powers were not accepted with complete enthusiasm. Serious questions were raised about the ethical consequences of using the new systems and about how they would affect the relationship between individuals and the social order. In England and Scotland these ethical issues were particularly pressing. By the middle of the eighteenth century British society was changing dramatically as a result of industrialization and urbanization. It could not be denied that the new technologies and plans were effective; in cities, towns, and increasingly in the countryside, the impact of development on work and life was clear. But also becoming obvious was that to seek progress, development, or other benefits through technology and conscious planning did not necessarily mean that they would be gained; that merely because people intend to improve their lives through rational effort does not assure that something better will automatically be achieved.

Mandeville and the Scottish Moralists: Progress and Reversals

With the understanding that man's lot was no longer the same as man's fate, that people now had substantial control over nature, came the sober recognition that things could still be awry. In this spirit, two related themes came to dominate English and Scottish social thought in the period: first, that the results of collective effort, what society as a whole accomplishes, may turn out to be quite different from what all of the individuals involved in the effort had intended; second, that the morality of individual acts, whether a person is "bad" or "good," may be a different matter from the morality of the society made up of individuals, whether the group is "bad" or "good." At issue here was the actual extent of the newly-discovered power of the individual. Would the well-intended application of science bring progress, or were there other factors that made the path between intent and outcome more complicated? This question was basic to the foundation of modern social science; it remains fundamental in current social thought about technology as well.

One of the first English writers to discuss these issues formally was Dutch-born physician Bernard de Mandeville (1670-1733). In his *Fable of the Bees* (1934) and other essays, Mandeville focused on the ways in which people can act in good conscience to benefit themselves and each other and yet create with such acts a corrupt, disordered society (Horne 1978; Schneider 1975:42-56). Several lessons were meant to be drawn from these studies. Most directly, Mandeville was offering a philosophical argument against altruism, one that was later to be incorporated into the economic theories of Adam Smith and his contemporaries. More generally, these observations and arguments were meant to show that social behavior is always subject to unintended and unanticipated consequences, that individual acts and social outcomes proceed by two different sets of laws (for more recent views see Lerner 1963 and Lange 1965).

With the understanding that, as the German sociologist Savigny put it, "the outcome of the general will is no one's will in particular," Mandeville and his successors established that over and above individual motives and behavior, an objective, "emergent," social order exists that can be studied and manipulated in its own right. These observers of the early industrial revolution were no longer content with merely arguing that one type of behavior is morally preferable to another, or with seeking to discover the types of ethical principles and laws that would assure the proper conduct of human affairs — as had been the main concern of most earlier religious and philosophical thought. They sought to study ethics, morality, and laws as historical phenomena and with the same detached interest that is applied to any other object of scientific study.

This shift in the way people viewed their own social relations expresses a new freedom from the theological and metaphysical concepts of law and morals — as somehow above scrutiny, questioning, and not subject to principles of cause and effect. Just as the new technologies had secularized the human prospect, the experience of development, industrialization, and their consequences challenged the older values and secularized our way of thinking about this prospect. But as Mandeville's writings illustrate well, this new concept of freedom was not absolute. Along with the discovery that good, bad, right, and wrong are relative matters that change with conditions, came the constant reminder of the potential reversals and ironies that can arise when people try to take their destiny into their own hands. At the same time that technology's impact on the moral order, freedom, laws, etc., was first recognized, it was also noted that the application of technology in a conscious effort to achieve progress can always produce unwanted results.

Mandeville's insights into the complex nature of progress and ethics in industrial society were developed further in Scotland, later in the eighteenth century, by the Scottish moralist philosophers: Adam Ferguson (1723-1816), Adam Smith (1723-90), John Millar (1735-1816), and others (see Lehmann 1930). The ideas and arguments of this group continue to influence law, philosophy, and social science in England, the United States, and elsewhere (Schneider 1971, 1976:18ff), in large part because of their keen understanding of the moral dilemmas of industrial society. In their writings on ancient and recent history and their studies of contemporary economics and social relations, the Scottish moralists had an explicit interest in discovering the best ways to achieve progress. But like Mandeville, they also emphasized the many ways in which willful attempts to progress can lead to a deterioration of conditions.

Ferguson's (1852, 1979) views on the decadence and glut that material prosperity brought to the Roman Empire and other "advanced societies," as well as his "strong streak of voluntarism" (Schneider 1976:20), illustrate these concerns well (Schneider 1979). John Millar's (1930) studies of British constitutional history also stress the often paradoxical causes and effects of conscious self-improvement, providing an early perspective on the limits of planned change. Ferguson and Millar used historical examples to show how the aims of people can differ substantially from the eventual outcomes of their actions. But in reviving our memories of such differences, as in Millar's study of the *Magna Carta* as an instrument of aristocracy which became a foundation for democracy, these early social scientists were equally reflecting on the effects of the new technologies and ideals in their own time.

These concepts are perhaps most familiar in the writings of the central figure among the Scots, Adam Smith. In his well-known *Wealth of Nations* (1776), *Theory of Moral Sentiments* (1759), and other works, Smith summarized effectively earlier observations such as those of Mandeville and applied them to current social, economic, and scientific trends. Like many of his contemporaries, Smith was both an inventor and sharp moral philosopher. In seeking to understand the factors that had contributed to England's newly acquired industrial prosperity and in attempting to provide a rational framework and body of techniques for ensuring that such prosperity would continue, he created the science of political economy. His wide-ranging interests are apparent in the scope of this new field. The theories and proposals of Smith and his students were based on the application of principles such as controlled observation, hypothesis-testing, and quantification to the study of market operation and production processes, psychological motives, morality, demographic trends, political behavior, and the rate of technological innovation. Smith's new science is generally recognized as the immediate predecessor of formal economics, but by virtue of its scope, it is also the field from which sociology, political science, and several other contemporary social science disciplines arose.

In the context of late-eighteenth-century urban Scotland, the birth of modern social science occurred as the result of dramatic and perplexing social changes and the attempt on the part of Adam Smith and others to make sense of them. The main principles of this new perspective reflect the events and leading concerns of the time: detailed analyses of the causes of material progress, a central focus on the role of the individual, his freedoms and constraints, a detached, relativistic view of morality, and a new concept of the forces that emerge unconsciously from collective acts, what Smith called "the invisible hand." The latter principle is important as a sociological aspect of late-eighteenth-century British thought. The idea of an invisible hand expresses a basic and uniquely modern appreciation of how the technologies of industrial society operate in human affairs. Within the first scientifically based program for the improvement of the human condition, the operation of the invisible hand served to warn of the ironic nature of coordinated planning. In opening the way for the application of science to economic, political, and social relations, Smith, Millar, and Ferguson viewed progress sympathetically but skeptically. What was clear to them, and what makes their work so relevant today, was their recognition that in the wake of the industrial revolution morality and technology had entered into a new and complex relationship.

The first birth of sociology was identical to the first birth of modern economics, political science, geography, and other fields (under the labels of moral philosophy and political economy); as such, it was very much affected by the first birth of modern engineering, planning, architecture, and management. Its occurence in mid- to late-nineteenth-century England and Scotland is significant, for in this revolutionary era the old ethical systems and morality in general had become problematic and thus the proper subject for rational inquiry. Partly as cause and partly as consequence, science was proposed as the basis not only of the invention of new machines and processes, but also of the study and practice of social change. From the moment of its birth, sociology (nee social science) began to inquire into the causes and effects of technological innovation, to seek a scientifically sound basis for human development and to moderate the enthusiasm that often accompanied this pioneering spirit.

Second Birth: Development Institutionalized

While the connection between social science, social change, and development is often only adumbrated or implicit in the work of Mandeville and the Scots (in part because these distinctions had not yet made their way into the vocabulary of formal thought), it is an explicit and central theme in the events surrounding the second birth of sociology, in France in the late eighteenth and early nineteenth centuries.

Society and Philosophy in Revolutionary France

At the time of the English industrial revolution, France was a leading colonial power. But at least temporarily, its level of development and position in European affairs were secondary to England's. Its course of domestic, economic, and political change generally followed that of other modernizing societies, and it benefited earlier and in some ways more fundamentally than imperial Germany from the "penalty" of England's leadership (Veblen 1939). By the 1780s, France had accumulated sufficient resources within as well as outside of its formal feudal system to support its own industrial revolution.

Though at first political institutions constrained (rather than facilitated) the French industrial revolution and localized it in Paris and a few large population centers, it was as sweeping as that of England. Indeed, in scientific, social scientific, and technological pursuits, France quickly superseded England as the world leader (Rose and Rose 1969,ch.2).

> France was the center of the scientific world [after the mid-eighteenth century] . . . . During the first three decades of the nineteenth century, however, French scientific supremacy became much more unequivocally established. In spite of the brillance of some British scientists such as Dalton, Davy, Faraday, and Young, neither in Britain nor anywhere else were there first-rate scientists covering all the then existing fields of science. Only in France, or more precisely, in Paris, were all fields of science pursued at an advanced level (Ben-David 1971:88-89; see also Crosland 1967).

In part because of the country's lack of economic development and its sudden rise to prominence in intellectual realms, the key social and technological issues of revolutionary France were not directly related to industrialization; by the late eighteenth century, it was no longer meaningful to ask if the country should industrialize but only how to do so and how to deal with related social impacts. The French Revolution — and France's "rehearsal" for it through its participation in the American

Revolution — centered not so much on economic transformations as on the proper methods of accommodating these transformations within new political and social forms. Like other European societies (and, unlike today's Third World countries), France was to shape its concept of modern statehood around the newly-emerging industrial-capitalist economy. Seen in this light, the revolutionary era that extended from before 1789 to about 1830 was a struggle not to initiate but to institutionalize development, and thus establish a place in the social order for technology and social science (cf. Nisbet 1971).

The French birth of sociology and technology placed explicit emphasis on the active pursuit of development — though the Scots had more than an inkling of this possibility:

> The French Revolution thus established a new view of social realities. History was no longer written in terms of dynastic empires, individual genius, or self-proclaimed spiritual leaders, but in terms of the flow and thrust of mass man. By conceiving of development as a secular process, historians like Michelet and, earlier, Condorcet, fashioned a theory of change which was simultaneously "scientific" and "moral" (Horowitz 1972:43).

The French Fathers: Saint-Simon and Comte

Henri Saint-Simon (1760-1825) and his colleague Auguste Comte (1798-1857) are generally held to be the French "Fathers" of social science and sociology in particular. In Comte's case, this is related to the fact that he coined the term, first *social physics,* then *sociology,* and because of his *Positive Polity* (1896) and *Positive Philosophy* (1912) he extended Smith's program for a comprehensive science of society.[2] The sociology that Saint-Simon, Comte, and their contemporaries had in mind was far more encompassing than what we refer to as sociology today; it was to include all other fields of study (scientific, humanistic, and technical) deemed legitimate in the new scientific era. Comte explicitly assigned to sociology the role of "science of sciences" and charged it with replacing theological and metaphysical systems as the foundation of the moral order in modern society (Comte 1896:150ff).

These views were part of the general project, begun by Saint-Simon, to employ science and technology in reorganizing postrevolutionary France (Manuel 1956).[3] The main principles of this project were positivism — the belief in empirical science as the ultimate method of determining truth — and socialism — the belief in an egalitarian society as the ultimate solution to problems, such as those raised earlier in England and Scotland, of individual and group morality. For Saint-Simon and his contemporaries, positivism and socialism were the means whereby society could resolve the major dilemmas of the industrial age then beginning in France: the most advanced forms of thought and social organization were to become established as the mark of a truly developed nation.

Saint-Simon envisioned a highly socialized economic order, democratically governed — politically and morally — by men of learning dedicated to positivist principles (Saint-Simon 1952). This system, while ancient in concept, is one of the first modern models of technocracy, government by scientists and engineers (Winner 1977, ch. 4). With it, the pursuit of social and nonsocial science was linked to the application of technology for the improvement of man. This doctrine was issued in response to the unprecedented development then occurring. It was meant to provide a structure for making progress in the economy, science, and technology a regular and integral part of the ongoing business of the state.

Comte expanded upon these ideas through an historical examination of the ways in which change in dominant types of social organization — from tribal, to feudal, to contemporary — corresponded with shifts in dominant modes of thought — from theological, to metaphysical, to scientific. This "law of three stages" was the basis of the first modern sociology of knowledge, in which the characteristics of knowledge systems are explained in terms of social factors rather than simply judged as true or false. Like the treatment of morality as historically and socially relative that had been pioneered by the Scots, Comte's approach sought to view truth itself as an appropriate object of scientific study, subject to the same causes and conditions that affect any other aspect of human behavior. From this standpoint, Comte argued that sociology offered the correct method for understanding and controlling social relations in the industrial age; that not only was social science superior to the old religious and philosophical systems on ethical grounds — it could better promote rational and moral conduct — but that the type of truth it yielded was more consonant with the times, compared to which the old systems were obsolete (Schneider 1976:24ff).

In addition to his attempt to provide details and a concrete justification for Saint-Simon's program, Comte also gave early French sociology a skeptical view of progress. Through his experiences with the Terror, the Napoleonic era, and reconstruction, he had seen the result of what he deemed unscientific pseudoexperimentation with the social order. He reacted with a new criticism of individualism and revolution similar to that of Mandeville and the Scots. For Comte, "liberty of thought is an illogical idea and 'justly deserves the charge of anarchy brought against it by the ablest defender of the theological school'" (Nisbet 1971:33 quoting Comte).

Many of the ideas of Saint-Simon and Comte can be traced to their Enlightenment predecessors the *philosophes,* Condorcet, and Turgot. Their writings exhibit the characteristic concern for material and spiritual progress that distinguishes the second birth of sociology and technology.

> The stress on the necessary linkages between the ages of mankind, the emphasis on the inevitable increase in the cultural inheritance of humanity, the belief in the powers of science — these and many other elements clearly are the major ingredients in Comte's synthesis and make him a continuator of the tradition that was started by Turgot (Coser 1977:22).

But for Comte especially, the influence of the conservative critics of progress such as de Bonald and de Maistre is equally strong. His views were not unconditionally optimistic; they were often grimly ironic, and for their time even hard-headed.

Despite these reservations, Comte held to a doctrine of rule by science in which social science was to have a predominant role. He believed that with modernization "the right of free inquiry will abide within its natural and permanent limits; that is, men will discuss, under appropriate intellectual conditions, the real connections of various consequences with fundamental rules universally respected" (Comte 1896, II:25; see also Coser 1977:8ff). Comte and his contemporaries were self-consciously committed to the integration of all the social sciences and the technical fields. Social theorists such as Prosper Enfantin, Victor Considerant, Le Play, Comte, and "hundreds of Saint-Simonians" sought out the company of leading scientists and inventors such as Lagrange, Fourier, Poinsot, Ampere, Gay-Lussac, and Cauchy at the Ecole Polytechnique of Paris (Hayek 1952). In this and other ways, the first sociologists attempted to establish their field as one branch (the "highest," to be sure) of a general system that encompassed all the sciences and engineering as well. It was based upon the "dissimilar scientific traditions of Bacon and Descartes [but also upon] Bossuet's Catholic Vision. Montesquieu and Hume, de Condillac, and the *ideologues,* as well as the natural scientists from Newton's day until [their] own, all exercised considerable influence" (Coser 1977:21).

The extent to which the educational and political programs of the first French sociologists were realized has not been great. But they did give new meaning and a new vocabulary to social inquiry. Subsequent events have tended to cast the French founders as idealists and even "Messianic Bohemians" (Salamon 1960). Yet their vision was well-anchored in the needs of their time — and perhaps of our time as well. Their root commitment was to the orderly and controlled development of humanity in an industrial era. To accomplish this they had to draw on diverse intellectual resources and invent sociology. The field was formulated not merely as a formal discipline whose own development was foremost (although Saint-Simon, Comte, and the others appreciated the purer orientation in science), but as a knowledge system inextricably tied to technology and government. This is a major source of the appeal of the French sociological vision but, as we shall discuss, it is also a cause of its failure to be fully put into practice then or yet.

Participants in the second birth of sociology, though men of learning, were not quietist scientists (as we now understand the role),

> who created a system in the silence of an office, but activitists who thought they were called upon to rebuild society on a new basis; and it is only too obvious that such a construction could not be done with a scientific method. In truth, the needs, the aspirations of all kinds which were causing the ferment in French society, guided the statesmen of the time and determined the main course of the simultaneously destructive and restorative work which they undertook (Durkheim 1890, 1971:36).

Saint-Simon, Comte, and their colleagues attempted to demonstrate the fundamental reasonableness of development through science and socialism. In forging this argument, they established the terms of a continuing search for a reconciled sociology and technology. But, as Durkheim (1971:38) observed, their perspective also constituted "a religion with its martyrs and apostles." In attempting to apply with religious fervor these principles of scientific development, the French founders created a modern social science and at the same time a doctrine that was later to be characterized as empty "utopian socialism."

Third Birth: Development Demystified

The influence of Karl Marx (1818-83) and Marxism on modern views about society and technological change is great and growing. As a nineteenth-century European intellectual, Marx was immersed in the social and philosophical currents of his time. As a German-born, well-traveled resident of England, he had the opportunity to observe and participate in several stages of the industrial revolution. As a student of the works of the Scottish moralists and French sociologists, he was conversant with the reversals and ironies inherent in development, with the program for a comprehensive social science, and with the prospects for a socialist society. In synthesizing these experiences and observations, and in reacting to the dominant romanticist orientation in European thought of his time, Marx — with Friedrich Engels and their "young Marxist" followers — invented, for the third time, a modern scientific perspective on social relations. Like his predecessors, he also formulated a body of theory and techniques, a technology, for what he held to be the correct application of scientific principles to promote human development.

Society and Social Thought in Mid-Nineteenth-Century Europe

The years between 1848 and 1865 are decisive in European (and world) history because they encompass an era of widespread domestic social unrest, including revolution, civil war, and their threat, and because they mark the beginnings of the modern communist movement. A chief cause of this social turbulence, and of the communist response to it, was once more the industrial revolution. By 1848, however, the major questions raised about industrialization were neither focused on existence, i.e., "is it to be?" — as had been the case in England a century earlier, nor on how the revolution was to be politically accomodated. The dominance of industry and the hegemony of the middle classes — including a bourgeois king in Paris — were established facts in the developed parts of Europe Marx knew. Instead, this "Age of Democratic Revolution" (Palmer 1962) was characterized by struggle over the distribution of economic and political benefits associated with the largely institutionalized capitalist-industrial system. It was the problem of the distribution of the fruits of the industrial revolution between city and countryside, classes, and nations that ignited Europe (and North America) during this era. Correspondingly, it is also the main problem addressed in Marx's *Communist Manifesto* of 1848, and, in one way or another, in all of his work.

The focus of Marx and Engels on distribution has led to the understanding that they are among the first "technology assessors" (Rothman 1978). *Capital* and other writings provide a sweeping assessment of the impact of the new industrial (as well as ancient and prehistoric) technologies on the organization of the economy, the state, the family, and culture. Marx assigned a master role to technology — what he referred to as the "forces of production"[4] — as a cause of social change. His analyses and arguments pointedly trace the connection between capitalist-industrial technologies — particularly, the distribution of wealth and power — and the nature of work, political relations, and philosophy in a contemporary society. In extending and criticizing the views of Smith and Saint-Simon, Marx provided a degree of detail and explicitness concerning the relationship between technology and society that continues to influence our understanding today.

Through his study of social conditions in industrial Europe, Marx had an early opportunity to see the changes that the new system could and might bring. But in straining to apply the observations of the Scots, the French sociologists, and especially of Hegel (1770-1831) to the current situation, he found it necessary to create a new, more adequate science of society. In the place of Hegel's "spirit" as the prime mover of history, Marx inserted the forces of production, as Smith and Saint-Simon had suggested. In so doing, technology was once again made a major subject of what we now call sociological inquiry.

The power of technology to transform human as well as nonhuman nature had been observed before Marx. In Marx's case and earlier, this observation was accompanied by a commitment to use knowledge in consciously effecting such transformations. But events in mid-nineteenth-century Europe did not allow Marx to assume, as did the Enlightenment philosophers, that the technological possibility of a good, just, or socialist society was sufficient (in the hands of scientific men of good will) to realize this potential in concrete social relations. It was on this basis that Saint-Simon, Proudhon (1809-65), and other French and English social theorists were categorized as utopians (Engels 1892). While the concept of development remained central to Marx's thought, it was demystified. By 1848, it was clear that development would not inevitably proceed in an orderly, rational fashion merely because Europe had entered the age of science and technology. For Marx the dialectical nature of productive forces, marked principally by the perennial struggle between the means and relations of production, assured that the development spurred by modern technology must occur in a conflictual and uneven manner. Like the other founders of social science, Marx was certain that the industrial revolution had secularized the human prospect. Like the Scots, he focused on the unintended and ironic side of progress, on the "darker side" of Enlightenment. But like the French Fathers, he also believed in the power of man, aided by technology, eventually to mitigate negative impacts.

According to Engels, what makes Marx's theory of mitigation scientific and that of predecessors utopian is his belief in political action, class struggle in particular, as a necessary condition for the kind of society that Saint-Simon and Comte had tried to argue or legislate into being. Marx's theories on the impact of technology, particularly its future impact, place a strong emphasis on the active, collective attempt on the part of society to intervene and direct technology to achieve various ends — including, ultimately, a socialist order. Marx made no sharp distinctions between social scientific theory, technology, and political praxis; rather, he made their connection essential. The communist intellectual was viewed as an "engineer" of revolution whose actions were to be based on scientific socialism. In a more-than-metaphorical sense, Marx regarded the doctrine of dialectical materialism as the new technology of the working class, with which it could take control of and redirect the older capitalist technologies. In a similar vein, he countered Saint-Simon's vision of technocracy with a far more grim description of the new industrial order. It was a view of society ruled, not by scientists and technicians for the good of all, but by the dominant middle classes for their own interests; an order in which technology was constantly being used for specifically political and economic ends.[5]

Marx and Technology's Stratified Impact

One of the lasting contributions to contemporary social science of the third, Marxist, birth is its framework for understanding social inequality or stratification.[6] Marx did not deny the tremendous economic and social benefits provided by the industrial revolution. But these benefits were analyzed in the light of a more detailed societal perspective, from which it is necessary to ask who benefits and who loses from such changes. Marx's work offered an early statement on the differential impact of technological innovation. This rendered futile the search for "the" social, economic, or demographic effects of a change in the forces of production. The nature of society assures that technology's effects will be stratified; each class or other aggregation such as ethnic group, city, or rural population will be affected differently. In crucial instances, different groups may even be affected in opposite ways. The specific, though (by contemporary standards) overly simple, proposition that steam power or another innovation might bring wealth to one group and misery to another became with Marx a basic premise in research on technology.

Stratification, as Marx knew, has crucial bearing not only on sociological theory and research, but also affects the way in which technological innovation is conducted: who conducts it, in whose interests, and for what purposes. These considerations identify an important way in which social factors shape technology. Yet at this point, perhaps more than at any other in Marx's work, the current social science disciplines — some of which have embraced the stress on stratification — diverge from one another and from engineering and other technical fields — which have generally ignored it. In the 1848 era, there was a third attempt to formulate a systematic, scientific explanation of the moral and social consequences of industrialization. By this time the economic and political effects of the industrial revolution had spread to the masses. The results were, by Marx's account, less than entirely progressive, liberating, or enlightening.

This recognition gave a new, critical dimension to European social thought. Marx's characteristic call for political action in the face of the unequal and unjust consequences of industrialization turned such criticism into a powerful weapon. But for these and other reasons, to be discussed in the following chapter, the connection between technology and inequality was not universally accepted. The political implications of the connection and of Marx's views became instead the objects of intense partisanship. A general result of this invention of social science was a great fission between social and technical approaches to technology, one that is only now being attended to as the approaches grow to depend on one another once more.

Summary: The Three Births and Development

Despite the differences in their generations, social backgrounds, and philosophical orientations, participants in the three births of sociology shared an interest in understanding and promoting social and economic development. This identifies them as specifically modern thinkers and sets them apart from earlier theologians, philosophers, and inventors. Commitment to development provided them with a view of technology as conditioned and limited in every respect by human relationships. The social sciences as a whole were to be treated, like other sciences, as a source of knowledge and principles for what are today the fields of engineering, planning, architecture, and management; and by the very nature of their activities and aims, these technical fields were felt to rely on social conditions in equally critical ways.[7]

In light of the development concept, technology was seen as knowledge to transform humanity and nature for the better, to free man from the limitations on his powers once accepted as inevitable. Social science was to be an adjunct to technology because it was required to help understand these objects: humanity and nature, knowledge, and freedom. It became clear with Marx's theories that social science must be used in understanding the interests and behavior of participants in technological activity: owners, technicians, workers, etc.; that these too are consequential and scientifically comprehensible parts of the innovation process. From these observations it followed that social science and technology are mutually dependent means to achieve a common end: development and progress through the application of scientific principles to human affairs. Though Adam Smith, Saint-Simon, and Marx explained these matters in different and sometimes contradictory ways, their shared understanding of the basic roles that social science, technology, and development play in industrial society remains a major contribution of the 1750-1850 era as a whole.

The three births occurred in idealistic times, in the midst of first economic, then political, then social revolution. The established ethical and legal principles and knowledge systems were called into question along with the old economies and political institutions (Merton 1970). Ethics, law, and knowledge were made problematic, objects that can be studied and reformed. This was a primary condition for the invention of social science as it occurred in England, France, and Germany. Revolutionary conditions caused people to look at their own lives in a new way. But because of the unprecedented effects of development ideals and realities, people were provided with a unique set of answers to the questions raised by revolution. For Smith, Comte, and the others, these answers lay in the pursuit of social science and technology.

Sociology (and other social sciences) arose as a result of the rapid and dramatic changes associated with industrialization. Its original overt purpose was to help ensure that these changes were for the best.

> In this context, then, sociologists seem to be not only the product of this process of dissolution but also a natural attempt to assist in the reorganization of human society [a most momentous technological undertaking], to help in the reorganization and readaptation of the individual himself. What we read in textbooks and treatises on sociology is very often nothing else than a retrospective collection of new insights which were gained during periods of social unrest. In the work of Saint-Simon and Comte, for instance, the impact of such direct experiences is clearly visible (Mannheim 1953:210).

The characterization of sociology as a product of dissolution and as an attempt to aid in reorganization aptly summarizes the conditions under which it originated. But it remains appropriate even as we examine sociology and technology in other, less-than-revolutionary, times and under less than development-oriented conditions.

Notes

1. A discussion of the ways in which Marx summarized, criticized, and provided an alternative to the optimism of the Enlightenment is contained in Zeitlin (1968, pt.3).
2. "Comte had high praise for Adam Smith, though not for most of his successors . . . . But he was also critical of Smith, and particularly his successors, for their belief in the self-regulating character of the market [which] systematizes anarchy" (Coser 1977:25).
3. Significant differences existed between the views of Saint-Simon and Comte (and de Maistre, de Bonald, Helvetius, and the "Saint-Simonians, as well). Durkheim and more recently Gouldner (in Durkheim 1958), Zeitlin (1968), and others have drawn attention to many misunderstandings, reinterpretations, and revisions to which Saint-Simon's theories of history, the elite, knowledge, and social justice were subject at the hands of his followers. In this discussion the broad themes that characterized the French birth as a whole are emphasized.

4. The term is used extensively in *Capital* (1954, vol.1:370ff), *The German Ideology* (1964:40ff), and *A Contribution to the Critique of the Political Economy* (1904). Despite the fact that forces of production can be further divided into (a) means and (b) relations, the equivalence between technology and more general "forces" is underscored here. Neither means nor relations ever occur or have meaning without the other (no matter how poor the fit may at times seem). Technology proper (i.e., as knowledge to transform nature based on science) might be arguably assigned to the means of production. But as soon as the science of management of productive relations is formulated, it becomes technology and the relations it manages may be viewed as inventions (this is noted by Marx [1973] in the latter sections of the *Grundrisse*). The boundary between means and relations is permeable because of the intimate relationship between technological innovation and social process.
5. The theory of engineers and technicians as an independent class was developed later, beginning with Weber (1947). Gouldner (1975) discusses intellectuals as "technologists" of revolution and Edwin Layton (1971) has provided a detailed study of class conciousness among engineers and technicians per se. Burnham's (1941) discussion is also relevant as what C. Wright Mills called "a Marx for managers."
6. The importance of stratification in Marx's writing is well illustrated in the collection by Bottomore and Rubel (1961); Wax (1965) discusses the centrality of stratification in social science generally.
7. "In the past, sociologists and engineers have been most at home with one another when they have attempted to carve out a future that would provide both order and progress, to use Comte's motto" (Horowitz 1972:426).

2.

Sociology and Technology in the Wake of Fission: Max Weber and European Social Thought

These comments on social-scientific views of technology continue with an outline of some events of the academic era, the period between the late nineteenth and early twentieth centuries. The formulation of the social sciences up to the time of Marx occurred within a period of transition from amateur to academic science. By the last quarter of the nineteenth century, the sciences, technical fields, and the more traditional learned disciplines had been (or were rapidly becoming) institutionalized within an academic framework. Since that time, and largely within the physical confines of the university, college, and technical school, and the intellectual confines of departmental disciplines, the social sciences grew apart from one another and from engineering and other technical fields.

The academic system in Europe neither wholly initiated nor solely perpetuated this disintegration. But as the principal set of organizations through which industrial society assigned its development priorities, recruited and trained personnel for the professions, and carried on research, it played a major role. The evolution of academic science occurred as a subprocess of a more general movement in the industrialized countries which followed the third (Marxist) birth of sociology. By the late 1800s, these countries were occupied with the problems of distributing the benefits of industrialization; and in each, communist movements were challenging established authorities with their own revolutionary solution to these problems. In response, politically and economically powerfully groups committed social resources to the building of "the technological society" (e.g., Ellul 1964), largely — though not exclusively — through the academic system.

Specialization, professionalization, and bureaucratization were instituted universally: in the factory, in government, in the arts, and in education — especially higher education. By the turn of the twentieth century the West had grasped the lesson of the industrial revolution and had institutionalized development. This achievement was not based on a Smithian model of the perfectly competitive market, nor on a Marxist model in which a disenfranchised working class would seize power and institute socialism, nor even on a Saint-Simonian vision of technocracy. Instead, an R&D approach to development was initiated in which producers, sellers, buyers, and even the working class were transformed into highly trained specialists. In the universities, schools, and professions, functions were defined and graded in terms of the service they rendered to the development of industrial, and frequently this meant military, technology.

As it bore on the relationship between sociology and technology, this process contributed to the formal exclusion of the types of holistic, moralistic, technology-centered, and application-oriented social science dominant during the preceding formative years. In their place there arose the telling distinction between objective technical disciplines based on non-social sciences and subjective, ideological pursuits such as social science (and some of the humanities). The institutionalization of academic science was accompanied by a necessary process of selecting out of the university new disciplines recognized as nonscientific and of cleansing authorized, scientific disciplines of ideological elements. Technology came to be identified with practical and applied physics, chemistry, geology, etc., and social science was, in many respects, its antithesis.

This chapter discusses these features of the academic era. It begins with comments on the historical process of academic institutionalization, particularly as it related to the formation and relative status of modern disciplines. Because of the effects of political and economic changes, the societal context in which academic institutionalization occurred is also considered here. Of special note is early industrial society's definition and method of pursuing social and economic development. Finally, the effects of these trends on the substance and organization of social science, and of Weberian sociology in particular, are examined.

Development Domesticated

For the earliest modern social scientists, the human and technological aspects of industrial society were closely related. For Marx in particular, this is reflected in the multidimensional concept of productive forces, among others. Until the late nineteenth century, technology was commonly viewed not as something inhuman or alienated from man but as "an extension of man."[1] As Marx (1964:372) put it: "Technology discloses Man's mode of dealing with nature, the process of production by which he sustains his life, and thereby also lays bare the mode of formation of his social relations and of the mental conceptions that flow from them." Seen in this way, the appropriate use of technology requires no less than a theory of human freedom and its limits; a theory which had been the major preoccupation of social thought since long before the industrial revolution (Wax 1965). From this perspective, the process of learning, applying, and innovating technologies is an inherently social process, influenced by values, group norms, interests, opportunity structures, stratification, ethnicity, etc.: the inventor and technician were seen as social actors who required a "reflexive" orientation, a way of understanding their own actions and motives in relation to the technologies with which they worked, in order to be effective.[2]

Societal response to this view of social science — especially its Marxist version — during the academic era was complex and ambiguous. It was clear that a new cultural force would henceforth play a prominent role in political and economic life. With Marxism, industrial society was provided with a systematic critique of progress and a coherent strategy for ensuring further "genuine" (i.e., socialist) development. But Saint-Simonianism, Marxism, and amateur social science generally were not formally treated as a practical adjunct to development in late-nineteenth-century Europe. The implications of these approaches, their truths, and their mistakes were studied and debated, not as regular part of academic activity, but in unofficial ways, and in the case of Marxism, in the often-illegal world of radical politics. In this sense, the evolution of formal academic disciplines, including sociology, and the evolution of amateur styles of social science — holistic and moralistic "technology assessment" — occurred in relative mutual isolation and even conflict.

European Development in an International Context

The manner in which late-nineteenth-century Europe sought to solve the problems of industrialization and distribution was incompatible with the institutionalization of Marxism — or of the kind of social science that had preceded it. This incompatibility arose from a difference over the central issue of what constitutes development. The industrialized countries were in a position, in relation to world economy, that enabled them to treat development as an economic entity, i.e., capital accumulation.[3] Due not only to industrialization but equally to the structure of their colonial empires and the consequent international division of labor

— in which free labor existed in some areas, serfdom in others, and slavery in others (Wallerstein 1974, ch. 1) — Europe and later North America measured development in terms of increments in money and machines. This was possible not because development is merely capital accumulation but because, as emerging international powers, the industrializing countries were able either to sustain the other dimensions of development with domestic growth or effectively ignore them.

European development between the mid-eighteenth and mid-nineteenth centuries was not an isolated process. It took place in conjunction with the expansion and consolidation of a worldwide economic, political, and cultural system which England, France, and Germany largely controlled and of which they were the main beneficiaries. The existence of this system provided European markets with abundant and increasingly (but never fully) dependable foreign sources of raw materials and labor. These inputs could be used in the relative absence of social disruption (in Europe, that is). It is true that European society underwent great internal changes during the early industrial era, including urban/rural conflict, economic depression, working-class agitation, and massive population migrations and dislocations. But in comparison to the revolutionary events of the preceding century and the disruptions in native societies in the far-flung colonies, this was an era of peace and prosperity in industrial Europe. As George Orwell (1968a:397) observed, "the overwhelming bulk of the British proletariat [did] not live in Britain, but in Asia and Africa."

This situation allowed the industrializing countries to concentrate on one end of a productive process whose other parts were widely dispersed over the colonies of Asia, Africa, and the New World, and in the less-developed areas of Europe. The economies of England and other European countries were geared to it in the sense that domestic skilled labor could be treated as scarce and expensive and other factors — unskilled labor, raw materials, etc. — as plentiful and cheap. Development (economic growth) in these countries coincided with increases in innovations which used nonhuman — and nonrenewable — forms of energy and which replaced men with machines in other ways. The characteristic features of modern industrial — labor-saving, high-energy, capital-intensive — technology were the result of economic opportunity as much as any other factor.[4] While European workers and farmers did suffer from unemployment, low wages, and poor living and working conditions because of these economic changes, freedom from manual labor, material prosperity, and time for cultural pursuits that Europe as a whole gained were less costly because they were largely paid for by the "underdeveloping" Third World countries (Rodney 1972; Frank 1969).

In this social economic context it seemed that industrial technology — if correctly employed — had no serious social or political impact that could not be mitigated (or encouraged) with additional innovations in industrial technology. One result of this was the incapacity of most Europeans to see the need for a socially oriented approach to technological innovation. As long as negative social effects were only being felt in remote areas of Asia, Africa, and the Western Hemisphere, increasing innovation ensured that capital accumulation could pay its own way in material benefits at home. Development was thus equated with development in a singularly capital-intensive sector of the world economic system which happened to be virtually the exclusive property of industrializing countries. It was and still is a type of development which is in the interest of some at the expense of others.

Development, Science, and Ideology

One by-product of this domestication of development, with its reliance on a relatively narrow concept of technological innovation, has bearing both on the general relationship between sociology and technology and on their reconciliation today. The "fission" between the two refers to an event of the post-1848 era which sociologists of knowledge such as Scheler (1960) and Mannheim (1968) have regarded as the most significant cultural effect of Marxism in Europe. It is the conscious, institutionalized attempt to separate knowledge, including social knowledge, into two broad categories: scientific and ideological (Gouldner 1976,pt.2).

During the amateur era, the word *ideology* (coined in France in the late eighteenth century) was merely a descriptive term for a system of ideas (Zeitlin 1968). It was assumed by Saint-Simon and Comte, for example, that an ideology (particularly theirs) could be based on science and thus be "true." Somewhat later, others — including Marx (Marx and Engels 1960) — began to use the word pejoratively to indicate distortion and bias. In current usage an ideology, in the second sense, is a distorted, interest-serving account; it is to the benefit of a group or groups to have the account believed. It contains a description of the world, or a diagnosis, and a prescription for action, a therapy. Significantly, ideologies distort by exclusion of certain facts, narrowness of focus, and overgeneralization. By the late nineteenth century, this sense of *ideology* — in word or concept — had become an antonym for *science* (Moore 1969).[5]

Because of their association with radical politics, but also because their concept of development differed from that to which industrial society was committed, Marxist and related holistic types of social sciences were classed as ideology. They were not recognized either as academic sciences, classics, nor humanities (which in the German system maintained considerable influence).[6] The fission affected the growth of academic science in support of the general societal commitment to development via capital accumulation. Holistic social science about and as a part of technology had no or very low status in official circles. Non-social sciences were presumably purified of all ideological tendencies and increasingly prospered as "neutral" servants of the industrial order.

From the 1880s through World War II, this trend continued and grew. The university, with its disciplinary departments, came to be viewed as the natural home of intellectual activity. The departmental system of dividing the curriculum was elevated to an ontology, a theory of reality, and formal logical positivism gave science an epistemology, a theory of knowledge, with which to see reality in terms of the structure of this disciplinary system. The relationship of mutual support between non-social science and engineering was strengthened, although their relative prestige was subject to frequent debate (and science usually won; see Kranzberg 1979). Engineering and other technical fields were consciously organized to serve industrial purposes, principally capital accumulation. They acted as intermediaries between the fully enfranchised academic world of non-social science and the directly economic world of industry. Thus arose the characterization of contemporary engineering and related fields as having "one foot in the university and one foot in business." This applied to individual engineers, to their professional organizations, and to their new institutes of technology (see Brittain and McMath 1976).

Split off from non-social science and cast as ideology, the applied, social scientific study of technology, its sources, and its consequences, had no established place in the new academic milieux. This meant that in the midst of the large-scale institutionalization that occurred in the academic era, there existed no institutional means for determining what social ends technology was to serve, what were its social and cultural effects, and how it was conducted — a rather serious omission.

Social science began to be taught at German and other European universities in the name of other disciplines, such as one of the humanities, law, and later following the example of non-social sciences, under several different names: first, economics and psychology — often under medicine — then sociology, political science, anthropology, linguistics, geography, communications, demography, etc. (and the specialization continues to proliferate; see Chubin 1976). As the academic era progressed, the type of social science deemed most acceptable was also that which most resembled, in method and scope, the non-social sciences. The results of this emulation include an accumulation of some formal theory, much atheoretical empirical research, and the achievement of a certain degree of credibility among non-social scientists and the public for the more quantitative social sciences: economics, demography, survey-oriented sociology, and experimental psychology. But the costs of this limited acceptance were also great. Greatest perhaps is the loss, as the principal motivation of social research, of the amateur commitment to seek scientific knowledge about the human condition and to use that knowledge for human development.

Institutionalization served to fragment the social sciences into seve.al overlapping but competing disciplines. It created a division of labor and a hierarchy in which sociology was second, or third, in order of disciplinary recognition, material resources, size, and prestige after economics and/or psychology.[7] In addition, it served to establish disciplines like physics, chemistry, and biology as models of scientific professionalism.

Following the German academic model — and in the context of the threefold splitting of science from ideology, social science from non-social science, and sociology from economics, psychology, etc., late-nineteenth-century sociology was left with a relatively obscure professional domain, remote from technology. At the center of the discipline there emerged a dominant natural-science orientation which, especially as sociologists came under the influence of logical positivism, promoted the proliferation of technically precise, narrowly focused empirical studies. But this kind of academic sociology was never the only type practiced. Had it won the unchallenged support of most sociologists (or of all sociologists most of the time), the history of the field would be less complex. Instead, since its earliest days as an established discipline, sociology has hedged its commitment to "social physics" with a substantial investment in phenomenological, activist, and other models (discussed in part IV).[8]

The ideal of a social science modeled after pure physics, as attractive as it may be on other grounds, projects a mechanical and ethically insensitive view of man. There was some enthusiasm for this view among early academic sociologists, but it was moderated by their knowledge of the very different conceptions shared by their amateur predecesors. The founders of academic sociology in Europe, Emile Durkheim (1858-1917), Max Weber (1864-1920), Vilfredo Pareto (1848-1923), and others, strove to be faithful to principles such as accuracy and logical consistency, and to abide by scientific standards in their work.[9] Yet they could not wholly ignore or entirely avoid integrating into their analyses an "unscientific" sensitivity to moral purpose and an awareness of the ethical and political implications of even their most quantitatively oriented descriptive work.

As a result, early European academic sociology was further weakened as a source of theory about and especially as a basis of technology. Because there was ambiguity, even within individual sociologists, about the extent to which pure description and explanation were desirable or even possible, the discipline was, from the beginning, unable to make a whole-hearted commitment either to accepting or altering the course of innovation. Specialization and the application of scientific principles made it difficult for sociology to say anything unique or meaningful about technology; at most, technology could be treated as part of the environment in which social relations take place. At the same time, the broad backgrounds in law, economics, and engineering of early academic sociologists caused them to give technology a major though implicit role in their theories. Typically, other strictly sociological concepts were featured, such as the division of labor, the industrial order, and bureaucracy, but technology was thought of as an important cause of social change and, from an ethical point of view, as a source of concern. The pure physics model suggested that technology, if considered at all, was to be studied, described, and explained by — but certainly not equated with — sociology. The interest in morality that characterized amateur social science and continued to affect the approaches of Durkheim, Weber, and their contemporaries, promoted intervention through political means or other activities to affect technological innovation.

While technology was not an overt factor in early academic sociology as it was for Saint-Simon, Marx, Comte, and Smith, it remained a force in the social relations which this sociology was intended to explain. Beyond their roles as pure scientists and disciplinary sociologists, Weber and the others were aware of this and saw the need to do something about it. But what? Generally the answer was to lament the partly explained effects of technology but to be basically passive; to keep separate one's political actions and the pursuit of sociology and leave to non-social scientists, technicians, and industrialists the conduct of technological design and innovation.

Weber and Bureaucratic Society

The views of Max Weber, one of the most influential early European academic sociologists, illustrate well the sources of this ambiguity (Weber 1946a, b) and other aspects of academic social thought of the period.[10] His understanding was that technology and economic factors had increasingly forced society away from primary, familial, and communal relations and toward rational, impersonal, secondary forms. The accumulation of conscious decisions on the part of political and economic leaders along with the inheritance of unconscious cultural predispositions such as the Protestant ethic, set industrial societies of the late nineteenth century on a course of development which required a distinctive resolution to the problem of integrating society and technology. This resolution was, as Weber pointedly argues, the bureaucratization of society: increasing dependence on rational administration and on highly organized and impersonal bureaus and corporation in all areas of life (Weber 1947).

Bureaucracy, R&D, and Development

The domestication of development, whereby it was equated with the accumulation of industrial capital, ensured that successful technological innovation would be judged in relation to its value over older technologies in achieving certain industrial ends. An "advance" was accomplished when a more profitable, labor- and time-saving way of doing work replaced a less profitable, less efficient way. The international division of labor that prevailed during the era of European industrialization permitted these societies to invest heavily in such technological advances and at the same time to achieve real improvements in wages, living standards, and social conditions (for most). The development of natural and human resources was presumably the automatic outcome of technologies which contributed to the accumulation of industrial capital. Weber and academic sociologists after him stressed that human relations were not left to develop on their own accord. They too became factors of production to be organized according to the requirements of industrial progress.

In part as a reaction to Marxism, schools and professions in Europe in the late nineteenth and early twentieth centuries chose to achieve an "unalienated" relationship between society and machine by excluding older types of social theory as part of the knowledge base of technology. Instead of making the humanization of mechanical technology a central program, early industrial society, through its state and corporate administrative structures, concentrated on the mechanization of human relations. The bureaucratic society of which Weber spoke is one in which social organization is increasingly geared to the key technologies and technological R&D systems. It is a society in which the satisfaction of traditional but irrational needs is postponed or excluded to ensure that the larger system functions effectively.

When its social and economic organizations are functioning effectively, bureaucratic society has also provided material benefits which, though not always sufficient to compensate for what is sacrificed, can be called "progress." But, as Weber and his contemporaries observed, nothing can guarantee that bureaucratic society is or will ever be organized to serve rational ends in terms of the interests of members of the society or human interest generally. There appear to be serious moral and economic dilemmas associated with bureaucratic society. On one hand, such a society necessarily follows from the requirements of continued technological advance; to argue against it is to argue against progress. On the other hand, technological advance may be exacting a price (in the loss of traditional human relations and values) which it is not always able to repay with material prosperity.

Europe's material advances after the industrial revolution were often accompanied by the impoverishment of its own traditional and/or lower classes as well as that of the populations in the colonies, and by a growing sense of powerlessness and insecurity among all classes (Mannheim 1954,pt.4,viii). Bureaucratic society was thus always less than entirely rational in economic and psychological ways. In addition, in times of economic crisis or when bureaucracy is combined with the "command economy" (Horowitz 1972:205ff) type of political structure prevalent in contemporary Eastern Europe, benefits for most members of bureaucratic society are slight.[11] Whatever scarce economic and political resources are available, under conditions of economic crisis or of the command economy, are required to ensure political stability (typically, it is argued that stability is necessary for recovery or continued progress). Under such conditions, bureaucratic society is irrational except in the most narrowly functional sense that things work in an orderly fashion.

Economic interests and policies in Europe supported the innovation of specific types of industrial technologies, i.e., those which operate at scales and with labor/capital and labor/nonhuman-energy ratios most conducive to capital accumulation. These technologies were designed by men who worked with the book of natural science in one hand and the double-entry accounts book in the other, and with social science nowhere in sight. Everywhere the machine intruded into traditional social relations: restructuring the labor force and reorienting the laborer to his work. It affected his family life, religious views and participation, work/leisure rhythms, and the entire concept of sociability (Kranzberg and Gies 1975), yet these factors were not considered aspects of technology!

Both bureaucracy's tendency to promote order at the expense of progress and the lack of social scientific orientation in technological activity were causes of distress to Weber and other observers of European society during the late nineteenth century. Unwilling or unable merely to describe and analyze, Weber treated these as moral concerns. Here is a source of ambiguity in early academic sociology about the use of a pure physics model. Because they could not accept the rationality of bureaucratic society at face value or the low priority assigned to social and cultural factors in technology, Weber and many who have been influenced by him have found themselves simultaneously accepting and at odds with prevailing assumptions of the industrial order.

New industrial technologies brought about a general reorganization so that society as a whole grew to resemble a gigantic R&D system in which every institution and occupation became a specialized servant of the perpetual accumulation of technological advances. The traditionalist side of Weberian sociology — a strain matched by traditionalism in the work of Durkheim, Ferdinand Tönnies (1887-1963), and even Marx — laments the instrusion of the impersonal, efficient, and universal into the sacred realms of home, trade, and faith. Its liberal side is prepared to accept bureaucratization as long as the benefits, such as improvement in the living standard of the population, accrue. But when faced with the real prospect that development under an industrial bureaucratic order can be uneven, asymmetric, or — in bad times — reversed, Weberian sociology becomes critical of "the new world of rationalized efficiency [which] has turned into a monster that threatens to dehumanize its creators" (Coser 1977:232).

The alternative — for Weber, to the extent that he discussed one, and for many later sociologists — could not be a Marxist or utopian socialist society; for even with the elimination of capitalism or with other specifically economic changes, bureaucratization remained inexorable. The appropriate solution to the excesses and irrationalities of bureaucratic society could not be the removal of the rationalizing tendencies of the technologies which induce them; nor would a movement from private to collective ownership necessarily shake the foundations of the bureaucratic order. For Weber, bureaucracy transcends property norms, class interests, and ideological differences. These too — norms, interests, and ideologies — are formal (*Wert,* in German) matters with which bureaucracy cannot be concerned. Forsaking both reactionary and radical critiques of technology's main social effects in industrialized Europe, the Weberian palliative was to seek to establish an environment in which the social impact of technological innovation is moderated by cultural and leisure pursuits and, if necessary, by psychotherapy (as if cultural and leisure pursuits and psychotherapy were somehow immune from bureaucratization).

The Impact of Weber's System

Rather than develop the critical side of his theory of technology (in the face of the apparent inevitability of bureaucratization), Weber chose to elaborate on other issues, such as the economic, religious, stratification, and other institutional dimensions of bureaucratic society, and the methodological implications of attempting to study such a society scientifically. With this work, Weber created what is widely recognized as the first systematic sociology of the academic era (Gerth and Mills 1946, Introduction; Weber 1947). While technology is a central component in this system, in large part because it was so central for Marx and the other founders, the analysis is too comprehensive to be a sociology of technology. The focus on technology is deflected because of the dominance of bureaucracy. Weber strongly endorsed the attempt to humanize the social environment to mitigate bureaucracy's effects, but he did not approach the mitigation problem from the other direction: to humanize technology by including social knowledge as an element in its design and aims. As a result, a certain type of technology-centered perspective originated by Smith, Saint-Simon, and Marx was neglected in academic sociology in proportion to the significant influence of Weber's systematization.[12]

Weberian sociology is both an extension and a critique of preacademic social science. It is infused with concern about the ways in which technological change is intertwined with social change, and how these in turn affect development. Like his predecessors, Weber approached the study of technology and society from an interdisciplinary point of view. His methods and sources included elements of economics, political science, law, history, literature, comparative religion, and other fields. He left to subsequent generations of sociologists a set of key questions concerning method, the validity of social knowledge, and the proper political role for social scientists. While most of these questions have not yet been satisfactorily resolved, the debate which their attempted solution has generated continues to influence social scientific research and theory. Weber's work gave impetus to and provided the framework for much later academic sociological research on technology.

The point at which the limitations of Weber's approach are most evident is in his program (or lack thereof) of an applied social science. Weber treated the matters of taking sides and making policy in an elusive way (e.g., Gouldner 1962; Riley 1974). In his personal politics, he had both a conservative and later a social-democratic orientation. But nowhere in his work is a clear connection between conservative or social democratic policy and his theories of bureaucracy, technology, class, etc. Instead, though Weber had a strong inclination toward political activism (Bendix 1960), his sociological research on technology was relatively insulated from his political life (a characteristic of many subsequent academic sociologists).

Technology, like stratification and religion, was at most to be treated as one phenomenon among many in bureaucratic society: to be studied, to be accurately related to its true causes and effects, and viewed as a variable factor in history and across cultures. This type of treatment continues to be important in social science; it defines the distinctly sociological way of looking at technology. At the same time, it allows the questions of technology's impact to remain academic: technology is viewed as an object but not as a vehicle in which the "wisdom" of social science, as opposed to the prevailing social "ignorance," might be incorporated.[13]

This is partly due to Weber's pessimism about the ultimately antihuman character of technology. At the turn of the twentieth century, technology was understood as both a cause and an effect of bureaucracy's dehumanizing tendencies. The process of technological innovation appeared to require bureaucratization, the technologies themselves were adjusted to it, and their impact reinforced it. It seemed impossible to reorient technology and the innovation process so that it did not encourage bureaucratization or, beyond this, so that the growth of bureaucracy might even be slowed or reversed with technology. Under these circumstances, it makes little sense to consider "reforming" technology with social science.

In a historical and comparative perspective, these are very special and variable conditions. It is now known that technology, in itself, does not dehumanize in the way that Weber felt was inevitable, nor are the applied human sciences necessarily irrelevant to the design and innovation of technology. This is obvious, for example, in recent work in solar and "appropriate" technologies in the Third World (Bota, Weinstein, and Walton 1980). In the specific context of Europe's industrial revolution, when technological innovation was equated with development which in turn was identified with accumulation, the effects that Weber abhorred were a distinct possibility. Perhaps, as he believed, they were even impossible to withstand. But such pessimism as Weber's says more about the inapplicability of a certain type of social science in a certain type of society than about the futility of attempting to humanize technology.

While Weber distinguished between capitalist and bureaucratic society, he did accept as valid the general equation between development and growth in capital and capital-intensive technologies. This premise was central to the organization of production (and increasingly to other areas of social life) in industrializing society, yet it is false. It ignores the fact that such growth in the R&D sector requires a traditionally extensive manual labor supply and raw material inputs from somewhere, some underdevelopment in other parts of the world. Weber studied the societies of China, India, and the Mid-East in great detail, but he did not discuss China or India's role, or that of other non-European nations, as a factor in Europe's development equation: a factor which we now know to have been as important as — if not more important than — the innovation of energy-using/labor-saving technologies.

Weber's approach to technology fails to stress the earlier, broader, nondomesticated definition of development. The optimistic Enlightenment outlook — even the ironic optimism of Smith or the long-run optimism of Marx — concerning technology's impact is also absent in this approach. Also, we do not get a sense of the profound international interdependence that is now known to have characterized the era of European industrial development. Nor, finally, can we distinguish a role for the application of sociology to technology, as opposed to the sociological study of technology.

In these respects, the Weberian approach is a product of its time, the culture and structure of its society, and of the academic milieu in which the approach was formulated. Weber's influence in academic sociology is great. It has also contributed to the general subordination of an explicit technology-centered orientation to other interests. In combination with the tendency in industrializing Europe toward specialization of professional activities, the delegitimation of Marxism, and the drive to separate the non-social from the social sciences in the design of technology, the Weberian approach has served to keep the sociological focus on technology a secondary or derivative one in relation to other themes such as social organization and methodology. As the social and technical fields are discovering today, this approach, though it may have been appropriate for the academic era, is now more of an impediment than an aid to understanding.

Technology and Early-European Academic Sociology

The institutionalization of disciplinary sociology in the universities of Europe contributed to a general estrangement between social science and institutionalized technology. The efforts of early sociology's proponents helped the discipline achieve a certain degree of acceptance as an academic field. But the fact that it was so specialized and segregated from the other social sciences, its ambiguity concerning objectivity, and — especially as the academic era continued — the ever-present pessimism about technology's inevitable effects on human relations, made it difficult for European sociologists to communicate about or share in the general societal commitment to development via the innovation of capital-intensive technologies.

Interest in the technological dimensions of modern society was implicit during this era. But it persisted only as a marginal interdisciplinary orientation or was identified with an activist and thus less than entirely legitimate type of academic activity. The revolutionary programs of amateur social scientists inspired the creation of academic sociology and kept other than pure science models viable. As the discipline evolved, these programs were not enacted as integral to research or teaching. Thus the possibilities for an applied sociology to aid in social reorganization were not well explored.

As we seek to trace the effects of the early academic era on later developments in social science and sociology, this mixed heritage will again be evident. The heterogeneity, the central and pure-science model, the more peripheral but still influential moralistic and applied orientations, a generally secondary and objective focus on technology, and related characteristics of contemporary sociology all have roots in the events and observations of Weber and his era. As such, they are major sources of the differences between academic sociology and institutionalized technology. But, as will be discussed, in this heritage too are resources for helping to resolve these differences

Notes

1. Marshall McLuhan (1964) and especially Harold Innes (1951) have forcefully brought this view back into fashion. Significantly, as we discuss in relation to the Chicago School, it is the study of communications that most impresses us with the deeply human character of technology. McLuhan (1964:317) refers to Innes' research as "pioneer work in exploring the social consequences of the extensions of man."
2. "Reflexivity," the ability to comprehend one's motives, thought, and behavior as social phenomena, is a deeply rooted component of social science (Mills 1963, Sec.4; Merton 1949, chs. 10, 11; Wax 1965; Gouldner 1976, pt. 3). In part because it is a deeply humbling (humanizing) orientation, it may be the most potent political and intellectual tool in the sociologist's "methodological" repertoire.
3. For a succinct statement of this thesis, see Baran (1957).
4. David Noble's (1977) recent study discusses economic and other factors that shaped technology, especially in the United States.
5. Scientism, which had been recognized as a potentially dangerous doctrine in the seventeenth century, could now be viewed as one ideology among many (Ben-David 1971:126-28).
6. As Ben-David (1971, ch. 7) notes, it is significant that the German model came to dominate Europe (and the United States) despite the fact that England and France had begun new science and technology centers earlier and were, by all accounts, more developed countries. As he and others have suggested, this may be due to the fact that a feudal institution, such as Germany's educational system, best satisfied the need for the patronage arrange-

ment through which amateur science thrived. In any case, "by the middle of the nineteenth century, practically all scientists in Europe were teachers or students, and they worked more and more in groups consisting of a master and disciples" (108).
7. The struggles of Weber in Germany, Durkheim in France, and Herbert Spencer in England for the institutionalization of sociology are significant in many ways; but of special interest are their goals in relation to the professionalization of economics and psychology. The accounts of Coser (1977), Zeitlin (1968), Timasheff (1967), and Gouldner (Durkheim 1958) discuss sociology's battle for legitimacy vis à vis the more established social sciences. Perhaps nowhere is this tension more evident than in Durkheim's *Rules of Sociological Method* (1964). This book is ostensibly about method, but it is also a polemic in favor of sociology. In it, Durkheim argues that social facts require distinctive social explanations and he criticizes contemporary evolutionists, including the British sociologist Spencer, for their lack of attention to this specifically social realm.
8. The question of what determines a (proper) public arose in the work of John Dewey and the Chicago School and later of C. Wright Mills. One concern is whether or not public issues should determine problem selection in sociology (e.g., Mills 1943; Raushenbush 1979).
9. Both Durkheim and Weber began refereed sociological journals around the turn of the century: Durkheim's *Anée Sociologique* in 1897; and *Archiv fur Sozialwissenschaft und Socialpolitik*, of which Weber was the founding associate editor, in 1904 (Tiryakian 1971:429).
10. Sources used in this section include Gerth and Mills (1946), Weber (1947, 1958a,b), Bendix (1960), Coser (1977), and Eldridge (1980). The vast primary and secondary literature on Weber attests to his continuing influence as a principal founder of academic social science.
11. In the command economy, continued development is categorically equated with the continuation of the rule of the party or leader as the embodiment of the true way. In a democratic state, the appearance of separation of polity and economy and the priority of political ends are maintained even when state and industry must cooperate in ensuring their continued hegemony in the face of scarcity.
12. Part of the reason for this lies beyond the logic of Weber's system and in the general ethos of the early industrial era. Sociology did not concern itself with technology beyond a certain limited — albeit important — degree, and technology did not concern itself with sociology, in part because of the specific approach to R&D then prevalant: that the technician must know and obey the laws of nature and economic necessity but is exempt from having to consider the social norms which might conflict with technological advance. These were treated either as superstition and irrational conservatism deserving to be eliminated, or as capable of being compensated for by material prosperity. So long as bureaucracy continued to bring prosperity, other social factors need not be taken into account.
13. Other directions are possible. Here one must consider the human relations in industry or human factors engineering perspectives as they have developed. These approaches, though obviously related to the types of technological applications of sociology under discussion here, are meant to be applied at the level of management of enterprises rather than in the technological innovation

process per se. These two types of applications, as Veblen (1919) noted, differ significantly in the degree to which they require the exercise of reflectiveness. The second application, closer to the spirit of amateur sociology of technology though less frequently practiced, requires that the application of social knowledge to technology be formally as well as functionally rational. The first asks only for knowledge that will ensure that management goals (formally rational or not) will be achieved.

3.

Early Academic Approaches in the United States: The Sociology of Technology of Veblen and Ogburn

This chapter discusses some features of academic sociology in the United States between the late nineteenth century and World War II.[1] It contains a survey of the major trends in the discipline as it evolved from a social-gospel-oriented college course at the turn of this century to the highly organized scholarly profession it is today. The focus is on the ways in which U.S. sociology of this period dealt with technology and on the role it conceived for itself in the innovation process. As in earlier chapters, we will also observe how the culture and institutions of technological society have affected the way in which the social scientific study of technology has been conducted.

From its beginnings sociology in the United States was concerned with the impact on social relations and morality of economic growth, automation, and urbanization. At the time of the formal founding of the profession during the 1890-1910 decades, sociologists and the country as a whole were very much involved in progressive social reform. It was a period of great change, of a high degree of participation in reform activity by the middle classes and professionals, and also of general prosperity. These and other factors gave early U.S. sociology its characteristic concern with public issues, with social problems felt to be of immediate importance and to demand prompt solution. Through the World War I era, the depression, and up to the late 1930s, this orientation prevailed. With it, there developed a substantial interest in the study of the effects of modern technology on family life, education, population growth, political relations, and religion, and in the formulation of policies and programs for applying social science to alter, mitigate, or intensify these effects.

In the face of other concerns, many of these projects received less attention than was warranted, and they were left incomplete or were unsatisfactory in other ways. This type of sociology also had shortcomings that reflected the academic settings in which it developed. Yet this work kept open a line of inquiry and a definition of the purpose of social science that had originated with Smith, Saint-Simon, and Marx, and that became more significant as the academic era drew to a close. Despite its limitations, a large share of the work in the discipline during its formative years in the United States was directed to understanding industrialization and introducing (religiously inspired) social reforms. In the work of the influential Chicago School of the 1920s and 30s, these interests were refined and applied extensively to many urban, ethnic, and political problems. During the years preceding World War II, the social reform and Chicago School orientations were criticized and eventually replaced or combined with several new approaches, theories, and research specialties. The general effect was that the discipline moved away from whatever serious applied interest in technology it had.

One source of this change was an increasing commitment to a pure-physics model of social science. In place of the older styles, the new approaches selected problems and research methods on the basis of criteria such as testability, quantifiability, and statistical reliability. Since the late 1930s, U.S. sociology has been distinguished as much — if not more — by this empiricism as by its social problems orientation. Since that time these two styles have often come into conflict. Some U.S. sociologists continue to practice a type of "macrojournalism," complete with in-depth interviewing, a preference for studying current events, and a focus on public opinion, while others chose to concentrate on the perfection of the techniques of empirical social research. While in retrospect it is clear that these differences were often exaggerated for political reasons, at various points it appeared that a choice had to be made by individual sociologists and the profession between relevance to the concerns of contemporary society and faithfulness to the canons of scientific method. During much of this century, these differences served to divide U.S. sociology and to divert the energies of its practitioners.

Another factor that altered sociology's treatment of technology was the growing influence of Max Weber. Long before Weber became as important as he now is, the theories of Herbert Spencer were a major force in U.S. sociology. In contrast to Germany and other parts of Europe, the United States of the late nineteenth century was not yet prepared to accept as inevitable the "icy cold polar nights ahead" that Weber's theories forecast. Men like William Graham Sumner (1840-1910), the leading U.S. disciple of Spencer, saw sociology as the field that studies and promotes the causes of progress. Spencer's social evolutionary views, his program for academic sociology, and his specific observations on industrial progress articulated well with these values; and Sumner and his colleagues exploited the situation effectively in bringing Spencer to the fore. The approaches of Spencer, Sumner, their followers, and critics were dominant in U.S. sociology until well after World War I. Then, through the translation and interpretations of Frank Knight (Weber 1923), Theodore Abel (1929), Edward Shills, Talcott Parsons, and Hans Gerth, Weber was introduced to English-speaking audiences. His work was well received in the universities and particularly in sociology departments in the United States. Eventually, the Weberian focus on bureaucracy, organization, and method came to rule problem selection in the discipline. As a result, a great deal of research and teaching has been devoted to these subjects, and relatively little to a specific sociology of technology or, by the same token, to a type of sociology that can be readily applied to the conduct of technological design and innovation.

The rise of empiricism, the shift from a Spencerian to a Weberian type of theory, and a third factor, increasing specialization, served to create a distance between sociology and fields like engineering and the non-social sciences that were responsible for technological R&D. Because the main issues in U.S. sociology during much of the academic era were internal, it was always difficult for the discipline to contribute meaningfully to such extra academic concerns as engineering design and evaluation; most of sociology's resources were expended in an effort to establish an identity for itself and to legitimize itself as a science rather than provide service and support for technical fields. With the growing dominance of the pure-physics model, sociologists became less interested in attempting to understand something as global, elusive, and application-oriented as technology.

In each period in the history of academic sociology in the United States, important exceptions to this trend can be found. Several individual social scientists and groups attempted to extend the views first suggested during the amateur era in Europe. Three men, one representing pre-Chicago School sociology, one a member of the Chicago School, and one a student at the school, are especially noteworthy because of their work in exploring the implications that Sumner and Spencer's evolutionary theories, Weber's perspective, and other ideas have for the social-scientific study of technology. The last of the three, C. Wright Mills (1919-62), will be discussed in chapter 5; the first two, Thorstein Veblen (1857-1929) and William F. Ogburn (1886-1959) are featured in this chapter. Each of them, perhaps more than any of their respective contemporaries, stressed how values and social relations are systematically affected by — and affect — technological innovation, and how the type of innovation peculiar to U.S. and European industrial society has produced a distinctive R&D order. In these ways, their contributions are especially effective in illustrating the use of social science to clarify the distinction between the conscious shaping of natural and man-made materials, so that industrial profits and the power of ruling groups will be maximized as well as the other things that technology might be and do.

The Rise of Sociology in the United States

The effects of Americanization on academic sociology served both to reinforce and moderate those of institutionalization. In the United States, sociology became more pragmatic and eventually a quantitative field; and it developed a new focus on relatively short-run issues. These tendencies were in marked contrast — and often consciously so — with the speculative and theoretical character of European academic sociology (Mills 1969; Mannheim 1953). In time, these gave the discipline the appearance of a pure science, but especially in the earlier years, they made it susceptible to certain types of application as well.

In the period that many consider the golden age of U.S. sociology, the post-World War I era of the Chicago School, the discipline made some significant inroads, not only in liberal arts colleges but also in scientific and technical curricula and professional activities. Yet this was also a period of ethnic conflict, economic depression, and worker unrest; a time in which fundamental weaknesses in the capital accumulation model of development were evident to U.S. academics and publics.[2] As U.S. political and economic institutions recovered, this same pragmatist and issue-centered approach was subjected to criticism on grounds of being "soft" and/or partisan. The ensuing debate reopened the questions that had surrounded institutionalization of social science in Europe, such as Marx's "Is sociology science or ideology?" and Weber's "Can sociology be value neutral?" This in turn increased the segregation of the discipline from technical fields and reinforced the ambivalence of sociologists about their own role in technological society.

Dominance in academic sociology shifted, as it did in other fields, from Europe to North America prior to World War I (Hughes 1958). The first department, the first American association,[3] and the first American journal of sociology were founded at the University of Chicago between 1895 and 1905. By that time, sociology had already been offered as a college-level course (often in economics departments) in the Northeast, in California, at Johns Hopkins in Baltimore, and a few Midwestern universities.

By the time of its institutionalization in the United States, sociology had already adapted to the concerns of the country, particularly in its commitment to the study of democratic process and ethical issues. Lester Ward, Sumner, Frank Giddings, Albion Small, and E.A. Ross — the first five presidents of the American Sociological Society — though a diverse group, all had interests or personal backgrounds in the social science movement (Bernard and Bernard 1943), populism, progressivism, or the social gospel. The nineteen presidents of the society born before 1880 were "almost without exception fundamentally concerned with ethical issues. . . . Often their reformism was a secular version of the Christian concern with salvation and redemption and was a direct outgrowth of religious antecedents in their personal lives" (Hinkle and Hinkle 1954:3; see also Bramson 1961, ch. 4).[4]

In this cultural climate, the naturalism that had come to dominate academic science in Europe, particularly England, was readily incorporated into U.S. sociology. Sumner and others expressed their reformism through biological metaphors, evolutionary concepts, and outright social Darwinism (Hoffstader 1955). These early sociologists were especially receptive to the way in which Herbert Spencer dealt with the connection between social change and moral progress and with his characterization of industrial society as a higher stage of civilization, although there was considerable criticism of his specific political views (e.g., Spencer 1896, vol. 1, ch. 10; also Coser 1977:89-127). This combination of naturalism, Spencerian evolutionism, and social reform gave the discipline a holistic and moralistic perspective on the study of industry, mechanization, and urban growth. Through the influence of Spencer, his disciples such as Sumner, and his critics such as Small and Veblen, early academic sociology in this country formulated and attempted to apply a developmental orientation in which the field was to serve as a knowledge base for social change. Though not necessarily antitechnology, as moralists their vision of progress differed from the machine- and capital-accumulation developmental models that prevailed among the powerful upper middle classes of the era. There was, in much of their work, a distinct preference for the small town, the small business, and the intimate way of life which industrial growth seemed to be destroying (Mills 1943). For many early U.S. sociologists — with the important exception of Sumner (Schneider 1975:315-18) — the business classes were often blamed for promoting a general deterioration of morals. And as a group, the early sociologists were disdainful of the rising cultural materialism industrialization was producing.

One result of this lack of acceptance by sociologists of the dominant model of development was the low social status of the discipline (Small 1903; Bramson 1961).[5] In comparison to the non-social sciences, the technical fields, and even the better established social sciences such as economics and psychology — which had no concern with criticizing the morality of the times, sociology was often identified as a reformist, "concerned," but impotent fringe of the academic community. That it was accepted and tolerated at all attests to the political acumen of early U.S. sociologists — and perhaps also to the "fringe" character of the University of Chicago where most of the early activities began (Rhoades 1980a). Yet the effect of the field on society as a whole and, with the exception of professional education and social work, on other academic fields was slight (Faris 1945).

Despite these adversities, during the years between 1895 and the end of World War I, sociology departments were established throughout the country. A distinctly U.S. style was developed, and with it the number of sociological publications, subdisciplines, and advanced degrees granted grew. There also emerged in the work of one of the best-known writers of the period, Thorstein Veblen, a conscious and explicit sociology of technology. The outstanding feature of this work, as it was in that of Ogburn and Mills in subsequent generations of U.S. sociology, is the renewed sense of technology as a human, culturally-bound system, and Enlightenment-inspired relativism both sensitive to the vast potential for good in modern technology and critical of the uses to which technology can be put in a less than perfect society.

Veblen's Approach to Technology

The activities, ideas, and even the much-discussed social marginality [6] of Thorstein Veblen illustrate well the ambiguous character of early U.S. sociology.[7] Veblen was strongly influenced by the studies of progressive social evolution of Spencer and Sumner. Yet his arguments were directed against important aspects of evolutionary doctrine and, even more acutely, against the underlying classic economic theories upon which the evolutionism of his teacher, Sumner, was based. Veblen was at root as much a social evolutionist as his predecessors. Yet his profound critiques of "economic science" (Veblen 1919) combined with his continental and — to varying degrees — Marxist [8] orientation often made his formulations the very antithesis of Spencer and Sumner's.

The Engineer versus Entrepreneur

Veblen's revision of evolutionism was based on a difference between the way in which he and other evolutionists viewed technology. Central is his assessment of the "instincts" of the engineer as more advanced than those of the entrepreneur, or as he called his ideal type, the "Captain of Industry." [9] Spencer had elevated the industrialists' "pecuniary" and "predatory" motives to the status of a progressive doctrine, to which Veblen was strongly opposed. He also took issue with Sumner's evaluation of America's business classes as progressive. For Veblen, industrialists were the main source of resistance to technological advance and the principal impediment to the engineers' truly progressive goals: the growth of industry had coincided with the decline of creative social change.

This distinction between technical versus business classes is a radical one. It is a distinction that had been partly obscured by industrial society's effort to equate development with capital accumulation. With the understanding that technological progress and the growth of business do not necessarily coincide, Veblen struck the core of the "problem with" technology in the modern era: that its practice and practitioners had been subordinated to the will of the business enterprise. Veblen was not prepared to believe that the shape and effects of technological innovation must inevitably be tied to the accumulation of capital, even if the effects of such a relationship in prosperous time is good; nor, like Weber, did he believe that technology inevitably contributes to increasing bureaucratization. On the contrary, he was concerned with freeing technology, with elevating it to a position where it would direct changes in industry and society as a whole. Anticipating later social critics such as the Frankfurt School by nearly one-half century, Veblen envisioned a contemporary political struggle for control of technology in which a great deal is at stake. This is a struggle not between workers and owners but between the "Engineers" driven by a creative, scientific instinct and the "Price System" driven by the need to accumulate wealth. Veblen's understanding of technology led him to pose a choice between reason and greed as alternative stimulants to development.

Veblen assigned a prominent political role to technicians and to the principles of science and technology. In the sense that Saint-Simon and other founders of sociology used the word, he was a technocrat. He was a proponent of their ideal of a society ruled by men of reason. But Veblen was not a technocrat in the sense of a supporter of a dehumanized bureaucratic order of robot-people serving The Machine. His interest in the promotion of any specific program that promotes these ideas, such as

Technocracy, Inc., was slight. Veblen's goal was to humanize the process of technological innovation through reforms in academic social science. He hoped that by building an institutional economics that incorporates knowledge of human relations and sentiments, people could see through the abstractions from human relations that characterize classic economics (Seckler 1975). For Veblen, the prevailing economic science amounted to no more than lies that helped business make more money if everyone believed them. He also stressed the need for a complete anthropology, one that would talk about motives, irrational customs, etc.,[10] to replace such a narrow and mechancial "economic science." Veblen hoped to reorient social science with his arguments: contrary to current practice, technological activity should follow the most logical, reasonable, and efficient course rather than the most profitable.

Veblen's progressivism was expressed in this educational program: to separate technological activity and teaching from subordination to narrow economic interests, and to free our understanding of economic interests from stilted and inhuman academic theory. Veblen's observations on technology were sharpened by his knowledge that, while factors such as the profit motive and the drive to accumulate everywhere inform technological activity, they are in no way identical to it. Veblen was a radical social theorist who substituted for Marx's proletariat an elite — the technician who is simultaneously an anthropologist — to struggle for economic power against the captains of industry, in the evolution toward a more developed society. This elite's vision of development is motivated by the "instinct of workmanship," thus it is certainly a workers' elite, rather than the instinct to "emulate," which is the instinct of a leisure class. But it is an educated group close to power — not a lower class, remote from power and invariably having to be "led" by a "vanguard" party. In stressing the elite nature of those who are to make the "next revolution," Veblen provided a theory of revolutionary intellectuals which preceded by decades the findings of current research: that the role of proletarists and other masses in history has been greatly overestimated (e.g., Gurr 1970; Kautsky 1969; Gouldner 1975). It is not the most oppressed groups that initiate successful social revolutions, but the groups educated in the latest ideas who are themselves frustrated and feeling oppressed. And, Veblen reasoned, these would be the engineers of his society.

Throughout his work, Veblen emphasized that social, cultural, and political factors shape technological activity and its effects. His *Theory of the Leisure Class* (1934), by far his most popular and ironic book,[11] profusely illustrates how social customs and economic pressures had perverted technology: how a supposedly rational technological system can be organized to produce waste; how modern industrial technology can be employed not for the benefit of all but as the weapon of a "predatory" class in a competition for booty; and how the establishment of a society-wide R&D system has served not to satisfy human needs, but to supply a decades-long round of potlatches among the ruling classes. In his study of *Imperial Germany* (1939) there is a classic analysis of the comparative impact of technology and how its effects can vary with social conditions. Here too his stress is on the interaction between technological innovation and diffusion, on the one hand, and the cultural and political environment into which innovations diffuse, on the other. The book introduced the concept of the "penalty of taking the lead" and the theory of stage-skipping in the evolutionary process; that is, one nation often pays for developments later incorporated at no cost by another, which allows the second nation to progress faster than the first. These ideas remain cornerstones of contemporary social scientific perspectives on technology transfer.

Critique of Academic Social Science

Another of Veblen's contributions to the sociology of technology that points beyond the academic era is his critique of higher education in the United States. Central to this critique is the contrast he underscores between the rational, humane pursuit of useful knowledge and the activities pursued in its name within the universities of his day. Once again, as in the organization of production generally, he sees the task of generating knowledge to promote development as having been subordinated to competitive interests and to the "predilection for 'practical efficiency' — that is to say, for pecuniary success — prevalent in the American community" (Veblen 1957:164).

Scholarship and genuine creativity — of any but the most "sterile" variety — cannot be tolerated long by universities whose ends are "chiefly . . . notoriety, prestige, advertising in all its branches, and bearings" (Veblen 1957:167). Though the university might be the source of knowledge upon which to base a more rational technological order, the growth of the academic disciplines has caused it to act in support of (and imitate) the general elevation of pecuniary interests above reason. The result is an educational establishment which, in Veblen's view, perpetuates an obsolete and antihuman theory of development; within university departments of economics, sociology, and political science, "the resulting academic staff had become the byword of nugatory intrigue and vacant pedantry" (Veblen 1957:164).

Veblen's commentary on academe was part of his framework for a critical, interdisciplinary sociology, complemented by anthropology, economics, and moral philosophy, in which technology as it is and as it might be has a central role. His critiques of economic theory and his own theory of social stratification challenged prevailing scholarly and societal assumptions concerning development and the role of technological innovation in the development process. His observations on what is today the Third World were limited, based largely upon the ethnographic study of small-scale societies then undertaken by Franz Boas. He was aware of the international roots of modern technology, and he was among the first social scientists to point out how the industrial prosperity of one nation is as often due to the penalty paid by other nations as it is to the genius of the people (or to other "wholly internal" factors). Veblen attempted to integrate these observations with a theory of the role of higher education as an institution that sustains the dependence of potentially creative intellectuals on the requirements of the prevailing R&D system.

Despite Veblen's critical views and his attempt at playing the role, for which Sumner was noted, of *homme de lettres* as editor of the influential *Dial* magazine, he never achieved a complete break with the academic world he so despised. A man before his time in many ways (Dorfman 1973), Veblen was "trapped" with a vision of a postacademic social science in an era in which the major preoccupations of sociology were legitimacy as a discipline and garnering academic honors. From his student days at Carleton, Yale, and Cornell, to his checkered professional career at the universities of Chicago, Stanford, Missouri, and the New School for Social Research, Veblen was uncomfortable to the point of distraction with academic propriety and ritual. Yet each time he lost a job, he turned back to the academy for material support and redemption.[12] "He never could adjust to or even make peace with the genteel culture of American academia, and yet he was sufficiently attracted to it to want to make his mark in it" (Coser 1977:296).

Veblen was marginal to professional, disciplinary sociology as he was to U.S. society in general. For a man with his training, temperament, and unorthodox views, a career as a university professor was perhaps the only possible course. Yet because his views led him to see the inappropriateness of the university system for promoting the serious restructuring of our approach to technology he favored, he could never be happy with such a career. This factor, perhaps more than his Norwegian immigrant background or radical political orientation (inspired as much by the philosophy of Edward Bellamy as by the works of Marx), made it difficult for Veblen to adjust to his own role. He understood the power of technology, and was well acquainted with the irrationalities to which technological thinking and innovation had to respond. But he saw no alternative beyond the university post — because none existed — whereby one might put such understanding into practice. Notwithstanding some faddish attempts to make a political program out of Veblen's theories, his contribution did not transcend a relatively passive academic study of technology. This step could not be taken until long after his death when, at last, a real alternative to academic support for the social scientist with Veblen's type of interests began to emerge.

The Era of the Chicago School

Amidst the growth of academic sociology during the progressive era, the University of Chicago and its style of social gospel remained prominent. Beginning with the work of Charles H. Cooley (1864-1929) at the University of Michigan, Robert E. Park (1864-1944), W.I. Thomas (1863-1947), and W.F. Ogburn (the eighth, fifteenth, seventeenth, and nineteenth presidents of the American Sociological Society, respectively), the Chicago style was transformed into the Chicago School. This school was to dominate U.S. sociology unchallenged until 1933 and as a major current through the end of the academic era. With the emergence of the Chicago School, U.S. sociology became a major international force, which also meant the de-Europeanization of the discipline.[13] Much of the Chicago School approach to sociology came from an infusion of European (non-Spencerian) thought into the United States. The main personal link between the two continents was Robert Park, who studied in Germany and later taught at the University of Chicago (Matthews 1977). From Europe, Park brought the influential ideas of Weber and Georg Simmel (1858-1918), as well as the naturalistic, ecological perspective of the biologist of Ernst Häckel. Park and his colleagues applied and refined these perspectives within a specifically American social and intellectual context. The result was a prolific production of sociological research and researchers, particularly on urban life and communications (due in no small measure to the fact that Park had a background as a journalist). This work quickly dwarfed, at least in quantity, the output of European departments.

In comparison to their social-gospel oriented predecessors, "the United States sociologists of the second generation were often 'unconscious' liberals who were concerned over the 'disorganizing' aspects of American life under the impact of industrialism and urbanism" (Bramson 1971:77). The social and intellectual factors that contributed to this shift from overt progressivism to "unconscious" liberalism are various and even contradictory. But three elements might be pointed to as decisive (Mills 1969). One of these factors is the influence of Häckel's ecological perspective. Sociologists in the United States were receptive to such a biological model because naturalism had dominated during the pre-Chicago School generation. With this stress on ecology, the concept of system and the idea that the social order could be viewed, if not as an organism, then as a biological system, assumed importance in American social science. The ecological model introduced the notions of equilibrium/disequilibrium, organization, dominance, and succession into the sociological vocabulary (Hawley 1968; Berry and Kasarda 1976, ch. 1).

A second factor in sociology's shift from overt progressivism was the impact of social conditions as they were experienced in rapidly growing cities such as Chicago and through media such as Chicago newspapers. Here serious crises and problems in the systematic aspects of social relations were evident: social disorganization and diversity, maladjustment and personal disequilibrium. Cities like Chicago had swelled to two or three times their prewar numbers and were largely populated by diverse ethnic groups now forced to live together and communicate with one another. In connection with their ecological orientation, Chicago School sociologists took their cues for problem formulation from the social problems that lay before them. They sought to establish logical and social order amidst urban chaos: in general, through a focus on the city as a human "biosphere" and as a vast, complex communication network; in particular, the Chicago School was "intent on discovering the laws of social life in order to be able to predict and ultimately control the problems of urban existence" (Bramson 1971:75).

The third major factor that shaped the Chicago School approach was its pragmatist background — through the influence of William James and, at Chicago, John Dewey. This philosophical orientation led sociologists to the pursuit of social facts (in contrast to European interest in theory) in support of their observations on disorganization and as a basis for their practical suggestions for corrective action. The second generation of U.S. sociology was not content with what it considered speculative moralizing about the effects of industrialization. Its members required documentation in the laboratory of the real world. Which "real" world? Not coincidentally, this was the real world of the tough-minded, "cash value for ideas"-oriented journalist. In the Chicago School of sociology, the pragmatist view that correct thinking is hypothesis testing was made the foundation of a social science based on in-depth journalism. Chicago School sociologists "hit the streets," gathering detailed information about the lives, neighborhoods, and values of "newsmakers": the poor, the immigrant, the criminal, the superrich, the political bosses, the "man on the street," etc. (R. Park 1969).

The Chicago School's ecological orientation, its attempt to understand

and resolve the crises of urban growth, and its pragmatism linked with journalism gave U.S. sociology a rich and distinctive perspective on the problems of industrial society. in the work of Louis Wirth, Park, and the others, the domesticated and more limited concepts of development were criticized. Partly as a response to the human tragedies the depression had produced, these men promoted their own program for an applied sociology. Their aim was to aid people in adjusting to the rapid and baffling changes that had come in the aftermath of World War I. The work of W.F. Ogburn stands out in this regard and, by way of illustrating these points, will be discussed in some detail.

With the exception of Robert Redfield and some of his students, members of the Chicago School were generally unconcerned with the development process from a comparative international point of view. To this extent, they were not as attentive as they might have been to the special conditions that made it possible for the United States and industrialized countries of Europe to pursue development through machine and capital accumulation. In addition, though the Chicago School sociologists were aware of the liberating possibilities of technology and urban life, they more often expressed distress at the disruption they viewed. Rather than remaining open to a descriptive approach to technology's impact, they were more inclined to condemn it in favor of an organic, folk-oriented, communal ideal (Wirth 1969).[14]

Nevertheless, the Chicago School was often in touch with broader development issues, and thus sympathetic to technology studies in a preacademic sense. Even when this sympathy was expressed through "muckraking" journalism, the relevant social problems of the Chicago School were close in concept to those of the founders of social science. Park, his colleagues E.A. Burgess, Wirth, and others consciously attempted to apply these principles within their pragmatist research program. They were no longer in the position of fighting for bare academic recognition for sociology, as were Sumner and his contemporaries, but they did respond to the mission of showing that sociology was scientific and relevant to public concerns. They stood "midway between the sociology of the library and learned meditation on the one hand and the increasingly circumspect research techniques of the present day on the other hand" (Shills, quoted by Bramson 1971:74).

W.F. Ogburn's Focus on Technology

William Fielding Ogburn was sensitive to these problems and to many of the same issues that concerned Veblen as well. Trained in economics, political science, and history, he too was distressed by the way in which his society remained ruled by those reluctant to understand and/or come to terms with its own technology. As an educator, Ogburn, like Veblen, believed that the academic fields had failed to fulfill their potentially valuable roles in preparing people to live fruitful and creative lives in technological society. Motivated by these interests, he sought to contribute to the humanization of technical fields and also incorporate the study of technology into disciplinary sociology.

The Study of Technology and Applied Sociology

In the foreword to his book on the social antecedents and consequences of aeronautic technology, Ogburn (1946:iii) observes flatly that

> social science has profited less from the study of technology than it has from the study of geography, biology, or psychology, for few social scientists have written on the influence of technology. However, society has been revolutionized several times by technology, for most modern social changes are precipitated by mechanical invention and scientific discovery.

Ogburn's (1957:9) assessment of the technical fields points to the

> teachers in technological schools [who] instruct their students in how to make this and construct that; and though these fabrications are to be used by society and have an effect upon social life, such matters appear to be of no concern to technologists. It is as if there were a great wall separating technology and sociology.

While he always displayed a keen awareness of the moral and political implications of this "great wall," his main orientation — unlike Veblen — was not as social critic but as an architect determined to turn the wall into a two-way conduit. Through his own work and that of his students, such as S.C. Gilfillan, Ogburn sought to explore all facets of the relationship between social change and technology: from the premise that "technological work does not take place in a vacuum but generally in response to a social demand. So the origin is sociological" (1957:9) to the impact of technology on population, the family, cities, religion, health, recreation, crime, education, transportation, public administration, international relations, and international politics (these are categories employed in Ogburn 1946:313ff, 1934a:141ff). By means of such comprehensive exploration, Ogburn contributed perhaps more than any other academic sociologist to the reconciliation of sociology and technology.

Ogburn's interest ranged widely over traditional sociological topics, but he always took pains to relate problems of ethnicity, stratification, migration, etc., to technological factors, especially the innovation and diffusion of new technologies. Even his once widely used introductory text (Ogburn and Nimkoff 1940) maintains a persistent focus on the ways in which social forces, groups, and institutions affect and are affected by inventions, design, and discovery. Although he did not always succeed in providing adequate and systematic explanations of specific sociotechnical relationships — most of his best analyses are sage observations and insights — he contributed much to the reinterpretation of the social science tradition as technology-centered. In this regard he was also in accord with Louis Wirth and his other Chicago School colleagues (and opposed to the views of later spokesmen of pure sociology) in the belief that sociology was meant to be applied to improving man's lot. He believed that the study of sociology should help people "to get a balanced perspective on social life and social issues. If they know how the social order came to be what it is, they will be better prepared to direct the social changes ahead" (*Current Biography* 1955:460).

Ogburn was born in 1886 in Butler, Georgia. He was on the faculty at Columbia University between 1919 and 1927, and then at the University of Chicago for the majority of his productive years (1927-51). Like Veblen, who was at Chicago from 1892 to 1906, he was influenced by the dominant evolutionary perspectives of Sumner and Spencer and by the pragmatist philosophy of James and Dewey. Ogburn's analysis, too, is rooted in an evolutionary anthropology in which a concern for social progress is a dominant theme.[15] Intervening between Veblen's era and Ogburn's is the earlier work of Park, Burgess, W.I. Thomas, and Wirth. Though a relative latecomer, Ogburn did identify and benefit from the interaction with a body of colleagues who self-consciously constituted a school. This placed him much closer than Veblen to an academic mainstream (and exposed him more to the benefits and costs associated with such a status). But because of the type of academic sociology which his Chicago School colleagues were doing, it also moderated his progressive evolutionary orientation in favor of a more empirical and issue-centered approach.

The Chicago School's characteristic interest in the city, communication and community, democratic values, and publicly relevant social problems, as well as its journalistic style of gathering and reporting data, are easily discerned in Ogburn's work. Professionally as well, Ogburn shared the orientation of his colleagues. A strong supporter of the *American Journal of Sociology* and the American Sociological Society, he followed Park and W.I. Thomas and preceded E.W. Burgess and Louis Wirth as president of the society (in 1929). Because of his position of intellectual and political leadership within academic sociology, he did much to foster the development of a sociology of technology as a legitimate specialty, although one could hardly say that the specialty has achieved the prominence Ogburn felt it deserved.

Ogburn's sociological approach to technology was inspired by many of the same intellectual and social currents that influenced Veblen. Yet perhaps by virtue of his more established personal background and affiliation with the academy, and also because of the nature of the academic environment in which he worked, he was in some respects able to advance the field beyond the point to which Veblen had taken it. In particular, not only did Ogburn seek to understand and counter the irrational forces that oppose technological innovation or that narrow and distort its effects, but he also sought to understand in detail the broad range of social effects (and sources) associated with specific technologies. In his studies of the railroad, communications, the automobile, and the airplane, Ogburn performed the first recognizable social impact assessments. These are not narrow inventories or mere lists of possible effects. They are analyses of the social, environmental, economic, and political effects of these innovations based on often sound (and often very speculative) systematic interdisciplinary sociological theory.

In his popular work, as well, Ogburn had one clear and simple message: society must learn better mutual adjustment between itself and technology. Perhaps nowhere are these views better illustrated than in the little book *You and Machines* which Ogburn (1934b:5) wrote for the American Council on Education for use in adult education classes and workers' study groups. In a section entitled "Is the Machine an Enemy or a Friend?" he concludes:

> The machine does us both harm and good. It saves our lives, but maims our limbs. It brings us comforts, but causes us unhappiness. The problem before the human race is to see if we can increase our friendly relations with the machine and diminish its hostility.

Who is the real "enemy"? For Ogburn, like Veblen, the enemy was antiprogressive groups and ideologies which make it appear as if technology were at fault in order to mask real interests. Although Ogburn had some thoughts on the conflict between what Veblen called the instinct of workmanship and the instinct for emulation, these are not his terms. Betraying the not-always underlying liberalism in the Chicago School world view, his attack was directed at conservatism:

> People who do not want to change are called conservatives. They want to bring back the Good Old Days. But we know that sort of thing is foolish. We can't bring back the Good Old Days, no matter how much the old men want them. Some of the conservative old men try to pass laws to stop change. One might as well brush back the tides with a broom. Passing laws will never do it. If they want to stop change, they will have to break up the machine or, better still, poison all inventors (Ogburn 1934a:52).

Ogburn's contributions to the sociological approach to technology, though perhaps not as politically astute as those of Veblen, were more firmly anchored in pragmatism (both in its philosophical and sociological variants), and thus more susceptible to application. In this, as well as in the other sense noted, Ogburn was far less a victim of alienation between theory and praxis, far less a victim of what Veblen considered a sterile, elitist academic establishment. In addition, where Veblen sought to build whatever nonacademic career he could as a freelance intellectual (having managed to hold his job with the Food Administration only five months; Coser 1977:285), Ogburn had extensive "opportunity to apply many of the findings of sociology when he served in various Federal posts. He was examiner and head of the Cost of Living Department of the National War Labor Board in 1918 and 1919, and director of research for the President's Research Committee on Social Trends from 1930 to 1933" (*Current Biography* 1955:461). [16] In several respects therefore, but especially in nonacademic experience, Ogburn came closer than most of his contemporaries or academic predecessors to playing — or at least to delineating — the role of postacademic social scientist. For this reason alone he was a direct precursor and theorist of current activity in this field.

Assessment of Ogburn's Perspective

Despite the evident advances made in academic sociology of technology between Veblen's era and Ogburn's (not to discount some equally evident losses in political acumen and literary style), Ogburn remained an academic sociologist. While he had a genuine concern for developing an applied sociology program, this was as much a function of the general position of the Chicago School as the logical outcome of the applied, sociological study of technology per se. His personal and programmatic interests in application were ultimately subordinated to traditional academic matters: publishing scholarly books and articles, being active in professional organizations, and advancing the discipline. Because its major audience and purpose were academic, Ogburn's sociology of technology remains distorted by a selective focus on those aspects of the society/technology interface whose discussion and analysis would satisfy these professional audiences and purposes. In drawing on Ogburn's important work, the postacademic social scientist must also attend to its limitations resulting from the setting in which it was formulated.

An important case in point is Ogburn's tendency to assign an overly independent role to technological factors in social change. Throughout his work, but especially in his earlier, more evolutionary writings, he identifies and attempts to verify various instances in which technology appears to have definite, unmediated effects on social organization and nonmaterial culture. "Technology brings social changes . . . inventions such as the railroad, the automobile, or the airplane make certain changes in customs and social institutions" (Ogburn 1946:3). While he was aware (especially as his work matured) that technology is but one factor in a complex set of causes and social change, his work often has a technologically determinist bias.

Louis Schneider's comments call attention to this tendency:

> . . .much more than technology is involved. New technology comes into an already existing social — and technological — framework. Obviously, the advent of the automobile can reduce railroad traffic only if railroad traffic already exists to be reduced. Ogburn has a tendency to write as if technology had direct effects, practically without human choices, where this kind of view was quite unjustified. The *automobile* does not implacably dictate that it shall be used as a major instrument to work out a new kind of vacation. The *automobile* never increased petting or sexual intercourse. It merely offered itself as an instrument which *might* be used to carry on sexual activities otherwise inconvenient or impossible to arrange. To use it in such ways is clearly a human choice. What we really often mean by "the social consequences of technology" is "the social consequences of technology that men have chosen to use in particular ways." The knife can heal as well as kill. The explosive can remove an undesirable rock formation as well as blow up people. The airplane can drop packages of food as well as lethal bombs. We make the choices, and in this whole sphere there is much tragic irony about the choices we make (1975:50, fn. 25).

Was Ogburn a technocrat, a technological determinist, or both? On the first point, if by "technocrat" we mean a theorist who favors an order in which technology and "the instinct of workmanship" rule over pecuniary interests, then he, like Veblen, surely was a technocrat. Ogburn was as opposed to the rule of the industrial and business classes as Veblen. His "chosen class" of the (perhaps mythical) past was also a worker class; not the proletariat of Marx, but the elite "laborer[s] who owned the tools they worked with." It is also in this context that Ogburn argued for more power for the technical classes — the modern laborer, alienated from his tools because "now capital owns the machines." "And so," relates Ogburn, "machines became the property of people who had money to invest . . . [but the owner] doesn't pay any attention to labor conditions; he has never seen the people who are working for him. He is chiefly interested in getting dividends, his share of the profits earned" (1934b:18-19). On the second point, if we are referring to a theorist who believes that the effects of technological activity and innovation are beyond the will of human actors, then unlike, e.g., Jacques

Ellul, Ogburn was not a technological determinist. Because he was engaged in the struggle for his version of technocracy as an academic, it was an unfortunate but nevertheless necessary device to exaggerate the importance of technology. This rhetorical exaggeration, unlike the views of contemporary technological determinists, was not rooted in a deep-seated, quasi-religious Luddism; it was part of a partisan political struggle for the legitimacy of technology studies.

Academic sociology of Ogburn's (and every other) era has been involved in a twofold search for the truth about social relations and for legitimacy as a science. The classic response to the second priority, to become scientistic in pursuing the first, was a common feature of Ogburn's contemporaries (and some of his predecessors and successors as well). In overstating the independence of technology as the independent variable and in oversimplifying its effect on social life, Ogburn was conceding to the pressures of respectability for his discipline and his specialty. While this was an important battle to be fought, and we should be grateful for Ogburn's efforts, it is not the battle of the postacademic social scientist. Under the influence of its era, Ogburn's sociology of technology was "for" the promotion of itself, the Chicago School, sociology in general, and, not least, William Fielding Ogburn. As the academic era comes to an end (coincident with organized sociology's victory in its fight for academic legitimacy), the narrowly-technocratic, technologically determinist, or otherwise distorted tendencies to which the sociology of technology may be prone is less useful and less necessary.

Positivist Challenge to the Chicago School

With the growing influence of the Chicago School through the 1930s also came increasing pressure for an even more empirical approach.[17] The result was a gradual loss of the dominant focus on public issues and disorganization and its replacement by a methodology-propelled, neutralist orientation. In place of the rule of the Chicago School there arose in the mid-1930s a series of factionalizations within the discipline that persisted, in one form or another, throughout the academic era. U.S. sociology became less capable of pursuing an unambiguous relationship with the technical fields; when it did speak as a profession, it increasingly spoke the language of an uncommitted, highly specialized empirical discipline.

The principal source of this evolution away from the Chicago School approach in U.S. sociology "does not derive from a mistaken development within the science itself" (Frankfurt Institute 1972:127). Rather, it arose from "the nature of [its] subject matter and the position assigned to sociology" in U.S. society during the late academic era. The Chicago School was prominent during an era in which to be empirical and responsive to social currents could, and often did, lead to broad, application-oriented sociology. At that time the problems of development were themselves open to public debate. But between the late 1930s and the World War II era, the "development = accumulation" formula had regained its validity. The classic Chicago School focus on disorganization appeared to be impertinent, radical, and unscientific. As had occurred in Europe one-half century earlier, this counterposition did not reign unchallenged. It took its place alongside and often in conflict with, the Chicago School and other perspectives. This recreated within U.S. sociology of the later academic era the schisms over objectivity, phenomenology, and Marxism that had surfaced decades earlier in Europe.

Struggle for Legitimacy

While for some purposes it is useful to divide the history of U.S. academic sociology into progressive, Chicago School, and positivist eras, it is equally fruitful to note the continuous and persistent features of the field that have endured from Marx's time until our own. These include a characteristic two-front battle which academic sociologists in the United States have fought: one against the forces that stand in the way of attaining social knowledge — the battle for truth; the other against the forces that stand in the way of legitimacy for the sociological enterprise — the battle for status. It would be an overstatement and an oversimplification to suggest that to win the second battle the first was consistently forfeited by academic sociologists [18] Yet the tradeoffs (indeed, the dialectic) between the two have, in a cumulative sense, played a major role in shaping the field, especially in relation to technology.

It would be wrong to see as epochal the shifts in orientation between the founders, the Chicago School, and the post- (and anti-) Chicago School of U.S. sociology. Similarly, one must be wary of overstating the differences — though there are many important ones — between U.S. and European (or specifically national) sociologies during the academic era. Beyond the continual specialization and increasing heterogeneity among and within these specific styles, the discipline was united in its twofold search for scientific knowledge and "a place" in academe. Every sociologist so-far mentioned was involved in the legitimacy struggle, [19] used his own research to illustrate and argue for a legitimate social science. The bearing of this fact on the question of sociology's relation to technology is crucial. The accumulated product of academic sociology, in its theoretical or empiricist versions, its conservative, progressive, or "unconscious liberal" manifestations, is, after all, a broad, encompassing ideology. It is an interest-serving account that distorts by exclusion, by a limiting of focus. It is a therapy-centered ideology (which paradoxically has often used empiricism and positivism — evident descriptive, antiactivist systems — to legitimize itself): the principal cure being "grant sociology a status equivalent to other, more 'scientific' fields." Could it be otherwise? Could a sociology which does not distort, by exclusion or in other ways, exist under any conditions? These questions must be answered in several steps, of which only one relates directly to the fate of academic sociology per se. It seems apparent that under the conditions that prevailed during the academic era it could not have been otherwise.

The fission itself raised the question of whether a particular body of social thought is scientific or ideological. The work of social scientists was, from the beginning of the academic era, subject to scrutiny on this ground, including self-scrutiny. Specialization among the social sciences left sociology proper with a limited repertoire of (legitimate) concepts to confront the monumental complexities of human relations; the sociological vision (per se) could do little but distort through selectivity. Societal commitment to machine and capital-oriented development defined legitimacy in terms of willingness and ability to share this definition and related assumptions. Despite a strong tendency by individual sociologists to accept these terms, the discipline as a whole has more often been factionalized than it has come to a consensus on the issue, precisely because such a commitment serves interests which (many have argued) are countersociological. Questions of neutrality, objectivity, and partisanship became entangled with the question of disciplinary identity. While in other fields (because their subject matter is not inherently normative) the neutrality question was resolved in favor of neutrality as a condition of professional identity.

These factors produced a situation in which sociology was fated to develop in relative isolation from the disciplines that made it comprehensive enough to be meaningful, and left it without the capacity to resolve the issue (because it is a sociological issue) of its proper relationship to established science and industry. Debates over which is "correct," the European theoretical style or the American empiricist style, Small's progressivism or George Lundberg's (et al.1968) positivism, etc., can be exaggerated out of proportion unless we remember that at least one-half of their "correctness" (in time and place) was as the "correct" way to do legitimate sociology.

The fact that academic sociology is ideological and legitimacy-centered does not invalidate any of its findings or larger truths. It merely provides perspective on why sociologists (of all hues) seem to speak such a "strange" language, report their results in such a "contorted" fashion, and choose to study such an "obscure" range of phenomena. Part of the reason — and it relates deeply to the alienation between sociology and technology — is that the language, results, and range of phenomena of sociology have characteristically been oriented (1) to answering questions about social reality and (2) to arguing that sociology deserves to be considered a science (in terms of many different standards).

The period between the mid-1930s and World War II was one in which an increasingly technical and logical positivist orientation moved to the center of the sociological profession. This change has often been considered significant because it brought sociology closer to its goal of being scientific and because it (incidentally) improved its social status. But this has been the claim of every innovation or change in orientation within the discipline since the beginning of the academic era (and earlier). The dominance of abstract empiricism could not, in light of the centrality of the debate on whether or not social science requires a separate method, resolve for the discipline the question of what it means to be "socially" scientific. It is difficult to agree with those who see a progressive shift in the waning of the Chicago School and the waxing of the "Eastern" school in American sociology, or in the more recent Americanization of European sociology (Jay 1973:282).

The passing of the Chicago School era was a loss for the neoclassical development orientation. The change also served to emphasize the now-prevalent sociological tendency to narrow and trivialize, against which the Chicago School argued. In this sense, the shift does represent a move away from integration between sociology and technology [20]— especially in view of the fact that the Chicago School orientation did not disappear, but along with various other orthodoxies, continued to vie for a central role in the discipline. The seriousness of this change, its character, and the fact that it was contested were succinctly summarized by Louis Wirth (1948:274).

> The development of sociology from 1915 to the present seems to follow successive phases of a cycle. Beginning with an interest in the practical problems of everyday life, sociology, under pressure of the quest for respectability and academic legitimation, moved into an era of excessive concern with building up a technical vocabulary and finding rationalizations for systems of classifications and other abstract categories of thought. This led to the emergence of a cult of unintelligibility and increasing remoteness from the concrete reality of the social world. As a result of widespread and often justified bitter criticism from academic and especially journalistic quarters, as well as a growing sense of irrelevance and futility on the part of

sociologists themselves, there followed a period of fact-gathering and intensive, but more or less aimless, study of small and often disconnected "problems" and the immersion into the development of super-refined techniques for ordering and summarizing the crude data thus gathered. While this accumulation of mountains of authentic but meaningless facts and the invention of complicated scientific gadgets for processing these crude data in a more or less mechanical fashion lent a certain aura of pseudo-scientific glamor to the sociologists engaged in it, it obviously lacked the sense of values and hence of direction of the older, philosophically more sophisticated, speculative sociology, while at the same time it yielded a minimum of either practically useful or scientifically generalizable conclusions.

Politics Within the Profession

Despite the methodological and moral consequences of the post-Chicago School movement, the most significant outcome of the mid-1930s break was political. Prior to about 1935, U.S. sociology and the American Sociological Society existed in a state of close support, if not absolute identity. The organization — in the spirit of progress — was dedicated to promoting the well-being of the discipline and practitioners. For many years, the *American Journal of Sociology* was the only U.S. sociological periodical. Later, regional journals were added, but in close connection with the society. Because of the small size and simplicity of the sociological community in the United State, the profession, and the craft of sociology (Horowitz 1971) were equally of concern in the discipline's battle for legitimacy. But with the challenge by the positivists, it became apparent that the profession did not nor could it any longer speak for all sociologists. This was certified with the establishment of the *American Sociological Review (ASR)* as the official journal. The situation intensified under the pressure of growth, increasing specialization, and (renewed) ideological heterogeneity.

Among the effects of the mid-1930s break was the emergence of an acute awareness among sociologists of the importance of the professional organization — a matter which had not been of serious concern when the profession comfortably coincided with the field. Because technological society had become the organizational society, because sociologists study organization and because the discipline underwent a struggle for power within the professional organization, the post-Chicago School era was as concerned about professionalization as it was about operationalization. The result is most significant.

The displacement of the Chicago School resulted in the establishment of an identity between positivist sociology and the profession. *ASR* sociologists of the 1930s sought to place their orientation at the center of the discipline through the power exercised by the profession. In view of the history of alternate visions of the sociological mission and the vast difference between the positivist commitment and sociology's classic development orientation, this move was most abrasive. It contributed to further factionalization within the field and to the expenditure of considerable energy on internal issues.

The attempt to equate sociology with positivism through the powers granted the professional organization was an audacious move. For some, perhaps, it was made in order to achieve a better relationship between sociology and technology though one suspects that for most, nothing could have been further from their minds. Yet in an ironic way it was successful. In very prosperous times, the profession could not keep up with the growth in the field. But under conditions of scarcity, the profession has been able to coordinate and orchestrate the content and direction of academic sociology. The irony is that just as the positivists achieved the establishment of their orientation as professional orthodoxy, academic sociology was beginning a period of significant decline.

Notes

1. U.S. sociology was, in comparison to its European counterpart, first and fundamentally academic. Sorokin (1929:57) observed that "American sociology has grown up as a child nursed by universities and colleges; while in Europe, its modern start since August Comte, and development in a considerable degree have taken place outside of the universities and colleges."
2. George Gurvitch (1971:61) has commented on this as follows: "American sociology developed rapidly in the period of the great Wall Street crisis, in the period of the New Deal of Roosevelt, in the midst of terrible worries when people wanted to foresee the most immediate future, when business circles, extremely upset, sought for a science capable of forecasting near-term that which economists and, all the more, politicians of that period had failed to envisage."
3. The American Sociological Society was founded in 1895 as a branch of the American Economics Association (founded in 1884). See Bramson (1961, ch. 4). For a recent historical review, see Rhoades (1980a).
4. Jane Addams, who was — significantly — a social worker, looms large in the early days of University of Chicago sociology (Lasch 1965). To find an important "left-wing" exception to (and critic of) this progressivism, one must turn to Veblen — the archetype of the academic maverick.
5. The following image of sociology at the turn of the twentieth century is reported by Albion Small (1903:16) founder of the *American Journal of Sociology* and fourth president of the American Sociological Society: "The distinguished president of an eastern college took occasion, in an address before leading Chicago citizens to associate the name 'sociologist' with the term 'freaks' and 'faddists,' and he is reported to have said that sociology seemed to him to have nothing to do except to gather up what is

Early Academic Approaches in the United States   65

left after political science and economics have done all that is important with the facts of society."

6. Veblen's marginality is a strong interest of his biographers: e.g., Coser (1977), Lerner (1948), Dowd (1958), Dorfman (1940), Diggons (1968), and Rosenberg (1956).
7. According to many of Veblen's critics (e.g., Dowd 1958; Dobiansky 1957), his work also illustrates a number of less desirable traits such as obfuscation, pomposity, and a streak of antiprogressivism. According to Layton (1971), Veblen was influenced by Gant, a colleague of F.W. Taylor. Through this association, Veblen's role as ideologue for antiprogressivist forces was strengthened.
8. The degree of Veblen's Marxism has been a matter of much debate among Marxists and others; see Coser (1977:289ff), Dorfman (1973), Rosenberg (1956, ch. 5), Aaron (1951).
9. Veblen's instinct theory, though often criticized as unscientific, remains a powerful metaphorical device.
10. Veblen was greatly influenced in this regard by Franz Boas and the newly-emerging Chicago School of Anthropology.
11. This book, like most of his other writings, is inherently ironic in terms of its subject matter. Most of his critics and biographers (e.g., Coser 1977; Lerner 1948) are certain that he is a very sarcastic man, that he could not possibly be serious when he speaks of "trained incapacity," "organized inefficiency," and similar paradoxes. It is more likely that Veblen was dead serious about the absurdity he observed in the world.
12. For a pertinent commentary on Veblen as a "failure," see C. Wright Mills' introduction to Veblen (1953).
13. This is not to say that European academic sociology did not bear much fruit, only that European social thought now had a serious competitor — just as European society now had to take the U.S. role in world affairs more seriously. For a comparison and evaluation of European versus U.S. sociology in the early-to-middle academic era, see Mannheim (1953). Mannheim felt that the theoretical orientation of European sociology helped retain its critical and reflexive edge which was eventually lost in the U.S. mainstream.
14. They often confused the effects of the novelty of the urban environment on immigrants with those of the city per se. Thus Wirth and others argued — wrongly — that urban life (rather than lack of stability in the city) made community solidarity difficult if not impossible to achieve (Berry and Kasarda 1976, ch. 2).
15. "Culture lag," introduced in 1922 — before the Chicago School experience moderated Ogburn's evolutionary orientation — is a classic example of a social evolutionary concept. Ogburn (1922) held that the material and nonmaterial aspects of culture follow their own, relatively independent, evolutionary course. Usually material culture advances more rapidly; thus most (but not all) lag occurs when society is not ready to accommodate the inventions it has wrought.
16. In addition, he was "director of the consumer advisory board, National (Industrial) Recovery Administration in 1933, special advisor of the Resettlement Administration, 1936, research consultant and member of the science committee of the National Resources Planning Board from 1935 to 1943,

and chairman of the Census Advisory Committee in 1941"*(Current Biography* 1955:461).
17. In 1937, the *American Sociological Review* (based in the East) replaced the *AJS* as the official journal of the society (Lengermann 1979).This victory is recorded (by Bramson, for example) as "only symbolic," however.
18. One reason is that such a characterization ignores the important role played by personal and other non- or subprofessional interests. Another is that it does not account for the many ways in which these two can support one another (and have done so).
19. It may be true that intellectual leaders in many fields are also active in professional affairs. But in sociology, the professional activity of those thus far mentioned centered on value issues rather than, as for most professionals, leadership and normative contributions to their disciplines.
20. Some might argue that the move toward positivist sociology brings the discipline closer to the nonsocial sciences and technical fields because the other fields have been positivist for many years. This argument misunderstands positivism, the nonsocial sciences, and their uses of scientific method.

4.

Karl Mannheim and the Postacademic Program

Philosophers and sociologists once thought that there was a tendency towards rational and moral progress inherent in the human mind. That this is untrue is clear to everyone who knows what is happening in the contemporary world, for it can be asserted with confidence that in the past decades, we have receded rather than advanced as far as moral and rational progress is concerned.

Karl Mannheim (1954:51).

The European university, with its departmental structure and emphasis on the individual scholar surrounded by disciples, was already becoming inadequate as a basis for scientific activity by the late nineteenth century (Ben-David 1971:129-38). As the number of subdisciplines and special fields, the cost and complexity of research, and the size of the scientific community increased, science became decreasingly academic and more of an organizational pursuit (Price 1963). Through World War I and the succeeding years, government and, especially in the United States, corporations gradually came to replace the schools as sources of financial support for physics, chemistry, and the biological sciences. By the middle of this century a distinct shift had occurred in the character of work and interpersonal relations, not only within the established natural sciences, but also in economics, sociology, political science, and the social sciences generally. These trends were intensified by the wholesale recruitment of scientists and other professionals to the war effort by both the Allied and Axis powers during the late 1930s and early 1940s.

One of the first European writers to recognize these changes and attempt to assess their bearing on the relationship between technology and society was the Hungarian-born sociologist Karl Mannheim (1893-1947). During and immediately after World War II, Mannheim (then working in England) began to revise his earlier observations on the sociology of knowledge (1968:1952)and culture (1956) in light of the rise of fascist and communist totalitarianism in Europe, the war itself, and the new and powerful role technology had begun to play in the conduct of war and other business of state. Inspired by the deep sense of uncertainty which characterized his generation and by his long-standing commitment to aid in the reconstruction of modern society through the application of social science, Mannheim (1950:291) posed a question that forces itself upon us today with equal urgency, and perhaps with an even greater sense of practicality: "On the plane of scientific thinking, there is neither tragedy nor comedy. There are facts and processses, neutral in themselves. But is it possible to experience human events in an atmosphere devoid of the tragic sense, where there is no climax, no one responsible, no hero but a lot of atoms that behave in a certain way?"

In an attempt to answer this question, a framework for a new postacademic social science was provided. Though largely unattended to in much academic sociology, it gave impetus to the theories of C. Wright Mills and several later sociologists in Europe and the United States. Many of the issues addressed in this book were closely considered by Mannheim in later writings. Mannheim's response to the question about scientific truth and moral commitment is contained in the program he began to develop in his last three works, *Man and Society in an Age of Reconstruction, Diagnosis of Our Time,* and especially *Freedom, Power, and Democratic Planning.* Though this work has a definite pedagogical purpose, Mannheim was aware that if such a task of educational reform was to be taken seriously, it was necessary to discuss the kinds of social change with which such reforms must articulate (Coser 1977:447-9). For him, the program for a postwar (i.e., postacademic) social science had little meaning except in relation to the events and personalities of the technological society which this new (but old) type of science was meant to understand and alter.

Mannheim explored the relationship between the formulation of a morally sensitive, interdisciplinary, and developmentalist social science and the rise of technology-aided planned social change with rigor and engagement. These explorations have been discussed in some detail by his students such as Adoph Lowe, Kurt Wolff, Hans Gerth, C. Wright Mills, and several sociologists and philosophers associated with the New School for Social Research. In this and later chapters, some of the leading themes in postacademic social science that originate in Mannheim's analysis are considered (Simonds 1978). While it is impossible to review here this entire comprehensive approach, it is evident, even from a general outline of his work and that of social scientists influenced by him, that postacademic social science is not a traditionless abstraction. It is a real activity, first worked out early in the present era by a truly modern social thinker. While we refer only briefly to Mannheim's analysis of the role of social science in technological society in this short chapter, the analysis offers a solution to the "crisis of our time."

Elements of the Program

Mannheim's program is based on the view that technological society is increasingly shaped not by forces such as a free market, public opinion, or egalitarian ideals, but by elite technicians and planners unresponsive to — insulated from — the effects of such forces. This condition of "massification" is the result of decades of technological innovation and political change. Because it differs so much from the type of society envisioned by earlier philosophers of both the Right and Left, it poses a special challenge to educators and citizens in even the most democratic societies (Mannheim 1956, pt. 2: 81-116). In the absence of effective means of public participation, people cannot affect the design and innovation process in industry, communication, education, advertising, science, and other "social techniques" such as welfare and health systems (Mannheim 1954, pt. 5). Yet planned technological innovation of one type or another has everywhere replaced both laissez faire and socialist approaches to change, approaches upon which effective democratic participation through consumption and voting are premised. Industrial, medical, and other "technological delivery systems" (Wenk 1979) increasingly shape our individual and collective lives; but they are increasingly specialized to the point at which rule by elites is inevitable. Machines and social techniques have been improved and extended since the earliest days of the industrial revolution, but most especially in the postwar, postacademic era "in order to solve the problem of mass organization. But the main thing" about these modern technologies, as they affect society and as they concern social scientists, "is not only their greater efficiency, but that such efficiency favors minority rule" (Mannheim 1950:7).

In Mannheim's later writings, various alternative responses, theoretical and actually instituted, to those tendencies are examined (Mannheim 1954). This assessment applies his earlier distinction between ideological, utopian, and scientific views of social change in some detail (Mannheim 1968,1956). As the earlier works had dealt with social knowledge, the later works sought to establish a "third way" of dealing with technological innovation. This approach recognizes that innovation cannot be left to proceed of its own accord, but it also opposes fascist (ideological) and communist (utopian) solutions: "The end of *laissez faire* and the necessity for planning are unavoidable consequences of the present situation and the nature of modern techniques . . . . The alternatives are no longer 'planning or *laissez faire?*' but 'planning for what?' and 'what kind of planning'" (Mannheim 1950:8).

In discussing specific features of this third way, Mannheim provided observations on the place of technology in modern political and social life and on the place of social knowledge in the design and innovation of technology. He stressed that postwar society requires not ideologies and utopias in education and public life generally, but an interdisciplinary scientific orientation to understanding technology and its impact. But understanding cannot be either the entire or the final purpose of this orientation. To so narrow the concepts of education and social science would be to deny their now crucial intervening position between the theory and practice of social change. In support of these observations, Mannheim called upon and further developed his analyses of elites, rationality, war and dictatorship, planning, freedom, and — with increasing importance in his very last writings — religion (especially 1950,ch. 13). The result of these explorations is an impressive theory of technological society: of its inherently dangerous tendencies toward disorganization and/or totalitarianism, its inherent liberating potential, and the role in it for a new, technology-centered, applied social science in combating disorganization and totalitarianism and promoting development.[1]

Mannheim's work has influenced present-day sociology and related fields in several ways. In the following two chapters, its formative effects on Mills' research on knowledge, power, and elites, and its role as a counterposition for the Frankfurt School will be noted. The very active period in the late 1950s and 1960s of work on mass society and pluralism in the United States (e.g., Kornhauser 1959; Riesman, Denny, and Glasser 1950; Vidich and Bensman 1958) was largely inspired by questions posed by Mannheim. In addition, several recent social scientists have continued this exploration of the society/technology interface in an attempt to produce social knowledge in the human interest. This work has special relevance for postacademic social science because, while it represents a continuation of the program begun by Mannheim, it also takes account of current conditions that make possible the program's realization. Because the time has come when the opportunities for technology and policy-focused research are relatively substantial, it is now possible to speak of developing the applied side of Mannheim's postwar social science — a side relatively neglected by him and much of the academic research he influenced.

Among the several features of Mannheim's program that have received attention since his death, some touch especially closely on types of problems that arose with persistence in the preacademic era, in the work of academic sociologists of technology such as Veblen and Ogburn, and have arisen once more in postacademic contexts. One of these is an interest in understanding the range of unintended, often ironic, outcomes of technological innovation. This focus on the unplanned side of planned change, a fundamental aspect of the first birth of sociology and technology and of Mannheim's analysis, has since been discussed by theorists in the United States such as Robert K. Merton and Louis Schneider. It is part of a sociological tradition that stresses the effects of technological innovation which do not stop when the overt purpose is achieved — e.g., to span a body of water, to travel rapidly, etc. — but which reverberate through the social order and extend beyond the expectations of planners and technicians.

This focus on the social ironies of technology (and the related idea of the sociologist as ironist) has recently provided some key theoretical and methodological links between sociology and applied fields such as social impact analysis, technology assessment, and technology policy formulation. Another feature of Mannheim's program was the attempt to understand, through the sociology of knowledge, the relationship between technological innovation, on one hand, and changes in ideas and idea systems, on the other. Of particular interest to Mannheim, as it was in Marx's era and earlier, was the relationship between the process of innovation and theoretical systems, such as social science and the major political ideologies, meant to account for innovation and its impact. Following Mannheim, Mills, the Frankfurt School, Alvin Gouldner, and others have developed this perspective to the point that it now serves as an important part of the conduct of sociotechnical research.

The ironic focus and the sociology of knowledge perspective are discussed at greater length in this and later chapters of this part and in part II. A third and complementary element, the application of social science to development, is highlighted in part III. Mannheim's program not only provided a central, practical response to classic concerns with technology, with its ironies, and with theories meant to understand them, but it was also formulated in a uniquely explicit awareness of the role of the Third World in European development, in the past, present, and in the future as far as it can be foreseen (Mannheim 1950). Mannheim combined the stress of preacademics (especially the French positivists) on the intimate relationship between sociology, technology, and development with observations about the impact on innovation and on our understanding of innovation in today's developing countries and their search for a "third way." This stress on development, as a complement to the other themes, constitutes a substantial basis for the transition to postacademic social science.

Technology, Irony, and Detachment

Since the days of Mandeville, modern writers have appreciated the awesome power of technology to transform our environments, populations, societies, and cultures.[2] Characteristically, this appreciation has been stimulated by the "mocking discrepancies" between intent and outcome so frequently observed in the wake of technological innovation. This sensitivity to the ironies of technology is a major component of Mannheim's approach that has continued to receive attention in later academic and postacademic research.

Technology is a very special kind of knowledge; it is purposive. The thing, often the only thing, that distinguishes technology from other kinds of knowledge is its anticipated effects. We cannot begin to comprehend the role of technology in society and history without considering the uses to which it is, and is meant to be, put. The ironic focus stresses that, beyond these rational considerations, there are more extensive effects unanticipated by technicians or by society at large. These effects greatly complicate the course of progress and development.

Irony and Technology in Mannheim's Work

The underlying sense of tragedy in Mannheim's theories and in the social science tradition of which these theories are a part derives from the understanding that ironies, great and small, have accompanied technological innovation since the industrial revolution. Chief among these is the contradiction between the Enlightenment intention to use technology to set people free from bondage to tradition and nature and the current situation in fascist, communist, and mass societies alike in which technology serves to keep most people in bondage to the few political and technical elites. Mannheim's approach and the postacademic program in general are attempts to develop and apply a social science that will redirect technological innovation to achieve the spirit as well as the letter of its original purpose. The very concept of technology today requires an expanded view of purposes — based on the recognition that social impacts, in all their complexity, are as important as technical effectiveness, efficiency, etc., in the achievement of the goals of innovation.

This theme is stressed at several points in Mannheim's writings. At times (e.g., 1954, pt. 1), his ironic focus is directed to the problem of the "two types" of rationality in technological society (closely related to

Weber's *Wert* and *Zweckrationale*): the "formal" rationality of means and ends in engineering and the "substantive" rationality of the ends to which the products of formal rationality are put. Failure to take account of substantive rationality, or the tendency to replace substantive rationality with interest and ideology (Mannheim 1952, pt. 4) has led to a general "victimization," a loss of control (despite democratic hopes) by most members of society over decisions that greatly influence them.

At times Mannheim's concern with the ironies of technology are related to his educational reforms (especially in the later sections of 1950, 1954). Here the education of technicians, planners, and their publics in a democratic society must include a sense of the extensive and often dissimulated outcomes of innovation. Without such awareness, the conduct of the state, economy, communications industry, schools, and social techniques will be based on a dangerously incomplete model of the world: in the absence of an ironic focus on technology, innovation is understood only as a technically rational process (plus "error" and "side-effects"); dangerously incomplete because, while "vast industrial empires could not be held together without these modern means . . .these very same techniques also make for dictatorship" (Mannheim 1950:7).

Another specific point at which an interest in the ironies of technology arises in Mannheim's work is in relation to "detachment" as a social science method. With the connection between technology and the unintended effects of innovation in view, Mannheim suggests that a sustained focus on the ironic aspects of planned social change can help to cultivate necessary intellectual distance between observer and observed (see Schneider 1975, chs. 1, 2; Brown 1973; and Stanley 1972, for later statements). In assuming such a focus, the social scientist — like the dramatist — frees himself from and can "see beyond" the interest-bound perspectives of technicians, elites, and publics. The ironist's role is associated with a specifically sociological state of mind, an objective perspective on goal-directed behavior relatively unaffected by actors' motivations.[3]

Detachment and Engineering Evaluation

Detachment from the interests, motivations, intentions, and anticipations — in brief, from a very substantial part of subjectivity — of actors has been a much-pursued and debated goal, in social science generally and in the social scientific study of technology in particular (e.g., Rudner 1966:66ff).[4] Mannheim (1958:253) devoted much attention to the problem of detachment in Marx, Scheler, and his own sociology of knowledge. For him,

> That which within a given group is accepted as absolute appears to the outsider conditioned by the group situation and recognized as partial . . . . This type of knowledge presupposes a detached perspective. . . . this detached perspective can be gained. . . [if] within the same society two or more socially determined modes of interpretation come into conflict and, in criticizing one another, render one another transparent and establish perspectives with reference to each other.

These and related observations are employed in Mannheim's theory of the *Freischwebende* (free floating) intellectual and in his equating lack of detachment with ideology. A contrast is drawn between interests and the limited, distorted perspective on planned change which interests entail, and freedom from interests and the less distorted perspective which detachment produces.

In his earlier work, Mannheim's analysis centered on the distinction between a detached, sociological account of social change and accounts distorted by conservative, utopian, and other specifically political interests. In his later writings and in the work of his successors, these concerns are extended to a consideration of other sources of distortion, especially those created as a result of the process of technological innovation itself. In this instance, the group situation of actors is seen to influence their perceptions and communication not only because of personal and class loyalties, world views, and private motives, but also by virtue of the constraints that come from responding to the requirements of technical rationality. Even in the most "neutral" approaches to technological change, "the subject limits his perspective in order to deal with a particular problem directly practical or theoretical" (Merton 1949:256).

Such limitation, though an essential part of engineering design and evaluation, acts to obscure all the possible effects of innovation which appear to be unrelated to specific technical purposes. As a result, control over impacts that lie outside of this range is not exercised by the planner and technician. These impacts are ignored until they either make themselves known as side-effects and ironies or become the domain of political and economic elites who thus use innovation for their own ends. As antidote to this, Mannheim's program proposes the incorporation into the planning process of a detached, ironic focus on innovation, one that "extends the observer's attention beyond the questions of whether or not the behavior attains its avowed purpose. Temporarily ignoring these explicit purposes," the focus on the unanticipated consequences of engineering and planned change "directs attention toward another range of consequences" (Merton 1949:64). The suggestion here is that the intended effects of innovation may not be the most important ones, that they can be overshadowed or even reversed by the "other" range of consequences. Interest in such ironies and near-ironies of planned change has recently been expressed in the engineering evaluation and policy-oriented fields of technology assessment and social impact assessment.

Technology and Ideology

Mannheim's concern with detachment has been extended not only in connection with the ironies of technology, but it has also figured prominently in the sociological study of knowledge. Technology is a type of knowledge, and as such it is susceptible to the same analysis that Mannheim and other sociologists have applied to political ideologies, literature, and even science. In this analysis, technology proves to be particularly prone to be shaped not only by the avowed purposes of the technician, but also by the myriad, often unanticipated, sociopsychological factors related to our awareness (or lack therof) of its impacts.

Sociological analysis of the relationship between knowledge and technological change began early in the industrial age. Earliest, perhaps, was Saint-Simon's observation that prescientific systems of thought had become obsolete. As noted in chapters 1 and 2, after the mid-nineteenth century and coincident with the rising influence of Marxism in Europe, this analysis experienced the fission that resulted in the separation of two familiar sets of perspectives on technology and culture, the scientific and the ideological. Each of these perspectives has been used by writers and scholars, many of whom have been discussed here, to characterize the factual and moral implications of technological revolution to the public. Social scientific and ideological accounts of technology have frequently been used to criticize each other for portraying a false characterization; yet they remain ironically dependent upon one another and upon the momentum of the technological revolution which they seek to explain.

Focus on Knowledge

The writings of Mannheim and later social scientists on this "dialectic of ideology and technology" (Mills 1940; Gouldner 1976) are of special interest to the postacademic social scientist. Influenced by the theories of Weber and Durkheim and especially by various schools of European Marxist theory, this perspective has provided a unique insight into the relationship between technology and other idea systems (social-scientific and ideological). Its main focus is on the body of thought, generated since the split, that has sought to reconcile the ideal and material aspects of modern social life: values, sentiments, and technology. This work demonstrates well how the light (and darkness) generated by technological revolution has been the principal source of illumination (and confusion) for social theory during the past two hundred years. According to this perspective, explanations of technology are two-sided. Ordinarily, there is a descriptive, diagnostic dimension (what Gouldner calls a "report") and a prescriptive, therapeutic dimension (a "command"). The model of the theorist that emerges is not that of a physicist but of a physician. This perspective also stresses the importance of the political dimension in all theories about technology; one's political orientation — among many other sources of motivation — is bound to affect the nature of his diagnosis and therapy.

The purpose of this perspective is not to discover if propositions deriving from theories about technology are "empirically" true. Examination of the noncognitive, moral implications of theories, which must be excluded by empiricism, are essential implications in social thought about technology. Limited by its focus on truth value, empiricism presumes a level of simplicity of technology and its impact that is wholly unjustified. Unlike observations made in sciences described by positivist philosophies (if any exist), those produced in the social-scientific and sociophilosophical study of technology are not "empirical" in a strict and simple sense. Such observations selectively focus on ("bracket," in the terms of phenomenology) a set of data from an imponderably vast universe of possible foci. The expression of ideas about technology in propositional form, upon which empiricism insists, reinforces selectivity, revealing only a small set of relationships among inconceivably many possible ones. What is not said about the social causes and effects of technology is often more important than what *is* said. The question that needs to be asked is not whether a particular claim about technology is true, for the answer is undoubtedly, "yes, given certain definitions, under certain circumstances." Rather, it is: (1) what does such a claim say about the "health" of the social order in light of the observer's (or some other) diagnostic criteria, and (2) what does it prescribe to improve the health of the social order?

Sociology of Knowledge and Sociology/Technology

The sociology of knowledge tradition of Marx, Durkheim, Weber, and Mannheim seeks to relate social-scientific (and other) observations to the social conditions that influence the framing, utterance, publication, and process of certification as "truth" or "falsity" of such observations. In application to the study of technology, this diagnostic-therapeutic formulation summarizes a significant connecting link between social thought and social conditions. Consciously or not, every social theorist is affected in his choice of questions, data, techniques for data accumulation and hypothesis testing,[5] mode of reporting, and choice of audience by his individual and group-related interests. There are obviously many interests that simultaneously affect any social philosopher or scientist: some relate to his willingness or lack thereof to accept particular aspects of the political status quo, others relate to his own professional goals and opportunity structures. Beyond the truth of falsity of their specific propositions, some kinds of knowledge provide grim, while others provide promising, diagnoses of the objects to which they refer. Some suggest specifically reactionary, reformist, or revolutionary "solutions" to problems revealed in their diagnoses, while others suggest (or imply) a therapy of benign neglect, etc. "The facts" of a complex and dynamic social reality are likely to support any and all possible diagnoses and therapies, depending on how these facts are gathered, interpreted, and communicated. The meaningfulness of any social observation (its truth) depends not upon the state of the raw facts but upon the ideological orientation in which it is cast and understood. This ideological orientation relates, in turn, to the individual and group interests of the observer and his audiences.

This perspective has many implications that bear on postacademic social science. Perhaps the most important of these is that a postacademic social scientist knows and acknowledges his selectiveness. Because the postacademic social scientist is not constrained to uphold the view, which Mannheim takes to be a myth, that social theory ought to be or even can be undistorted by the investigator's therapeutic and diagnostic commitments, he can begin to understand the true limits of his own thinking. More specifically, this perspective reminds us that our social theories about technological innovation, shaped as they are by our interests and other motivations, are themselves indirect products of technological innovation. Since technological changes affect the nature of work, they inevitably affect the nature of social-scientific (and ideological) work: new bureaucratic requirements create different disciplines and provide different opportunities for intellectuals.

Technology: Ideas and Realities

In characterizing the relationship between technology and idea systems as dialectical, the sociology of knowledge approach points to several things. Foremost, perhaps, is the view that the two realms, ideology in the broad sense, including social science, and technology as practical know-how are major components of what Marx called the superstructure and material base, the Yin and Yang of modern society. As such, they continually invite (and defy) synthesis, as indicated by the perpetual search for reconciled theory and praxis.

Academic Context

As a sociological perspective on knowledge, this approach presents a description and partial explanation of the ways in which systems of ideas, whether political ideologies, social sciences, or other myth structures, present distorted depictions of technology and its social consequences. According to this formulation, science can present a no more truthful picture of social change than any other idea system. Such systems differ because they serve different cognitive and social purposes. In particular, academic social science, in stressing (at most) the purely descriptive side of technology's impact, emphasizes the diagnostic and suppresses the therapeutic function of theory. But by virtue of is antitherapeutic (and antiideological) character, such social science systematically distorts the role played by its own diagnoses in the design, innovation, and evaluation of technology.

The type of distortion of social reality common to various styles of academic social science, on one hand, and to a range of ideological perspectives, on the other, is not mere falsification. Both can be highly rational, highly elaborated codes that distort by exclusion. They focus only on a sample of the myriad pertinent facts, unintended and intended outcomes, and moral consequences they must confront. The languages the academic social sciences use to discuss technology and social change fail to grasp reality in all its complexity because they stress description and ignore prescription. Political ideologies distort by deemphasizing the descriptive role of language and by stressing programs to change reality. Neither credits the other with its actual degree of rationality.

The apparent paradox of symbolic systems that are rational yet systematically distorting has led sociologists to the exploration of an important but relatively neglected field, linguistics. Mannheim's observations on this problem have recently articulated with the search of German sociologist Jurgen Habermas for an ideal speech situation in which undistorted communication can take place. Alvin Gouldner's investigation of the many ways in which communication can be distorted, and Basil Bernstein's (1972) work on elaborated and unelaborated codes have furthered this research. In suggesting how academic social science and ideology are alike (they are both highly elaborated, impersonal ways of discoursing about the world) and how they are both remote from ordinary discourse and the language of mass communications, this exploration into linguistics connects with other significant insights on democracy, communication, and education of Mannheim, the Chicago School, and the Frankfurt School.

Bureaucratic Context

As a study of the social context of technology, the sociology of knowledge perspective focuses on the social stratification system in modern bureacratic organizations: industrial, governmental, and especially communications and "the new consciousness industry" — or what Mills called the "cultural apparatus." In such organizations, technicians are seen as engaged in a power struggle with older-style bureaucratic officials. As part of their political platform, planners and engineers project a view of the world in which technology rules purely in the interest of progress. They uphold a social-scientific distortion —"the end of ideology" ideology — that suppresses the therapeutic function by negating the future, for the future is here and no therapy is necessary.

While it is clear that the means and ends of organizations in democratic society are no longer the sole domain of capitalists, bureaucrats, and political appointees, technicians distort the situation when they suggest that technical expertise has altogether replaced the partisanship of the old guard. Instead: "The technical staff is alienated from the ends, the officials from the means. While now developed more than ever before, technology remains subject to the ultimate control of nontechnicians, bureaucratic officials, and political appointees" (Gouldner 1976:255). To attack this status quo, the technician becomes a "technologue." His account is different in many respects from that of the old ideologue, for he exemplifies and reports the end of ideology; but his purpose is the same. The account, based in academic social science, with which technologues justify a wished-for status quo is a weapon in an ancient battle for control of the means of production.

The new ideology of the technocrat has already spawned a sort of antithesis in the ideology of environmental doom. The prophets of technology's destructive powers are like their technocratic opponents who cannot see an end to progress. In their zeal to reform, they both underestimate the consequences (for good and evil) of technology. The doomsayers have struck at a vulnerability in technology that makes allies of the most conservative and most liberal of ideologues under the label "environmentalist." [6] The ideology of environmental doom provides an eschatology (albeit not a happy one) and abrupt outcomes, where technology conceives of perpetuity; it commands, where technocracy stresses passivity; it asks for heroes and for control, where technocracy projects routinized and cybernetically monitored social change. The problem is that technology means both progress *and* destruction, and more (see chapters 8 and 9).

Aim of the Postacademic Program

Such an analysis of ideology and technology in a democratic society delineates the social implications of the technocratic, antitechnocratic, and old guard bureaucratic ideologies. As elaborated impersonal accounts, they offer alternative diagnoses and therapies of the role of practical knowledge in an industrial age. But the postacademic program seeks to embrace none of these. In placing technology and accounts of its impact in a social context and focusing on the ironies of planned action, its aim is to develop a critical, humanistic perspective on technology and science. Like Mannheim and his Frankfurt School critics, Mills, and other past and present students of technology as well, the postacademic social scientist seeks a perspective from which ideology, technology, their sources, and their consequences can be studied and criticized. But this is a perspective that can also be used to analyze itself and thus provide a relatively detached, undistorted account. Technology theory interested only in intended outcomes, uncritical of its own limitations, and unconcerned with its own role in technological society is not only obsolete but is also a dangerous and dehumanized type of knowledge.

In a continuation of these remarks in later chapters of this book, specific ideological accounts of technology's effects will be discussed and additional prescriptive remarks about the postacademic perspective will be offered. As in this chapter, our purpose is not to legislate the "true path" to the social study of technology — for such a goal is not only presumptuous and morally questionable, it is also utterly unrealistic even if we could claim possession of such a truth. The purpose of these and later comments is to identify in the writings of Mannheim and others a basis for the conduct of applied research on technology faithful to the ideals of the eighteenth- and early-nineteenth-century founders of sociology, responsive to the needs of our time, supportive of democratic principles, and which can serve as an alternative to the several types of approaches current in technological societies. This is an ambitious aim. But the depth of understanding that Mannheim's observations provide and the sense of urgency with which his program is infused suggest that the attempt might not be altogether futile. Because Mannheim's assessment opening this chapter is as accurate today as it was more than thirty years ago when it was written, it seems a virtual necessity that such an attempt be made.

Notes

1. Robert Lynd (1939) provided one of the most complete statements of this position from the perspective of U.S. academic sociology. Earlier, Albion Small (1903) put the same case in similar manner: "Sociology is good for nothing until it can enrich average life at least, our primary task is to work out correct statements of social problems and valid methods for solving them."
2. D.C. Muecke traces the term back to Plato's discussion of *Eironeia* (dissimulation) in *The Republic*. Aristotle used it in ethical and aesthetic contexts, and both Hindu and Hebrew myth history (as in *Bhagavad Gita* and *Torah)* show a profound awareness of irony and its implications. These early uses share a moral purpose akin to that served in technology assessment; that is, to warn of possible shocking consequences of even the best-laid plans. According to Muecke (1970:14ff) the word *irony* appeared in English only in the early sixteenth century.
3. See for example Mannheim (1968:250ff). Also relevant, though Mannheim used neither "irony" nor "latent function," is Mannheim (1956, esp. pt. 1; 1944:59-79; reprinted in Wolff 1971:367-84). Mannheim's "socially unattached intellectual" and our "sociological ironist" are closely related types (Coser 1977:435-36; Lyman and Scott 1970).
4. See Brown (1977:53ff) for a somewhat more detailed discussion of what he calls "distance." Despite Brown's suggestive comments, the question of standards or units by which this distance is to be measured remains unanswered. Clearly, it is to be measured along a multidimensional (and unbounded) scale. "Perfect" detachment is an ideal on two accounts: (1) one can always achieve "greater" distance; (2) one can achieve distance in an unlimited number of ways (contexts, foci).
5. In full: "Since no scientific hypothesis is ever completely verified, in accepting a hypothesis, the scientist must make the decision that the evidence is sufficiently strong or that the probability is sufficiently high to warrant the acceptance of the hypothesis" (Rudner 1953:2).
6. Norman Podhoretz (1971:7) has summarized this analysis: "Technology is destroying ecology and will end by destroying us all; so stated, in the most general terms, runs the formula which in not more than a year or two has won a most amazing degree of critical acceptance. . . [but] no proof exists that the end of the world is at hand. We do not even have persuasive evidence pointing to that conclusion. All we have, exactly as the men of 'yesterday' did are warnings and exhortations to the effect that we are doomed unless we repent and change our ways and return to the proper path. That such warnings and exhortations are often voiced by professional scientists and couched in the language of science does not in the least endow them with the authority of tested scientific statements. When they speak of these matters, the scientists in question speak not as scientists, but as moralists and ideologues, and no one ought to be fooled."

5.

C. Wright Mills and Pragmatist Sociology

Of the several academic sociologists under discussion in this section, C. Wright Mills is least likely to be identified with the sociological study of technology.[1] His work ranged over many familiar sociological concerns, such as stratification, social change, and the sociology of knowledge, and broke ground in several new areas as well. But his major focus was not on the interaction between people and machines as was Ogburn's; nor was the conflict between technician and entrepreneur, so important to Veblen, an explicit point of reference in his analysis. Nevertheless, Mill's often pungent observations on academic sociology and on the social role of the intellectual (or in his words "the man of knowledge") are of great value to those who seek to reshape the discipline in the present postacademic era. In these and several other ways, he well represents the social scientists whose contributions, though not formulated as social impact analyses or histories of specific inventions, often outweigh those of more readily identifiable sociologists of technology.

Mills was strongly influenced by the work of Mannheim, the Chicago School, and by the general American sociological tradition. He was a serious student of this tradition, and his observations on the works of U.S. (and European) social scientists and schools are among the best to be found. Of special interest are his interpretations and applications of the pragmatism of C.S. Pierce, William James, John Dewey, and the Chicago School sociologists and social philosophers. His major monograph on American social thought (Mills 1969) is a book that began as his Ph.D. dissertation (at the University of Wisconsin) and whose theses were elaborated throughout his brief and productive career. It is rich with insights about social evolutionism, the growth of pragmatist sociology, and the doctrine of applied social science. Its pointed critique, which matured and deepened with the years, justifies the book's Veblenesque subtitle: "The Higher Learning in America."

From this earlier work on the history of social theory to his last writings on Marxism, international relations, and knowledge, Mills argued with a sense of urgency for a sociology [2] in the human interest. At a time when the post- (and anti-) Chicago School entrenchment of the pure physics model had been virtually accomplished, he selected, out of the myriad things that a sociologist might do, those things that would make a difference in man's quest for democracy and social justice. He was exceptional among his contemporaries in his support of the developmental sociology program of Robert Lynd, Mannheim, and the European founders of social science.[3] Mills' contribution is perhaps best understood in comparison to the styles of academic sociology he sought to counter and transcend. Irving Louis Horowitz (1963:4) has formulated the contrast in this way:

> Mills' search for answers, no less than for the right questions, meant a conscious abandonment on his part of two established traditions in sociology: empiricism as an ideology limiting discourse to low-level generalities, small group studies, and an ethically unconcerned posture; and rationalism as an ideology committed to abstract solutions of concrete issues, and an ethical involvement having sources independent of and even alien to the findings of the behavioral and historical sciences.[4]

Stratification and Elites

In addition to his programmatic works and critical essays on sociology, Mills' commitments were expressed in his substantive research, especially on stratification and international relations. This research was designed to illustrate the advantages of the type of ethically engaged intellectual craftsmanship he favored, and reveal truths about aspects of modern society which might make people freer. Not as a matter of design but as a natural outcome of the scope of this inquiry, he also had much to say about the social antecedents and consequences of technology and the manner in which social science might be used as a knowledge base for technological innovation. Mills addressed a combination of theoretical and practical problems in his research on stratification. He sought to reconcile certain differences between the major European theories of class and class consciousness, particularly the Marxist and anti-Marxist types: To what extent does the working class feel hostile toward capitalists? Is economic inequality growing or diminishing in capitalist societies? Has a new bureaucratic system replaced old-style free enterprise capitalism? His main purpose in this work was not merely to decide which school of stratification theory was right or wrong. He wished to use knowledge about social inequality as a tool for political reform (Mills 1963, pt.3, ch. 4).

Like Marx, Mills believed that people are deceived about the extent to which the rewards of wealth, prestige, and power are distributed justly in democratic societies. He felt that if the average man could understand how impotent he is in comparison to members of the ruling elite, he would be encouraged to join with other average men in a political movement for the creation of a more egalitarian system. Unlike Marx, Mills did not accept the premise that the sole or major point of inequality in contemporary society is between owners and nonowners of the "means of production," nor that the source of conflict between elites and nonelites was solely or largely economic. Mills agreed with Max Weber and John Dewey that the society of Marx's era had been succeeded by another, less vulnerable one; that the main obstacles to social progress lay in antidemocratic tendencies of our political system, as opposed to antisocialistic tendencies in the economy.

Mills' specific stratification studies covered a wide range of topics, including labor union executives, the new middle classes (Mills 1951; 1963, pt.1, chs. 7, 8, pt. 3, chs. 2,3,10), and political, economic, and military elites (Mills 1956; 1963, pt. 1, ch. 7, pt. 2, ch. 5). This work was largely undertaken to discover if bourgeois/proletarian polarization or conflict could be discerned in contemporary United States and other capitalist countries. The results revealed how the tendency toward rationalization had greatly influenced worker, middle-class, and elite lifestyles and chances in creating a mass, but unequal, system incomprehensible in classic Marxists terms. He found Weber's stress on bureaucratic organization in modern society to the point. "It is no accident," Mills (1963:53) insists, "that Max Weber is more and more frequently quoted for his thesis that the historical drift may be seen as the bureaucratization of industrial societies, irrespective of their constitutional governments. It is this *form* of organization which is taken to be the substance of history, the more so as it is identified with a growing rationality of modern society."

While Mills employed a variety of techniques in his research, including interviews, the assembly of collective statistical biographies, and documentary analysis, his general method was that of the historical sociologist. With this method he sought to examine class and bureaucracy as variables — dynamic features of an ever-changing technological society — rather than as rigid frameworks to which all other historical forces would have to adjust. His theoretical work on class centered on the formulation of a pragmatist alternative not only to Marxist but also to Weberian orthodoxy.

This goal was expressed in his major empirical studies such as the *The Power Elite* and *White Collar*, and perhaps more so in his later critiques of radical and liberal theories of stratification (e.g., Mills 1962).[5] These studies and arguments remain useful guides to understanding the structure of modern society and as illustrations of the uses and limits of classic theory in social science.

In furthering his practical aim, Mills sought to examine those aspects of stratification in modern society which have direct bearing on the fate of individuals.[6] In *The Sociological Imagination,* he defined sociology as the field that focuses on "the intersection of biography and history," and his sense of relevance regarding the study of stratification is everywhere informed by this definition. One consequence of this commitment is his conclusion concerning the irrelevance of much of the debate between various schools of stratification theory, debates that he felt had little value in resolving pressing public and private issues. His stratification research, though hardly positivist, was empirically oriented. He discounted many of the subtle issues associated with Marx-Weber dialogue, such as whether or not the working class could be expected to achieve true class consciousness, in favor of the description of patterns of social inequality in modern society.

Mills used the study of social inequality neither as a vehicle for organizing working-class revolution nor as a source of pessimism and resigned passivity. In conscious application of the approach of John Dewey and the Chicago School sociologists, he viewed it as part of a critical program to identify and alter the forces that impinge upon democracy and/or limit the formation of well-informed democratic publics. From this pragmatist perspective, the individual is seen as a potential victim of impersonal institutions, as an object of an increasingly antagonistic social order in which freedom and dignity can be preserved only if the drift of history and elite planning are understood and countered.

Mills extended Mannheim's critique of elite rule in the democratic countries of Western Europe and North America by showing how partisan interests now dominate technological society and subordinate the values of progress and rationality which such a society might embody. In his view, those who stand at the top of corporate, military, and governmental hierarchies — the "power elite" — constitute an interacting group; they know one another, belong to the same social circles, marry into one another's families, send their children to the same few schools and colleges, and their jobs are more or less interchangeable. They tend to have a common world view and come to share a "domesticated" concept of development and of the role of technology in the development process. The power of the modern military, corporate, and governmental organizations is the result of control-extending innovations in communications and industry. These organizations and the strategic position of the power elite within them have served to transform the elite vision of technology and development into the prevailing vision (Mills 1963, pt.4, ch. 11).

According to this analysis, under conditions of "massification" — the absence of well-informed, active publics and a lack of democratic access to elites — the way in which society supports particular technologies and manages its development programs will only incidentally be in the interest of the people who make up that society. National R&D priorities are increasingly geared to innovations that best serve the interests of organizations and their leaders, even when these run counter to those of the citizenry or humanity. Formal democratic institutions such as the right to vote, multiparty systems, and representative government remain intact, but control over the course of planned social change becomes increasingly an elite prerogative.

Mills' theory of the power elite has been criticized on several grounds: for being inaccurate, atheoretical, conspiratorial, and for being vague or deterministic in answering central questions such as "Who rules America?" Some of these criticisms cancel each other out, others — such as the attack by communist critics that the theory ignores capitalism as a determining factor — are merely polemic, and others yet are perhaps warranted. The purpose of presenting the theory here is neither to endorse nor refute it, but to illustrate how Mills dealt with technology. The thing to be stressed is that — as in other sociological views discussed in this section — technology is seen as a major cause and effect of other social forces such as political relations, the growth of organizations, and stratification systems, but it is not attributed special power to act either for or against people. To the extent that loss of personal autonomy and control is an ever-increasing trend in technological society — a great extent in the opinion of Mills, Mannheim, the Chicago School, and Veblen — this is due to how technologies are used and envisioned, whose definition of appropriate uses prevail, and what is allowed to count as legitimate development; these factors are to "blame" and not the technologies themselves.

The theory of the power elite may not give a thoroughly true picture of the political structure of democratic societies. But it makes it easier to see technology in context and to conceive of policy (difficult as it may be to formulate in detail or implement) that might affect our uses of technology and visions of development so that personal freedom is increased by planned change. This general aim, to see technology as a social factor and direct it to the democratization of mass society, gives meaning to Mills' attempt to understand the role of elites. It makes the theory a valuable asset for the practice of postacademic social science (especially in the absence of an adequate alternative), even if it must be treated as highly tenuous, partly outdated, or even biased.

The Role of Craftsmanship

Mills' interest in the abuses and potential proper uses of technology not only provided the impetus for his work on stratification, but it also underlies another major part of his approach, the focus on craftsmanship. As with Veblen, this focus logically begins not at the macroscopic level of the relationship between elites and masses, but with an examination of the details of the structured conflict between political and economic elites and the technician. Despite the importance of technology as a source and instrument of power, at this level the man of practical knowledge — the designer, technician, and the engineer — is characteristically a "man in the middle" (Mills 1963, pt. 3, ch.10). He is not "in control," his fate is a variant (albeit an important one) of that of the white-collar worker, the new middle class (Mills 1951). The white-collar technician has as little autonomy as the layman; with the rest of his class, he is caught up in status ambiguities, mobility consciousness, and a political role through which policies may be enacted but never decided upon. However noble or mean the values and interests of technicians may be, they are bounded by the limitations of being in the middle: subordinate to the vision of a remote and powerful elite, yet "above" the masses and potential publics influenced by their work.

What was seen by Veblen as a deep-running antagonism between the engineer's instinct of workmanship and the capitalist's drive toward pecuniary emulation (and for Ogburn as a battle between youth and conservatism) was viewed by Mills as the struggle between a segment of the bureaucratized middle class and an elite of business, government, and military enterpreneurs. Like Mannheim, Mills believed that the results of the ascendancy of this elite in combination with urban growth, changes in the postwar international politicoeconomic order, and the nature of the new technologies, would be far-reaching. In the political sphere, this situation is identified by the increasing atrophy of democratic institutions and the increasing massification of what once were (or might have been) public forums. In industry, it is expressed in the situation in which, barring far-reaching structural reforms, creative people " will tend to be commerical stars or commerical hacks. And *human development will continue to be trivialized,* human sensibilities blunted, and the quality of life distorted or impoverished" (Mills 1963:386).[7]

Among the most significant effects of this evolution toward overdevelopment, for Mills, is the manner in which the instinct of workmanship, what he referred to simply as "craftsmanship,"[8] has been systematically replaced by passive routinization in every realm of life. This diagnosis is based on classic Marxist theories of alienation, on Weber's theory of bureaucratic society, and on Veblen's commentary. But key changes in the role and world view of technicians had occurred between Veblen's time and Mills', including the highly conservative "Revolt of the Engineers" (Layton 1971) and the emergence of "a Marx for the Managers" (Mills 1963, pt.1, ch. 3; Orwell 1946). For this reason, Mills (1963:53) could neither look to the technician class as a new proletariat nor could he be sanguine about the unbridled course of technological progress:

> Apart from the opaque line of technological rationality, social life is drift and habituation. The irrational institutions, particularly pecuniary ones, are in the main only permissive; all they do is occasionally hinder the spread of a mechanical rationality into all areas of life. It is the men who nurse the big machines, the industrial population, who implement that which makes history.

In this light, the events of the early twentieth century have made the engineer a relatively ineffective actor, incapable of combating the trivialization of development because his role, too, has been trivialized by decisions implemented by those who "nurse" — as opposed to those who invent and/or are affected by — "the big machines." The loss of a sense of craftsmanship, as a value and in practice, has become so thoroughgoing that it has affected those who would otherwise be especially likely to pursue craftsmanlike careers: the artist, the designer, and the creative inventor.

> As practiced, craftsmanship in America has largely been trivialized into pitiful hobbies: it is part of leisure, not of work. As ethic, it is largely confined to a privileged group of professionals and intellectuals. What I am suggesting to you is that designers ought to take the values of craftsmanship as the central values for which they stand; that in accordance with it, they ought to do their work; and that they ought to use its norms in their social and economic and political visions of what society ought to become (Mills 1963:386).[9]

Before political victory for the technician over the entrepreneur can be endorsed by Mills, it is imperative that the technician first recapture the spirit and practice of his craft. In short, the engineer cannot be expected to humanize society until engineering itself has become humanized.

The International Context and the Third World

Mills' diagnosis of elite rule and the loss of craftmanship characteristic of overdevelopment is supported by his comments on international relations and the problems of underdeveloped societies. Two related themes inform most of his observations on technological society in a world perspective, and are of particular interest to the postacademic social scientist. First is the recognition (again, traceable to Mannheim) that technologically developed countries have come to exert enormous and undue influence in the affairs of the less developed. Not only does this promote the loss of democracy on a universal scale, but when technological advance is combined with nuclear monopoly (as it was during Mills' life when only the United States and the Soviet Union were nuclear powers), such influence places the very survival of mankind at risk. The second theme is the charge that overdevelopment has occurred at the same time that, partly by design and partly by neglect, underdeveloped societies have been allowed to stagnate economically and have been subjected to a long line of political dictators — imported and domestic. Mills was not what was later called a "dependency" theorist — a proponent of the view that all the problems of the Third World can be traced to Western colonialism — for he was unwilling to so simplify what was evidently a multidimensional, multinational set of interdependencies. But he did point to the fact that technological societies are not alone in this world and that their trivialized concept of development would have to be examined in relation to more than merely "the national interest."

Mills' comments on the hegemony of technologically advanced societies were meant to provide an alternative to the cold-war mentality that dominated thinking in the United States during his career (and to a degree to this day). His basic argument, as expressed in greatest detail in *The Causes of World War Three* (Mills 1958), is moralistic rather than political or economic. He avoided the main assumptions of cold-war discourse by discounting certain differences (but certainly not all) between technological U.S. and technological U.S.S.R. In many areas in which it matters to the individual and to humanity, both these bureaucratic, accumulative, high-technology-oriented megastates are prone to abuse power. At the limits, both these societies, perhaps for different reasons and by different channels, had by the mid-1950s become warrior states, armed with a state-of-the-art military technology they aimed at each other and at everybody else. The entire political and technological apparatus in both countries had become organized to "improve nuclear capabilities," that is, to increase the capacity to destroy totally and extensively. This is not, of course, the only business of the state in such highly advanced technological societies, but it is both important and symbolic of the extent to which the ideal of technology for development can be distorted.

Mills' message in this regard, at its simplest, was that might does not make right. That a great culture lag exists between the technological acumen of a society that can develop an effective nuclear offensive and defensive capability — using the land, the surface of the oceans, the waters underground, the sky, outer space, and the moon — and the ethical acumen of these societies' elites. His prescriptions for correcting this situation, while not usually offered as flat solutions, centered on a reexamination, a humanization, of our technological priorities through the improvement of conditions in the United States and the Soviet Union that promote the formation of publics and democratic participation. While like the power elite theory, these prescriptions can be questioned, the important point is that Mills boldly defined the moral dilemmas associated with technological leadership in the second half of this century.

The rising Third World is an aspect of modern international relations which Mills had only begun to investigate before he died in 1962 (a key turning point in the Indo-China war and U.S. involvement in it). [10] In later work he undertook abroad (Mills 1960), he established a framework for integrating our newly acquired knowledge of social, economic, and political conditions in the Third World with his observations on stratification, elites, craftsmanship, and technology. He recognized the enormity of the commitment to their own visions of development which increasingly drove the leaders and masses in less-developed countries. This commitment, and the desire for political and economic independence which accompanied it in Asia, Africa, and Latin America, provided a sharp contrast to the triviality of the social aspirations current in Europe and North America.

Like many postwar social scientists, Mills was impressed by the vast popularity and ideological appeal of the Third World independence movements. His reflections emphasized the limitations of technologically advanced countries to meet this challenge. He recognized that while the ideals and economic goals of revolutionary America and Russia had become relevant again in the Third World, a lack of will and/or ability on the part of elites in contemporary United States and Soviet Union made it difficult for them to accept this as part of their own heritage; instead, the common response was hostililty, misunderstanding, and imperialist designs. Mills felt that these limitations should be exposed and that knowledge of them should be borne in mind as Third World societies continued their search for development models. While the Russian and American Revolutions bear important positive lessons for today's Third World, the lessons to be learned from overdeveloped societies that have succeeded are largely negative. Mills (1963:156) was an early proponent of Mannheim's "third way" to development, a broad philosophy of social change that does not, like the other two ways of the Western and Soviet systems, subordinate the human interest to that of economic or political elites. For him

> The problem of the underdeveloped society is to achieve a higher material development of the sort that avoids the sad features of the overdeveloped society, and, hence, makes possible a variety of human beings, of styles of life, perhaps never seen before in human history. . . . You [the people of the Third World] are really on your own: the answer for you is not available in historical Europe or in contemporary North America or in Soviet Russia.

Critique of Social Science

Mills' observations on the nature of technological society were, like those of Veblen and Ogburn, complemented by a critique of contemporary academic sociology. It is difficult to appreciate his contribution without viewing it in relation to the intellectual environment in which he worked and against which he struggled. His commentary on the role of the schools and especially of university sociology programs was a product of John Dewey's and his own concern about the forces that sustain or erode a democratic system. Like Veblen, these comments also reflect Mills' marginal relation to the academic sociological establishment. These conditions produced in his work a powerful sociological analysis of current knowledge institutions, especially but not exclusively those concerned with social knowledge.

Mills' critique combines a theoretical base in the sociology of knowledge (Mills 1963,pt. 4, chs. 2,3,4) with an application to current activities associated with the production and distribution of culture, the "cultural apparatus" (Mills 1963, pt. 4, ch. 1). His views on sociology are part of the general study of the production and consumption of social knowledge. He begins with the premise that human experiences are not self-defining but mediated by culture: experience is interpreted through words, symbols, symbol systems, values, and, simply, convention. In modern and, increasingly, less-modern societies, the functions of mediation and interpretation are assigned to specialized roles and organizations. Intellectuals and other cultural elites, schools, and "knowledge factories" — such as research laboratories — are occupied full-time in defining and evaluating events and disseminating their views to the public.

As with other crucial functions, interpreting reality for society has become increasingly rationalized and bureaucratized in response to the requirements of "those who nurse the big machines." "In a word, what has happened in the last two centuries to work in general is now rapidly happening to artistic, scientific, and intellectual endeavor: now these, too, become part of society as a set of bureaucracies and as a great salesroom" (Mills 1963:418).[11] Under these circumstances, what society certifies as knowledge is increasingly prone to a type (or types) of distortion. Like the effects of bureaucracy in other realms, in the knowledge business it serves to make more difficult the creation of informed publics and limits the development of other conditions required for effective democracy.

Implicit in this assessment is Mills' ideal of the craftsman; in this case, it is the "intellectual craftsman." The knowledge produced and certified in technological society ordinarily does not bear the stamp of craftsmanship, nor does it reflect the responsibility of the producer inherent in the craft ideal. The producer of knowledge is an occupant of a position in a hierarchical organization; he has no grasp of the whole process of which his work is a part. The values and the needs of elites, entrepreneurs of the First World and party bosses in the Second World, dictate what is produced, authorized, and sustained.

An immediate result of this lack of autonomy for the producer of culture — the intellectual, artist, and scientist — is the proliferation of uninspired, prosaic, and uncritical interpretations of social experience for mass consumption: fiction magazines, books on how to improve your powers over others in ten easy steps, prime-time television, and popular religion. This is work whose final purpose is not to serve the producer or consumer but to assure profits and/or the continued hegemony of the ruling elite. Interpretations that the market or the regime will not bear, including criticism of the true motives behind the creation of culture, cannot become knowledge: the public remains misinformed or only partly informed. Bureaucratization of the production of culture supports an orientation toward "technical virtuosity" and sustains an educational system that reinforces it. With this orientation, specialized training is stressed in which techniques are treated as essential while investigation of the reasons behind or the consequences of applying the techniques are neglected.

Mills saw in the loss of intellectual craftsmanship, or as Veblen put it, in the subordination of the instinct for workmanship, the imminent prospect of the dehumanization of education. Mills (1963:369) observed a general lack of interest by schools and other producers of culture in fostering a widespread "birth of sensibilities." This is the result of passivity among intellectuals and based on well-reasoned consideration of bureaucratic demands placed on them. But regardless of its source, it can only be combated by creation and support of an independent, craftsmanlike role. Barring such a change, those whom technological society charges with the production of culture and with instruction in the means of interpreting experience can only contribute to the further alienation of themselves and others and to continued massification of the social order.

This analysis is presented in greatest detail in Mills' comments on the production of and instruction in academic social science, especially sociology. Here we find his most concrete illustration of how conditions in technological society affect the intellectual and distort the product of his labor, and it contains his most direct discussion of the types of distortion that result. These observations on contemporary sociology are found throughout his writings (but especially in Mills 1959; 1963, pt. 4, chs. 6-9; 1962). They are of interest to postacademic social science because they so acutely delineate and criticize the field as an academic pursuit. This work should not be thought of as exempt from criticism; many of Mills' specific comments are obsolete, erroneous, or more emotive than cognitive. Yet he was perhaps the most perceptive critic of sociology during a crucial period when the field was beginning to lose its identity as an exclusively academic discipline. More than any other sociologist of the postwar era, Mills knew the field — its promise and inanities — and cared about how it was to be used in these dangerous times.

The Promise

Central to these observations is the promise of what sociology might be, a promise that the discipline has occasionally approached fulfilling. This ideal is patterned after the general model of the intellectual craft. Mills contrasted it to what he considered to be the distorted products of the sociological imagination: research and teaching deadened by bureaucratization and the pressures of an academic environment. Mills, like others, understood the purpose of sociology to be knowledge for development. Based on this, the main thing that kept the discipline as practiced from coming closer to the promised ideal was the nature of the practitioner's "larger" commitments; that is, the lack of current interest in doing sociology to improve people's lives. The major weakness of academic sociology for Mills was not epistemological, methodological, or even political — it was moral. He did not debate whether it was possible to have a neutral, pure science, whose main object is to perpetuate itself within a professionally approved set of ideas and methods. He rejected this very popular model as ethically inappropriate; for him it was the morally bankrupt product of the bureaucratization of the sociologist and the consequent separation of his work from that of other social scientists, technicians, and other intellectuals.

Mills was also disturbed by the manner in which contemporary academic sociology had systematically ignored its own heritage or had understated the commitment of earlier social thinkers to an engaged, practical, and interdisciplinary orientation. Instead of being associated with "the promise," many of the early sociologists had been enshrined as defenders of the academic pure science model of the late 1930s, including Spencer, E.A. Ross, Comte, Durkheim, Mannheim, Marx, Veblen, Schumpeter, W.E.H. Lecky, and Weber. Yet each of these men self-consciously pursued his work in order to "contribute, however minutely, to the shaping of this society and to the course of its history, even as [we are] made by society and by its historical push-and-shove" (Mills 1959:6). The questionable ethics of the reigning pure sociology was, for Mills, compounded by the lack of honesty or knowledge concerning the founding commitment of the field — to be relevant to people. Absent in contemporary academic sociology is a commitment to grasp "history and biography and the relations between the two within society." The result, in Mills' (1959:6) view, was a half-hearted pseudoscience; for "no social study that does not come back to the problem of biography, of history, and of their intersection within society has completed its intellectual journey."

Distortions in Academic Sociology

To pursue the type of social study that Mills had in mind requires social and intellectual conditions that have ceased to exist as the bureaucratization of the social sciences has progressed. To practice such an engaged — which is not to say dishonest or unscientific — sociology, a certain independence, free-floatingness[12] from bureaucratic constraints is necessary. Such a social study takes seriously the vision of itself as a service craft, one whose major services include helping people to "acquire new ways of thinking... a transvaluation of values" (Mills 1959:7). In the schools and universities of overdeveloped societies, this craftlike vision is not taken seriously, in part because it is so difficult to realize. Instead, one of three basic images has tended to shape academic sociology (in rough chronological order): (1) a theory of history; (2) a systematic theory of "the nature of man and society"; and (3) today, empirical studies of contemporary facts and social problems (Mills 1959:22; see also Mills 1963, pt. 4, ch. 9). Each of these tendencies represents a distorted interpretation of the social science tradition and promise (and can provide a negative example for the postacademic field).

The first two tendencies are largely of interest on historical or intellectual grounds, since both are rarely practiced, and the second has had less impact than it might deserve because it is generally unintelligible (Black 1961). All three tendencies deflect rather than focus attention on history, biography, and their intersection. In the case of the theory of history, this is accomplished through an attempt to discern meaning and connectedness from the rise and fall of great civilizations and in the epochal events studied by the classic historian. As a sole preoccupation, it is of little use to contemporary human beings because it "can all too readily become distorted into a trans-historical straight-jacket into which the materials of human history are forced and out of which issue prophetic views (usually gloomy) of the future" (Mills 1959:22-23).The second tendency, systematic or "Grand Theory," arose "perhaps in reaction to the distortion" of the first. Also unable, or unwilling, to examine the actions of living individuals and groups in all their complexity, "history," for the Grand Theorist, "can be altogether abandoned: the systematic theory of the nature of man and of society all too readily becomes an elaborate and empty formalism in which the splitting of concepts and their endless rearrangement becomes the central endeavor" (Mills 1959:23).

Because, as a lone-scholar type of pursuit, the theory of history has been made obsolete by the technology of intellectual productivity, it is not likely to have a great impact on sociology in the future. Because of its obscure language, Grand Theory is also inaccessible to most sociologists but it has exercised some influence in academic circles, especially through the work of Talcott Parsons. Interest in Grand Theory has sources in sociology's struggle to appear scientific and in the general social conditions of overdevelopment, although it cannot be easily explained as the direct product of bureaucratization. For Mills, its role in academic sociology can be understood in terms of its ideological content. "The ideological meaning of grand theory tends strongly to legitimate stable forms of domination" (Mills 1959:49, fn. 19). It can perhaps be seen as a doctrine serving the interests of those who also benefit from the bureaucratic order. "Yet," Mills cautioned (twenty years ago), "only if there should arise a much greater need for elaborate legitimations among conservative groups would Grand Theory have a chance to become politically relevant."

It is much easier to understand the political relevance of the third, and by far most popular, tendency of late academic sociology, what Mills called "abstracted empiricism." For this is bureaucratic sociology par excellence. Its origins can be traced to the relatively low esteem and late institutionalization of academic sociology in comparison to other disciplines, particularly economics, psychology, and political science. As a result of these factors, disciplinary sociology in the United States was forced to specialize in the study of events and phenomena excluded as nonessential by older disciplines. Post- and anti-Chicago School sociologists were left with an unsystematic hodge-podge of problems from unrelated contexts, which could not be treated "scientifically." They set about to systematize the hodge-podge at the level of a covering, positivist methodology, and to redefine "good sociology" as technical virtuosity. They attempted to make their studies susceptible to scientific approach by stressing the formalities of research and ignoring "messy" substantive interrelationships and contextual factors. These "sociologists have tended to become specialists in the technique of research into almost anything, among them methods have become methodology . . .these tendencies — to scatter one's attention and to cultivate method for its own sake — are fit companions although they do not necessarily occur together" (Mills 1959:24).

Abstracted empiricism fits well into the dominant bureaucratic framework of culture production. It is highly susceptible to organization, specialization, and the performance of intellectual tasks that the intellectual has had no part in formulating and for whose final uses he need have no concern. It fosters technical virtuosity at the expense of an interpretation of the world in which people might be more effective actors. The findings and observations of the abstracted empiricists point in no particular direction; if studied as an exclusive intellectual diet, they can induce moral paralysis. Abstracted empiricism (at least to a greater degree than the other two tendencies) has given many academic sociologists a sense of prominence in a field that continues to reward virtuosity for its own sake, that celebrates a stilted image of pure science and the scientific method. It has provided the profession with an orthodoxy whereby it can control the discipline, at least within academe. It is the style of sociology most compatible with the search for academic canons of quality control and a rigid definition of legitimacy. Because of this and its highly organized, specialized, and technical character, abstracted empiricism has been used as the basis upon which to build a sociology whose development corresponds to that of the society it supposedly studies. It can be defined in terms of the endless accumulation of products that contribute to the accumulation of more products, and so on. It is the antithesis of the type of intellectual craftsmanship neccessary in the postacademic era.

Mills as Academic

At several points it has been stressed that Mills' analysis and criticism can be traced to the sociological tradition of which he was a student and scion. His pessimistic views on stratification, craftsmanship, and academic sociology are consonant with his general theory of the erosion of democracy. But standing somewhere between Veblen and Ogburn in this regard, Mills was also responding to a deep personal ambivalence concerning his own role in relation to his sociological colleagues, the prevailing abstracted empiricists and grand theorists.

Mills' career at Columbia University was marked by a striking discrepancy between, on one hand, the wide acclaim he won outside of American sociological circles, among nonsociologists within academe, and among students and young professors of sociology, and, on the other hand, the (at best) cool reception he received among his colleagues in the department and profession. His department chairman at Columbia, Paul Lazarsfeld, was a leading figure among abstracted empiricists; and the "enemy," the grand theorists and other distorters, dominated his professional environment. This perhaps made it inevitable that he would be a marginal man in academe. Jealousy his popularity inspired (the underestimated prime mover in academic departments) exacerbated the situation. While he was a member of the faculty of one of the leading sociology departments in the United States, he was for many of his colleagues the very opposite of what a professional sociologist should be — and he returned the sentiment.

Mills was an effective and vastly popular writer, perhaps — had he the years to develop his talent — of the caliber of Sumner, Veblen, or even H.L. Mencken. By his own standards, his theories about society's need for a personally relevant sociology were proved true. But he was affected, and his work was shaped, by that other set of standards from which he — like Veblen and Ogburn — could never entirely free himself, that of the academy. Here the results are more difficult to interpret.

Mills' dramatic life and his death at an early age, his unabashed espousal of the cause of the powerless, and his academic marginality all support the view that he was a martyr. Whether or not he "deserves" this status, this view obscures a significant feature of Mills' work and times. Again very much like Veblen, Mills was so much a man of his age, so prolific, so attuned to American society, and so knowledgeable about the sociological tradition, that he could not be understood, let alone appreciated, by the academics of his era. Mills may not have received his due rewards from Columbia University or the American Sociological Association; but the larger tragedy is that members of the profession did not have it in their power to evaluate him, for they were not his peers. Nor did they have the capacity to reward him, even if they could have evaluated his work properly. Mills was the prototype of the postacademic social scientist, with one small (but significant) exception: he was first and foremost an academic. It was still an important part of his reality that his occupation was the "training" of sociology majors at an Ivy League university and the writing of books and scholarly articles. The chief contradiction of Mills' observations on the role of the man of knowledge is that it points away from the role and values in which the man himself was immersed.

Mills in Retrospect

C. Wright Mills' main contribution to the reintegration of sociology and technology was not in his theories of technological impact nor his studies of inventions, but rather in the manner in which he grasped the essence of the role of the postacademic social scientist (though he could not perform it as effectively as he might have wished). He was a strong proponent of interdisciplinary, applied sociology. He was sensitive to the role of sociology in bureaucratic society and to its antecedents. He was alert to the manner in which this society had trivialized development. This was combined with and informed by an understanding of the social forces that shape knowledge, the intimate relationship between underdeveloped and overdeveloped societies, and by deep insight into the forces that have shaped and distorted academic sociology. For these reasons and, to reiterate, despite the fact that he did not style himself a sociologist of technology, his contributions to this field were great.

What distinguishes this approach is a sense of moral responsibility on the part of the social scientist to the public, to his predecessors, and to the subjects of his work. These are things that the craft of social science — perhaps to a greater degree than non-social sciences — requires. This ethical orientation was very explicit in Mills' work, as is evident from the list of "dos and don'ts" with which he instructed his students — many of whom are, along with their peers and students, now being called upon to practice sociology outside of the classroom and library.

> Do not allow public issues as they are officially formulated or troubles as they are privately felt, to determine the problems that you take up for study. Above all, do not give up your moral and political autonomy by accepting in somebody else's terms the illiberal practicality of the bureaucratic ethos or the liberal practicality of the moral scatter. Know that many personal troubles cannot be solved merely as troubles, but must be undetrstood in terms of public issues — and in terms of the problems of history-making. Know that the human meaning of public issues must be revealed by relating them to personal troubles — and to the problems of the individual life. Know that the problems of social science, when adequately formulated, must include both troubles and issues, both biography and history, and the range of their intricate relations. Within that range the life

of the individual and the making of societies occur; and within that range the sociological imagination has its chance to make a difference in the quality of human life on our time (Mills 1959:226).

Notes

1. He is also the only one who had formal engineering training at a technical university, Texas A & M, where he spent an "unpleasant year" in the 1930s (Horowitz 1963:6).
2. Mills did not like the word *sociology* because of the narrowness it had come to connote. *Social science* was more compatible, but he preferred the more catholic and less pretentious *social studies* — though he also objected to the implications of a high-schoolish "guts" course this term has (Mills 1959:18-19, fn. 2).
3. Mills' most succinct views on development can be found in 1963, pt.1, ch.9.
4. On this analysis of sociology see Mills (1959, chs. 2-4; 1963, pt.4).
5. As opposed to the rigidly determinist and/or evolutionary views of "vulgar" and "sophisticated" Marxists and Weberians, Mills sought a "plain" approach. Essential to this plainness (as discussed in Mills 1962, pt. 1) are (1) the rendering of the "iron laws" and "irrevocable effects" of the European sociologists into more modest hypotheses and variables, but (2) the maintenance of the *engagé* orientation that gave rise to the postulation of these supposedly iron laws and irrevocable effects.
6. Horowitz (1963:3) quotes from a letter that Mills wrote to a "white-collar wife": "It is one thing to talk about general questions on a national level, and quite another to tell an individual what to do. Most 'experts' dodge that question. I do not want to."
7. Mills' concern with the trivialization of development is key to his type of critical perspective — which in this respect is shared by Veblen and others. If the sociologist can disassociate development as such from the current societal definition of it, his study of technology's social antecedents, uses, and effects take on a radical edge.
8. Mills' use of the craft ideal can be found throughout his work; but perhaps the most concise programmatic statement is in his essay "On Intellectual Craftsmanship" (Mills 1959:195ff).
9. Mills continues with another reflection on development as it relates to craftsmanship: "Human society, in brief, ought to be built around craftsmanship as the central experience of the unalienated human being and at the very root of free human development . . . For the highest human ideal is: to become a good craftsman" (Mills 1963:386).
10. Horowitz (1966, 1974) formulated the first explicit sociological theory of First/Second/Third World relations; but many of his basic themes were discussed by Mills and earlier by Mannheim.
11. The sentence that precedes this in Mills' text should settle any question about the degree of his debt to Veblen: "The pervasive mechanisms of the market have indeed penetrated every feature of life. . . and made them subject to [N.B.] the pecuniary emulation."
12. This concept of "free-floatingness" — introduced by Mannheim — and the theory of the intellectual of which it is a part, are not incompatible with "engaged" sociology. To the contrary, as Mannheim (1950) argues, this perspective and sociological engagement complement each other.

6.

The Frankfurt School: A Critical Theory of Technology

The economic insanity, which is interwoven into the technology, is what threatens the spirit and today even the material survival of mankind, and not technological progress itself.

The Frankfurt Institute (1972:95)

This chapter concludes our historical overview of social scientific approaches to technology with some comments on the much-discussed views of the Frankfurt Institute for Social Research — or Frankfurt School, as it is usually called today.[1] This school originated in Germany in 1923, when several young philosophers, economists, and sociologists — including Friedrich Pollack (1894-1970) and Max Horkheimer (1895-1973) — received some private funds for founding an Institute for Marxist Studies at the University of Frankfurt. As an officially recognized, yet Marxist, university program, this was one of the first ventures of its kind in Western Europe. By the late 1800s, academic social science had turned away from the preacademic stress on criticism and application in favor of a more limited, descriptive, pure-science approach. The very establishment of the Frankfurt School added a significant missing dimension to legitimate academic research on technology and social change.

With the collapse of the Weimar Republic and Nazi rise to power, most of the institute's members were forced to flee Germany for religious and political reasons. In 1933, with the help of groups and individuals in the United States, the institute was given a new temporary home on the campus of Columbia University in New York. There, the original members attracted other radical and antifascist scholars from throughout the world, and the school enjoyed an enormously productive period of seventeen years. Scores of their books, articles, journals, pamphlets, lectures, and other works were written and published, and their work began to acquire an international reputation that has continued to grow to this day.

In 1950, many members of the school returned to Frankfurt, where the institute was reopened with Max Horkheimer as director and professor. Since then, with an increase in the number of translations of their later writings and the appearance of several English-language works — especially by those like Erich Fromm (1900-80) and Herbert Marcuse (1898-1980) who stayed in the United States — the research of Theodore Adorno (1903-69), Horkheimer, Marcuse, and the other members of the Frankfurt School has had a great impact on contemporary sociology and on popular culture as well. In Germany today, the school retains its influence on academic sociology through the editorial work and writing of Jurgen Habermas, his colleagues, and their students.[2] In the United States, scores, if not hundreds, of social scientists educated during the post-World War II and especially New Left eras have attempted to understand, elaborate upon, or critically engage the concepts of authoritarian personality, eros and civilization, Nazism, and the other rich and evocative themes that identify the Frankfurt School approach. Everyone familiar with developments in social science during the 1950s, 60s, and 70s is aware not only of the continuing importance of Karl Marx and Sigmund Freud, but also of attempts to combine perspectives of Marx and Freud made by popular Frankfurt School members such as Fromm and Adorno.

The last word on the scope and influence of the Frankfurt School has certainly not been spoken; and it would be impossible to attempt anything so definitive here. The Frankfurt School (along with Mannheim, Mills, and others) performed a Janus-like role in relation to postacademic social science: summarizing the achievements and shortcomings of the academic era and insinuating itself deeply in what is to come. As the years pass, and social science and the technical fields continue to rediscover one another, it is likely that the work of the Frankfurt School and its followers and critics will increasingly become a point of reference. With such explorations, the writings of the school will be continually reexamined and reevaluated. This discussion is not meant to provide a full accounting of the school's theories or of its strengths and limitations but rather to suggest why the postacademic social scientist would be much poorer without its legacy.

Social Study of Technology[3]

Extensive commentary on technology, its social implications, and its impact is to be found in the writings of the Frankfurt School. As in Mills' case, much of this commentary is not consciously formulated as a sociology of technology. The innovation and evolution of technology are viewed as key variables which, in combination with economic and psychological forces, have acted to shape modern society and have in turn been shaped to society's requirements (Fromm 1968).

Friedrich Pollack, the school's leading economist, was one of the first to provide an account of the basic political role of technology in modern society. In his research on *The Economic and Social Consequences of Automation* (1957) and other works, Pollack studied the transformation of the state, the private sector, and their relationship in capitalist countries. In his view, these have been greatly changed since the mid-nineteenth century by the spread of labor-saving innovations. Like Mannheim, Pollack argued that free enterprise no longer existed; that instead, the state had increasingly come to perform a service function in relation to private industry. Rather than being an obstacle to economic activity, government was now geared to serving the interest of capitalism by aiding in the displacement of human labor and, in fiscal matters, by subsidizing and absorbing the losses of industrial advance (O'Connor 1976). In this analysis we find an early, explicit model of industrial society organized as a huge R&D system.

The state, the main institution through which democratic sentiments may legitimately be expressed in modern society, had been turned against the average worker and into a servant of capital accumulation. This observation became a major premise of the Frankfurt School's general program of understanding and combating political authoritarianism. In this regard, Pollack's observations, Franz Neumann's studies of the causes of Nazism (1944; 1957), and Karl Wittfogel's writings on absolute power (1957) can all be viewed as broad social impact analyses, as discusssions of the way in which modern technology has affected social relations in Europe. They show how economic and political elites have been able to direct the innovation process to serve their own aims — the cost-reducing aim of automation, profit-maximizing mass production, power-maximizing heavy investment in military R&D, and the control-extending aims of developments in transport and communication. In the process, democratic procedure is subverted or nullified by the new, extensive powers that control of technological R&D provides. The discipline of the machine is extended to all areas of life; and with it comes state control of psychological processes, culture, and social activities.

Such a tendency operates in all modern societies, but in many instances it has become national policy.

From this Frankfurt School perspective, technology is seen as having been directed to promote what the few define as progress and as having given the ruling elites unprecedented power to coordinate social relations and private lives. As in Mannheim's theory, the key tragedy is that because of the relative inaccessibility of official planning to normal political channels, such forms of authoritarianism can emerge even within a formally democratic system — though the final product may be, as in Nazi Germany and the Soviet Union, a totalitarian expansionist state. As George Orwell put it in 1944, "so long as the world tendency is towards nationalism and totalitarianism, scientific progress simply helps it along" (1978:175).

Herbert Marcuse is the Frankfurt School member best known for his direct comments on technology. His major writings on the subject include several essays in sociology and psychology (Marcuse 1941; 1955), and his most important book, *One Dimensional Man* (1964). The main focus of this work is on the relationship between the forces of production, as Marx used the term, and culture and personality, as Freud understood these. In incorporating the views of Pollack, Neumann, Fromm, and others on the reach and impact of the industrial R&D orientation, this work summarizes and applies the Frankfurt School theory of totalitarianism to contemporary democratic societies such as the United States. Marcuse concluded that technology has contributed significantly to a general dehumanization and trivialization of our lives, thoughts, and aspirations. The two essential dimensions, biography and history, have been taken from modern man for the sake of the one dimension of "progress," that is, accumulation and industrial advance — or for even less rational purposes (Wolff and Barrington Moore 1967).[4] These changes produced in turn widespread frustration which at times has been expressed as a longing for direction and a sense of security, a longing that authoritarian elites can easily exploit.

Marcuse provided a massive and devasting critique of technological society. The effects of modern, labor-saving and leisure-trivializing technologies, of nuclear armament, and of advanced techniques for exercising and extending state control play a large part in this critique.[5] There is in this work a definite antitechnology, or "Luddite" orientation. Marcuse displays an ironic aversion to carrying his own Marxist and Freudian analyses to their logical conclusions, or at least as far as did some of his colleagues like Horkheimer and Adorno. This is the conslusion that technology is not "at fault," but rather that industrial society has (more or less consciously) concentrated on the creation of technological designs and applications which lead to the destruction of the human impulse and/or of the species itself.

Yet it would be wrong to class Marcuse with less politically sensitive proponents of the "autonomous technology" thesis (see chapter 8) or other variants of Luddism. When we view *One Dimensional Man* and Marcuse's observatioins on technology in the context of his other work, it is clear that he was well aware of the distinctions between the effects of economic and psychological forces and those of technology *per se*. Though Marcuse (and the school) has to an extent earned the reputation of being "against" technology, he was also sensitive to the important distinction between the current societal definition of development and that stressed in our discussion of Marx, his predecessors, Veblen, Ogburn, Mills, and Mannheim. He generally treated technology as a secondary cause, and identified as the appropriate target for applied sociology "the [political-] economic insanity, which is interwoven into the technology."

Like Mills and the others, some of the most important contributions of the Frankfurt School to the social-scientific study of technology lie not in its research on specific technologies, nor even in its observations on structure and change in technological society as a whole. They come from its perspective, its sense of engagement, and its methods of study. Perhaps we cannot accept some or most political opinions or the often excessive pessimism of Marcuse and his associates. But it is still possible to recognize the significant effect that their work has had on the philosophical viewpoint of those formulating and practicing postacademic social science today (e.g., Gouldner 1976, ch. 6).

Moral Dimensions and the "Other" Society

One contribution of the Frankfurt School likely to influence our philosophy of society for some time is its method of dealing with the morality of contemporary social institutions. Central to this method is a pair of ideal types, fascism and the "other" — or "wholly other" — society: one, the ultimate perversion; the other, the ultimate perfection of technological development. As a theoretical tool, this contrast defines the "critical" orientation by which the school members chose to be identified. In practice, members relentlessly compared social relations and cultural products, such as the family, work, art, and music, to their negative and positive ideals. The result was a type of sociological insight that goes beyond positivism in describing not only "what is" in today's world, but also the essential ingredient of applied sociology, "what might be" in the worst and best of all possible worlds (Frankfurt Institute 1972, ch. 3).

This contrast between fascist and "other" societies is one of the few features that allows us to identify a specifically Frankfurt School orientation or system. In most other respects, their personal moral views, choice of research techniques, and choice of subject matter, the Frankfurt School attempted to avoid and even to disavow systematization: "At the very heart of critical theory was an aversion to closed philosphical systems" (Jay 1973:40). Even the coherence provided by the fascist/"other" contrast is at best very loose. Despite this looseness, or at another level because of it, we can speak of sociology pursued in the Frankfurt School spirit, which unlike positivist approaches, will have a moral purpose.

> A sociology which is committed to the "postive" is in danger of losing all critical consciousness whatsoever. Then anything that diverges from the positive [i.e., anything that speaks of fascist of "other" societal tendencies], that urges upon sociology questioning the legitimation of the social instead of merely ascertaining and classifying it, becomes open to suspicion. . . .But only a critical spirit can make science more than a mere duplication of reality by means of thought, and to explain means at all times to break the spell of this duplication (Frankfurt Institute 1972:11).

In place of promoting such duplication, this approach reminds the applied sociologist of technology to seek to locate the "positive" in relation to the (ever-changing and nebulous) model of good and evil social orders. To be effective, the postacademic social scientist must be prepared to commit certain types of value judgments as an integral part of his work.

The concept of "other" society provides an ethical foundation for critical theory, but it is a classically precarious one. It is similar to the Marxist ideal of communist society but equally so to other utopian concepts constructed through the ages in every culture. In much of the writings of the Frankfurt School, but especially in the work of Walter Benjamin (1892-1940), the "other" society has less of a Marxist and more of an explicitly Messianic, Judaeo-Christian meaning (Benjamin 1958). To this extent, it is an elusive ideal. In some ways, the "other" is merely the opposite of fascism (Jay 1973, ch.5), in other ways, it is more and/or less than the antithesis of fascist societies, such as Nazi Germany, that have already existed. Rather than being even this definite, the "other" is instead used as an ever-receding ideal, against which no product of post-Enlightenment society can possibly measure up. It is a promise whose character must remain partly unknown. In part, this is because the shape of the world to come depends upon events yet to occur, and in part it is because the psychological and cultural impact of the industrial revolution, its economic and political institutions, and its technologies have made it impossible for members of post-Enlightenment societies such as ours to dream in sufficient detail about what might be.

Part of the dehumanization of modern life comes from our loss of the ability to conceive of a better world. As Benjamin observed (see also Jay 1973:175-76, 200-201), the depiction of the "other" society and its undefined aspects in the work of the Frankfurt School must be understood in the Talmudic sense: to leave the "other" society unnamed is to keep it divine. The ancient Hebrews considered it a sin to name God; for thereby He would be classified, made finite, reified, and ultimately — the cardinal sin — idolized. Similarly, the Frankfurt School resisted naming or giving specific shape to the "other" society. Yet paradoxically, in their view we cannot adequately understand such things as the role of technology in society or the role of sociology in relation to technology (or our own actions in relation to these roles) without the concept. Without a positive ideal, we cannot understand technology as the source of freedom.

A similar Talmudic, even cabalistic orientation toward the "other" society is found in the work of Erich Fromm (originally in Fromm 1927, in German). Here the "other" society is identified with Sabbath *(Shabbat)*. For Jews, Sabbath means the seventh day of the week, the day of rest and worship; but it also stands for a state of mind and being after which one constantly strives, in Yiddish, the state of being *Shabosdik*. In this second sense, Fromm and his colleagues view the "other" society as the "day" yet to come: the day on which we would be one with our spiritual roots, free of toilsome labor, and according to tradition, wear our best clothes, be well scrubbed, carry no money or other "burdensome" objects, eat the best food and have a wonderful time. Thus underlying the Frankfurt School's concept of a nonpositive social science, for some at least, is a promise of perpetual Sabbath that the Messianic tradition believes was given with the setting of the sun on the sixth day, the day of The Fall — a promise so grossly contradicted by the frightening motto *Arbeit Macht Frei,* posted at the death camp.[6]

Critique of Positivist Social Science

The fascist/"other" contrast was developed by the Frankfurt School as an alternative to the prevailing descriptive emphasis of positivist social science. It is complemented by two additional aspects of the school's critique of contemporay education. One is the school's emphasis on the advantages of interdisciplinary studies, and the other is its views on the need for a "critical" as opposed to a "pure" science orientation for sociology and related fields.

Interdisciplinarity

Like Marx and the other preacademics, Horkheimer, Adorno, and their colleagues emphasized the complexity of the task of explaining social change (for extended comments see Rose 1980). They recognized the values of a specifically sociological outlook, although their definition of the field was closer to Comte's *sociologie* than to the prevailing one. But they were also aware that sociology cannot by itself make social change comprehensible; that economic, psychological, historical, linguistic, anthropological, and philosophical orientations are also required. Their best known and most fundamental demonstration of this interdisciplinarity is in the pioneering syntheses of history, economics (Marx's materialism), and psychoanalysis (e.g., Brown 1959).[7] This once audacious move has, in detail and form, become an integral part of much contemporary social theory. By combining a focus on the erotic dimension and on the rational, material side of social life, the Frankfurt School introduced a uniquely holistic model of man.[8] With this model, the traditional questions concerning technology's impact, the relationship between functional and substantive rationality, and the effects of bureaucratic organization could be addressed in a way at once deeply personal and politically consequential.

In addition to this theoretical program, the Frankfurt School shared a commitment to the practice of interdisciplinarity, that is, to team research. The school varied in composition between its first Frankfurt, New York, and second Frankfurt periods. But it was always composed of an exceptionally heterogenous disciplinary mixture of scholars working in close collaboration. Most major social-scientific and humanistic fields are found in the backgrounds of Frankfurt School members: philosophy, sociology, economics, psychology, literature, music and music criticism, political science, linguistics, and theology. The collected works of the members reflect this range: from highly speculative to highly empirical approaches to research, from a symbolic, hermeneutic treatment of aesthetic and scriptural material to the "hard-headed" analysis of economic indicators and in-depth psychological tests. At its best, the work of the school went beyond multidisciplinary ecclecticism and achieved a truly cumulative, transdisciplinary outlook on social life.

Some of the reasons for this are obvious. Beyond its heterogeneity, there existed for the Frankfurt School a unifying impetus to direct attention to the problem at hand: an analysis of fascist society, culture, and personality. This final aim was to be pursued regardless of how its "proper" study might ordinarily be classified — as an issue in political science, anthropology, or psychology. The power of this problem-centered approach to unify was complemented by the shared antipathy among school members to "merely positivist" types of research. Their commitment to understand a very real and clearly harmful social trend motivated and strengthened the interdisciplinary character of their approach. What demanded attention for the Frankfurt School was not the kind of issue that arises within disciplinary paradigms; rather, it was a problem that arose from experience. The focus on experience, with all its complexity and ambiguity, necessarily reveals society's many-sided character and obviates the value of narrowly economic, sociological, or psychological explanations. Perhaps this is to say no more than interdisciplinarity is not an accidental feature of the Frankfurt School's approach. The school's stress on the concrete problem and its critique of positivism are matched by an ambivalence about its own disciplinary identity. Its work if often identified collectively as sociology; yet Pollack was an economist, Fromm a psychiatrist, Benjamin a linguist, etc. The term used for their approach, *critical theory (Kritische Theorie)*, is broadly classed as a type of philosophical theory; and many school members considered themselves to be philosophers (e.g., *Telos* 1, no. 1). To say that interdisciplinarity is not accidental to the Frankfurt School approach is also to say that it is an essential ingredient in subsequent social science approaches.

Positivist versus Critical Sociology

The Frankfurt School's commitment to interdisciplinarity is not only a preference for a certain style, it is also part of its critique of disciplinary social science, particularly sociology. Members of the school held that just as no social formation or cultural product is above scrutiny or beyond comparison with fascist and "other" societal elements, the type of social thought produced since the Enlightenment, and particularly in more recent times, is not exempt from analysis and criticism. As Jay, Rose, Gouldner, and other historians of the Frankfurt School have suggested, the study of the production of knowledge about society is integral to the school's study of the forces of production within society (see, e.g., references in Jay 1973:368-70; Frankfurt Institute 1972:191ff).[9] This work exhibits the most direct applications of the school's social psychology of knowledge as well as its most explicit objections to the program of Karl Mannheim as not sufficiently radical. This critique focuses on the narrowing and distorting tendencies of positivist sociology. It rejects as irresponsible the search for purity and similar scientific pretensions to which academic sociology is prone. It highlights the trivialized, ahistorical treatment of scattered milieux and relationships, and questions sociologists' tendency to seek adequate explanations with sociological concepts alone. It provides an argument for why the applied sociologist of technology must seek to understand the nonsociological side of what he studies and seek the company of nonsociologists to aid him in formulating such an understanding.

> There exists no more pure sociology than a pure history, psychology, or economics: even that substratum of psychology, the individual, is a mere abstraction when removed from his societal conditions. The scientific division of labor cannot be ignored if intellectual chaos is not to arise; however, it is certain that its division into disciplines cannot be equated with the structure of the thing in itself. That all disciplines which concern themselves with man are linked and forced to refer to each other need hardly be stressed, specifically now that the concept of totality has come to be a cliché (Frankfurt Institute 1972:10-11).

In this aspect of its work the Frankfurt School can be classed with the other critics of sociology discussed in this section. Like Veblen, Wirth, and Mills, the school viewed domination of academic social science disciplines by the pure-physics model as a key reason why the majority of people in modern societies find it so difficult to understand and even more difficult to control the effects of their own technology. Disciplinary sociology, at least as it has developed in Europe and North America, has, in the hope of gaining legitimacy, defined itself out of the business of examining technology as a political, economic, and cultural phenomenon and as a potential source of freedom. Sociology has ignored a very important part of the story of man because *technology, economy, polity, culture,* and *freedom* are not properly sociological terms.

These features of academic sociology can be partly explained as a result of the need for industrializing society to keep technology unexamined (Frankfurt Institute 1972, ch. 12; also Smith 1970). The state and the private sector could not afford to support a radical approach to innovation in which questions of substantive rationality — "Why is this specific technology being developed?" "Why is a particular design principle superior to another?" "Why is efficiency a sine qua non in engineering?" "Why is labor saving always an advance?" — are raised, in which the application of social-scientific (i.e. normative) principles to engineering design, innovation, etc., is promoted. Those who have controlled educational resources have attempted to avoid the destabilizing effects of such an approach and to maintain direct authority over the work of technicians, without competition from possible opposing ideas about what innovations would be best. They have acted to legitimize and encourage pure, disciplinary social science. Theories and methods that underlie critical, applied approaches to technology were classed as "impurities" and thus deemphasized. The result is that sociology and other social science disciplines increasingly concentrated on matters — such as "building" theories and "reducing variance" — different from the substantive rationality of technology, while technological innovation remained largely responsive to elite priorities and, with these established, to technical rationality.

In this light, the dominance of pure sociology in the universities of Europe and the United States is part of the general evolution of technological society toward massification. One segment consists of a small but powerful elite which exercises unchallenged control over national R&D priorities; the other segment includes everyone else — scientists, technicians, social scientists, laymen — all involved in the production and consumption of that which will best satisfy these priorities. Between the elites and masses, decreasing opportunities exist for the development of an organized program to examine and criticize national priorities or to transform the mass into publics to which elites would be accountable. Such a program has not been provided by the academic social sciences; although, the founders of the field believed that one would emerge, and the Frankfurt School and other sociological critics believed that one should.

The Frankfurt School's legacy is extensive and significant: this applies to its specific research on technology, development of critical theory, critique of positivism, related use of the fascist/"other"society polarity, general stress on interdisciplinarity, and its hopes for social science. This is especially evident when we consider its work in conjunction with that of other academic sociologists like Veblen, Ogburn, and Mills. In this context, the Frankfurt School approach can be viewed not as a source of unimpeachable wisdom but as an important part of the effort to eliminate Ogburn's "great wall" separating sociology and technology. In keeping with the critical spirit of the approach, it is also necessary to mention a few negative lessons that the postacademic social scientist might gain from the Franfurt School's work.

Limitations of the Approach

Pessimism

Because the members of the Frankfurt School were so intensely intellectual, and because they applied their critical method so unsparingly at times, their analysis often degenerated into a kind of dyspeptic round of complaints about all the things, which turned out to be most things, in the modern world which they happened to dislike. Despite their insistence that they were neither pessimists nor optimists, school members shared a nasty, systematic inclination to be negative, not merely in the logical sense, but also in personal psychological terms. At times it appears that their "method" was to wake up on the "wrong side of the bed," disturbed by this radio program, that new jazz record, another new poem, book, etc., and to run down to the office to write an essay on this, that, or another product of bourgeois decadence. This can be an aesthetically displeasing part of studying the work of the Frankfurt School members. Their ill-feeling toward the popular, the ordinary, everyday life verges on an even more serious breach of the humanistic commitment that they and other sociologists rightly critical of contemporary society espouse. At times it appears that the Frankfurt School was simply misanthropic.

In the face of the awesome social and personal upheavals experienced by members of the school, it is easy to understand why it was difficult for these men to maintain a cheerful disposition — although similar trials were experienced by their peers, like Mannheim, Dahrendorf, and Einstein, who remained less dyspeptic. The deeper one delves into the social forces in technological society before, during, and after the fascist era in Europe, the easier it is to conclude that all is lost, that there is no evidence of man's humanity. But any such conclusion still requires a leap of faith concerning the next piece of evidence or the next reinterpretation of old evidence to be produced. In the work of the Frankfurt School, this is often a leap of bad faith. Their biographies and the facts of history make it a type of bad faith easily understood and forgiven. Yet despair, like hope, is a poor substitute for sociological analysis, especially because of — and not in spite of — the specter of the holocaust. Survivors must be capable of performing the albeit monumental task of seeking and finding the humanity that continues to exist; even in a lousy radio show, a too-loud jazz tune, a silly Hollywood fan magazine, and perhaps most ironically, in the self-destructive nature of fascism.

The postacademic sociologist is an applied sociologist (economist, historian, etc.), which means that he must have some faith in the efficacy of knowledgeable action. This is not to say that we can afford to overestimate the power of knowledge or underestimate the formidable obstacles that lie in the path of an unalienated relationship between sociology and technology. Despite the necessity of maintaining the critical, "other"-society-oriented perspective developed by the Frankfurt School, it is equally important that the postacademic sociologist not fall prey to a hatred of life and people. Such hate is the weapon of the enemy. Instead, ways, known or yet to be discovered, must be implemented to direct this kind of emotional energy to the constructive task ahead: to begin to shape the machine in society's image and reshape society where it has been transformed into an adjunct of the machine or an unwilling object of its side effects

The Frankfurt School as Academics

The type of applied sociological orientation the Frankfurt School appears to endorse is often quite remote from the personal concerns and practices of its members. This is evidence of another limitation of their approach to which the postacademic social scientist should attend. As participants in academic milieux, and particularly as former students of an intensely formal German establishment, the Frankfurt School remained bound by the prevailing canons of professional scholarship. This is not to ignore the very pertinent fact of the school's marginality to academe (at least until Horkheimer's postwar return to Germany), for this marginality acted as both a cause and effect of the innovative character of its work. It is difficult to understand how the school's research could have been so sweeping in scope, so critical, and yet — for some, at least — so widely celebrated outside of established academic circles had it not been banished first from its own country, then to the peripheries of the Columbia University system, and (until recently) from the English-speaking sociological community in general. Most members of the school were Marxists of one variety or another, and this contributed to their marginality. Despite this, their roles as intellectuals, their language, and their audiences were decidedly academic. Along with the commitment to defeat fascism, their main order of business was the academic business of publishing books, journals, and didactic material for the instruction of students and professors, and for their own recognition and fame. Their main reforms were directed not to society but to the knowledge base and social structure of academic social thought.

In this regard, the work of the Frankfurt School continues to shake the foundations of academic social science. But it does not appear to have explicitly contributed to a better reconciliation of theory and practice, of sociology and technology. An often fervent cultism has been generated around its work, indicating that the school did not provide an adequate model for systematic innovation in social science. The lives of Frankfurt School members suggest that in order to be a critical sociologist, one must be a university professor (or even a doctrinaire Marxist professor). This suggestion is untrue, and the school's work itself would argue against it. Like Veblen and Mills, the Frankfurt School's analysis of its own activity could produce little beyond ambivalence. Like Ogburn, the individual members had at least a school of like-minded scholars to serve as a social anchor and help in directing and enhancing their creative energy. But it appeared to the school, as it does to academics generally, that there is something less than legitimate about using sociology to analyze and reshape technology to improve man's lot, as well as to understand its general effects. It is this incapacity, luxury, myopia, or whatever else it might be labeled, that postacademic social scientists must, at the risk of undermining their own unique role, seek to avoid.[10]

Argument Against the Humanities: The Critical Impulse

A final feature of the Frankfurt School approach that might be viewed as a limitation should be noted. This is the members' relative lack of interest in fiction, art, poetry, or traditional historical studies, not merely as subject matter, but as parts of an expanded, interdisciplinary sociology. Despite their inclusive reliance on all the social sciences, the school's members did not assign a major role to the humanities as disciplinary bases for critical theory. Often the role was argued to be nonexistent. For a perspective that claims to be so centered on the loss of human values, this appears to be a serious omission. Has not the artist or poet as much, if not more, to say about social relations as the sociologist or economist?

The attempts to respond to these questions (for no flat answers have been offered) by the Frankfurt School, and by critical sociology generally, are reflected in the absence of a definite place for the humanities in the type of postacademic social science under discussion here. Many dilemmas remain unresolved in this area: Can valuable social insight be gained from the widely acclaimed poetry of a person with the political sympathies of Ezra Pound? Can the beauty of Gaugain's paintings be authentic in light of the human costs that went into this work? Can Rudyard Kipling's sociological observations on India be taken seriously (Orwell 1978:105-6;122-24)? It is important to avoid the Philistinism that neglect of the fine arts, literature, and other traditional humanities can nuture. As George Steiner has observed, "much of the light we possess on our essential, inward condition is still gathered by the poet." But it is equally difficult to argue with the conclusion shared by the Frankfurt School and Steiner (1970:7) that "many parts of the mirror are today cracked and blurred."

Though it has yet to be documented or analyzed in sufficient detail by social science (but see Mannheim 1956:231ff), the humanities appear to have failed significantly to humanize.[11] We would like to believe that sentiments such as sympathy, compassion, and even empathy are instilled in those who read good books and listen to great music, and that these arts nurture such sentiments as cultural relativism. But there is little basis for this belief. It appears that the fascist, the sadist, and every other type of scoundrel bent on domination and the denial of the humanity of others has known and appreciated great literature and art. In fact, these types seem to pride themselves on their mastery of the humanities.

At a time when most of the world still cannot read and write, it seems absurd to celebrate the humanizing qualities of fields that have satisfied the intellectual curiosity not only of many decent people, but also of the most rapacious members of a tiny literate elite. There is nothing wrong with the humanities, except the perverse uses to which they have been put. Unfortunately, such uses are not merely a minor by-product of the artistic or literary enterprise about which the artist or writer need not concern himself. We would not think of indicting food merely because it nourishes bad people as well as good people. But art and food, as products, have different relationships to conscious creative activity, and they are meant to perform different functions in our lives. The point is not that the humanities are to be "indicted," but that social science cannot afford to celebrate them or become dependent upon them to have an effect that they apparently cannot have. The type of criticism of the humanities proposed by the Frankfurt School, George Steiner, and — by omission, at least — in this book, suggests only that social scientists and humanists have before them the task of helping to improve our understanding of the uses and potential of art, literature, etc., as social products and social forces. This is a task that very few have been capable of performing. The exceptions, authors and critics such as George Orwell, Steiner, and James Joyce, are noteworthy.[12]

Despite the dangers in dismissing the sociological value of the humanities, we cannot condemn the Frankfurt School's antipathy toward the arts and literature as something, like its academic nature, to be avoided. The reader who is sympathetic with the approaches of Veblen, Mannheim, Mills, Horkheimer, Marcuse, Erich Fromm, etc., might also agree — with T.W. Adorno — that "to write poetry after Auschwitz is barbaric" (1967:34), or with Martin Jay (1973:299) that "to write social theory and conduct scientific research was more tolerable if its critical, negative impulse was maintained. For, so the Frankfurt School always insisted, it was only by the refusal to celebrate the present that the possibility might be preserved of a future in which writing poetry would no longer be an act of barbarism." The maintenance of this critical impulse and the refusal to celebrate the present in their treatment of the humanities and in general is perhaps the major feature of the Frankfurt School approach. It represents an incomplete and easily abused basis for the applied social scientific study of technology. But in combination with other themes under consideration here, it is an essential part of the foundation of postacademic approaches.

Notes

1. See Jay (1973:xv,n.) and Slater (1977, ch. 1, sec. 1) for a discussion of the distinction between school and institute stages in the life of the group.
2. See e.g. Habermas (1971, 1970, 1968). His normative view of postacademic sociology as "knowledge in the human interest" is used throughout this book. The concept and Habermas' analysis have roots in the Frankfurt School, and there are definite commonalities with the work of Robert Lynd, Mills, and other American sociologists. Reiteration by Habermas at this time, during the transition between academic and postacademic eras, is significant. No sociologist can presume to know what is in the human interest. However, we must perpetually seek to know and make the practice of postacademic social science as consistent as possible with our hopefully ever-improving understanding. Perhaps it is easier to understand what is *not* in the human interest. As Habermas, the Frankfurt School, and others discussed here have observed, this negative category would include (1) totalitarianism, (2) genocide, and (3) — more elusively — anything that contributes to these ever-present tendencies in the human condition. On the positive side, one would like to say that social development is clearly in the human interest. Yet the idea of developmemt is so normative that even it — or the term, at least — has been used to promote and justify every sort of inhuman action, including totalitarianism and genocide (Horowitz 1980; chs. 1-3).
3. References for this section include Jay (1973), Slater (1977), Rose (1980), and various primary and secondary articles and essays. The volume of this work indicates an enormous and ever-expanding body of literature on the Frankfurt School (including several journals, e.g., *Social Research, Telos,* and *Theory and Society),* likely to grow more rapidly in this country under the impetus of new translations of work in German, French, and other languages.
4. Other works of the Frankfurt School directly addressed to these issues include Benjamin (1958), Adorno (1941, 1945), Adorno et al. (1950), and Fromm (1947).
5. The focus on communications technology, an interest shared with the Chicago School, Inness, and McLuhan, is significant. See e.g. the Frankfurt Institute (1972:201-2).
6. The Messianic idea of the "other" society as Sabbath is conveyed in the story told by Rabbi Shlomo Carlebach, roughly translated here:

   At 3:00 in the afternoon on the sixth day of creation, the Lord created man and woman. At 3:10, He warned them not to eat of the fruit of the tree which gives knowledge of good and evil. By 3:30, they had eaten the fruit. By 4:00, man and woman had been ejected from the Garden of Eden. By 4:30, Cain killed Abel, his only brother. Thus at 5:00 in the evening of the sixth day of Creation, Adam, Eve, and Cain stood outside the gates of Paradise — which were being guarded by angels with flaming swords — praying to the Lord to be let back in. And at 5:30, as the sun set on the sixth day, God gave them His answer: *Shabbat.*

7. Brown (1959:17) makes a pointed observation that the intersection between Marxism and Freudianism lies in the fact that both are, at root,

"sociological." "Marxism is a system of sociology; the importance of the 'economic' factor is a sociological question to be settled by sociologists. Freud, himself speaking as a sociologist, can say that in imposing repression 'at bottom, society's motive is economic.'"
8. The Frankfurt School's perspective has been most vulgarized — by its own membership as well as its followers — in the expression of the view that Marxism and Freudianism are not only necessary but are also sufficient bases for postacademic sociology. The result is a new stultified orthodoxy formed from the ruins of two perspectives which had previously (and since) suffered from independent stultification.
9. It has been suggested that the work of the Frankfurt School was primarily a critique not of society but of social knowlege, through which other aspects of society could then be criticized. While the Frankfurt School was not uninterested in empirical research, its work, perhaps especially *The Authoritarian Personality* (Adorno et al. 1950), testifies otherwise. But its basic data were verbalized observations on society, e.g., the words of sociologists, other social theorists, and of "the average man" (of special interest is Marcuse 1972:217-51).
10. For an extended comment on the gap between the Frankfurt School's own theory and praxis, see Slater (1977). The school was not unsympathetic to an applied sociology program, though characteristically its members were critical of existing attempts at application such as Taylorism. Their social relations made it impossible for them to develop and enact such a program. The later work of Willhelm Reich, a man with many things in common with the school, illustrates one unsuccessful attempt to apply Marxist-Freudian theory. As brilliant as are his scholarly observations on repression (and especially on fascism), his work on orgasm therapy and the orgone box reflect a trivialized, commercialized, and ludicrous practical application of a fundamentally profound idea.
11. This general argument was crystallized by George Steiner (1970:95-109). His most provocative statement is the highly controversial essay on language "The Hollow Miracle." In an earlier section of the book, he summarizes his diagnosis of the humanities and prescription: "Because the community of traditional values is splintered, because words themselves have been twisted and cheapened, because the classic forms of statement and metaphor are yielding to complex, transitional modes, the art of reading, of true literacy, must be reconstituted. It is the task of literary criticism to help us read as total human beings, by example of precision, fear, and delight. Compared to the act of creation, that task is secondary. But it has never counted more. Without it, creation itself may fall upon silence" (1970:11).
12. "Joyce," observes McLuhan, "was probably the only man ever to discover that all social changes are the effect of new technologies (self-amputations of our own being) on the order of our sensory lives. . . every major technical innovation will so distort our inner lives that wars necessarily result as misbegotten efforts to recover the old images (McLuhan and Fiore 1968:5).

PART II

Society and the Study of Technology

Introduction

The preceding chapters have presented an inventory of the contributions of selected amateur and academic social scientists to the applied, critical study of technology. Throughout this discussion it was suggested that while such an orientation could play only a minor role in the past, it is now becoming central as social science, like the non-social sciences, enters the postacademic era. Today increasing opportunities exist for the conduct of the type of research in which the social antecedents and consequences of technological innovation are featured and for participation by sociologists and other social scientists as equal partners in the process of technical design, diffusion, and evaluation. One major condition contributing to this transition is the growth of interest in social science outside of traditional academic settings such as disciplinary Ph.D.-granting departments. These opportunities have arisen at a time when social science has achieved its place as a valuable, though rather unusual, modern scientific pursuit.

In this and the following part, the history of social-scientific approaches to technology is placed in the background in order to focus on specific features of the transition to the postacademic orientation. This discussion continues from the point at which our review of Mannheim's program for a post-World War II social science concluded (chapter 4). This program provides a framework for integrating several current trends in sociotechnical research and education to be considered here: the expansion of the role of social science beyond purely academic activities; the focus on the unintended, unplanned, and often ironic character of the effects of technological change; the sociology of knowledge approach to the study of technology; the stress on interdisciplinarity; and, in part III, the commitment to a developmentalist sociology.

In the chapters in this part our attention centers on the ideas that define that postacademic style and on the material sources of the transition in social science from a sociology of technology to sociology/technology. To accurately assess the character of this transition, it is necessary to reflect both on the obstacles and the advantages that attend an increased role for social science in engineering, planning, and policy application.

On the encouraging side, public and professional interest in nonacademic activities has grown rapidly in the past two or three decades and a firm material basis for their support is evolving. This is evident in the birth of a new transdisciplinary academic field — the Study of Science, Technology, and Society (SSTS)— especially in the policy and applied orientations of this and other new fields, and in the considerable growth of interest in applied interdisciplinary research. From the point of view of the development of the social sciences, this renewed, broad-gauged focus on the role of technology in human affairs can provide new insights as well as an impetus for the improvement of methods and theoretical orientations that have so often been criticized.

The applied social-scientific study of technology is also a pursuit rife with potential ideological and political traps and implications. Such a study provides special opportunities for the (conscious and unconscious) substitution of the investigator's own hopes, fears, and values for scientific analysis. A detached perspective on technological impact is especially precarious. What has often been produced instead is a range of theories about technology that, while presuming to be undistorted accounts, provide only a (systematically) partial view of an exceedingly complex reality. To illustrate this point, two very different types of ideologies of technological impact, mentioned briefly in earlier chapters, are selected for more detailed examination in chapters 8 and 9. One of these, the positivist-technocratic view, equates progress and the "good life" with increases in the value of per capita statistical indicators of technological innovations — thereby begging important questions about the meaning of such indicators, increases in their value, and even the quality of life itself. The other perspective focuses on "autonomous" technology. It argues that innovation leads not to the good life but to a growing loss of control by members of technological society over decisions that affect their lives. In stressing the "Frankenstein's monster" side of technology, this perspective not only underestimates the extent to which technology enhances man's ability to control his lot, but ignores the key fact that the impact of this autonomy affects the distribution of power among different groups in society.

These issues relate closely or directly to the various communication gaps that exist among sociologists, technicians, and the public. In some respects, it appears that a reconciliation between sociology and technology is now inevitable. But unless the sociologist, the technician, and society proceed with a clear idea of what sociology and related fields have to offer, the reconciliation may not be a happy one.

7.

Recent Landmarks in the Transition to Postacademic Social Science

> *We, as professionals, seem to have that detached interest about problems of overwhelming importance which will affect following generations (if there are any). We get much more excited about threats to our own jobs and about affronts to professional dignity (or ethical standards) than we do about the fate of the world. I think we are going to have to change or everything will become academic — in that quite different sense of the term.*
>
> Alan Grimshaw (1979)

Today for the first time in over one hundred years sociology and related social sciences are, to a significant degree, being practiced outside of normal academic settings. Social scientists have always been used in limited ways by business, government, and R&D laboratories, and some fields, economics and demography in particular, have played major roles in such nonacademic work. But until recently, most of the activity and resources in sociology and even in economics has been directed to the two main academic functions: teaching, especially in disciplinary Ph.D. programs, and research, especially pure disciplinary research. Almost without exception, the career outlook for a person holding an advanced degree in sociology, for example, has been to teach in a department of sociology at a college or university — and in the high schools of about one-third of the states in the United States (provided that one also has a teaching certificate or degree). This situation began to change after World War II, and by about 1970, a new trend was clear.

A large number of social scientists, perhaps most, are now employed either (1) outside of colleges and universities, (2) within them, but in the capacity of full-time, usually interdisciplinary researchers, or (3) to teach technical, professional, science, and liberal arts students in a "service" capacity, that is, in non-degree-granting and/or multidisciplinary programs. These are the only types of employment for social scientists in which the number of jobs and salaries have grown substantially since the early 1970s. There is every reason to believe that these trends will continue, at least till the end of this century.

Postacademic Style

The category "nonacademic social science" encompasses a diverse range of activities: advisory, report writing, team research, and "service" teaching. One important feature is common to most of them: the discipline, e.g., sociology, is used not to perpetuate itself, but rather for a practical, nonsociological purpose. In most of this work the goal is not to communicate with, influence, and be influenced by other sociologists or sociologists-to-be, but to put sociology into practice in conjunction with public policy, business planning, technical R&D, and professional education. The category includes technical advising, the many things that have been called applied sociology, as well as teaching and research. It differs from academic social science not in theory or method, but in orientation, the way in which the discipline is used.

In this sense, nonacademic social science is a kind of technology. It is meant to be used in conjunction with other fields as technology. Unlike its academic counterpart, it is knowledge that is intended to do something: to change, increase, or improve something other than the stock of sociological knowledge in research reports and in students' minds. Nonacademic activities require the ability to communicate about concrete issues with professionals, technicians, and practitioners. They require closure, some type of finality in the form of a decision, advice, or a line of action: the pat conclusion to academic discusssions, "further research is needed," cannot be the only or even the main finding in nonacademic work. These activities require a relatively overt normative perspective; the nonacademic social scientist must have a clear sense of what he wants to achieve in society, what effect he hopes to have on the world with his work.

When few social scientists did anything other than academic work, the difference between the two orientations was not a serious problem. The focus of the disciplines could remain on specialized training and pure research, and this would satisfy the needs of all practitioners. But as has occurred in the non-social sciences during the past several decades, the overall orientation of social science disciplines is now shifting from academic to postacademic. Since the transition is so recent, social scientists outside of academia have found that the way they need to relate to their field is different from that for which their academic training prepared them.

As the role of social science in nonacademic activities increases, these activities themselves are changing. Just as the sociologist who works at a business school or with development policymakers will have a different approach from that of his academic counterpart, the physician, manager, or engineer who works on a regular basis with sociologists will be affected by the experience. Whether these changes will benefit the sociologist and the engineer or, more important, whether the relationship will result in a better way of doing business, formulating policy, etc., depends upon how the principals approach the task at hand. It is here that a lack of knowledge about nonacademic work, a lack of interest in it by the proponents of pure sociology, and lack of adequate role models — people from whose experience one can learn — can limit the effectiveness of postacademic social science (see volumes 1-3 of *Sociological Practice*).

The overview presented in part I emphasizes the observations, theories, and perspectives in amateur and academic sociology that focus on postacademic concerns: interdisciplinarity, a normative action orientation, a sense of the limitations of pure social science, and an explicit interest in technology. In this and the following chapters of this part this connection between academic sociology and postacademic social science is discussed and illustrated further. To emphasize the contributions of Veblen, Ogburn, Mannheim, Mills, and the Frankfurt School in this manner is not to suggest that their writings alone constitute an adequate basis for the theory and practice of postacademic social science. The range of problems the postacademic social scientist is likely to confront requires an extremely broad knowledge base, not only in specific academic disciplines but also in the technical and professional fields and several other areas. The special value of the work of Veblen and the others is that it so effectively drew upon the diversity of sources, ideas, and theories — beyond narrow disciplinary approaches — necessary for an effective social-scientific understanding of technology. This and the following chapters present several additional and more current observations on the relationship between society and its technologies and on the postacademic role. A combination of historical and biographical factors, including the considerable influence of the sociology of technology, has by now produced a well-articulated style suited to postacademic social science, though its roots are fundamentally academic.

What characterizes this style? Although this question cannot be definitely answered at such an early point in the postacademic era, it is possible to discern from the work of Mannheim and the others certain general features. For example, it clearly includes an applied, interdisciplinary, and critical orientation to the study of the social antecedents and consequences of technological design and innovation. It treats technology as knowledge, and is interested in the sources of our knowledge about technological change. It sees knowledge as a tool for understanding and shaping technology. It is also action-oriented in that, as a humanist science, it has a distinct commitment to understand ing the moral dimensions of social reality and to applying this understanding. Most generally, it seeks to insert the human (and frequently this means the "uncertain") element into all considerations of social change in technological society.

Often this style has been expressed through criticism of current sociologically naive theories of change or through criticism of socially insensitive institutions from an antiauthoritarian perspective. At times it has been expressed through the positive commitment to the simple principle that it is people who make and use machines and social techniques. It is people's decisions about how technology is to be employed and designed and not the machines, the social techniques, or other inventions, in themselves, that make history — although the effects of such decisions and policies frequently differ in ironic ways from their intended effects. From this it also follows that the postacademic style carries a commitment to communicate with other people about social change and technology, and not only with social scientists (although intraprofessional communication is obviously important too). While these commitments express general and widely espoused ideals, it is significant that in the transition from academic to postacademic science, a concrete opportunity for the realization of such ideals now exists.

The remainder of this chapter discusses aspects of the transition to postacademic social science which indicate that the material and intellectual bases for reconciliation between sociology and technology can be emphasized and made to prevail, and that the associated communication gaps might be more easily closed in the coming years. In particular, we examine three types of changes now occuring in professional activities: one is the growth of opportunities for nonacademic sociotechnical work, another is the formation of a new sociotechnical cross-discipline: the Study of Science, Technology, and Society, and the third is the incorporation of classic sociological techniques such as the ironic focus into technology policy analysis. These trends, in combination with similar changes in technical fields like engineering, planning, and management, are shaping the future of postacademic social science.

Shift to Nonacademic Support

Evidence of the growth and institutionalization of these types of technology-centered, applied social science can be found not only in sociological writings but also in a great variety of other sources and events. For example, the number of graduate, undergraduate, and high-school courses on sociotechnical issues now being offered in the United States, Europe, and the Third World is large and growing rapidly (e.g., Heitowit 1977; Moravcsik 1975). The appeal of television programs such as the 1979 Public Broadcasting System production "Connections," and Courses by Newspaper on energy and society (Kranzberg, Hall, and Sheiber 1980) and related subjects indicate a widespread public interest and acceptance of the sociology of technology perspective. Finally, changes in occupational patterns and in the definition of fields in several professions — academic, research, and technical — suggest even more convincingly that the postacademic era has arrived for social science.

The institutional beginnings of postacademic social science, for which seeds were sown as early as the amateur era, has diffuse boundaries. In some respects, the postacademic era began in earnest (in all fields) with the general social and economic recovery from World War II.[1] In other respects, the postacademic era has not yet been embraced, at least by the thousands of social scientists and their students who continue to equate their calling with traditional scholarly pursuits, in the interests of the discipline, and within the confines of college and university disciplinary departments. But opportunities for social scientists to apply their craft in other than these traditional ways have never been better.

The disciplines have achieved sufficient maturity, and their pratitioners sufficient prestige and political acumen, to give them credibility and a degree of demonstrated merit among policymakers, technicians, non-social scientists, and educators of many types. It is no longer necessary for social scientists to believe that they can or are somehow fated to communicate only with one another. This is being demonstrated every day, with varying degrees of success depending on personal and social factors, by the many men and women who find themselves to be "one of the sociologists," or perhaps, "the sociologist," while working among engineers, physicians, policymakers, and even more frequently for the interdisciplinary teacher, among other social scientists.

Other signs of the transition from disciplines oriented principally to reproduction of themselves and their output to those oriented to other ends can be identified. Textbook structure increasingly resembles not that of introductory physics texts but of introductory engineering texts. The focus is once more on problems and solutions, replacing the focus on the mastery of concepts, and traditional literature (Horowitz 1969). Government agencies such as the the Department of Energy, Housing, and Urban Development, and USAID now require social impact statements — assessments of the likely effects of a particular policy on family, employment, housing, population. etc — to accompany proposals for administrative and technological innovations. Increasing proportions even of teaching sociologists are doing their teaching, not among other sociologists in sociology degree programs, but among political scientists, architects, and managers in programs granting interdisciplinary degrees that require a sociological component. The effect of these trends is a steady movement toward an era when even academic social science's principal role — in graduate education especially — will be to train interdisciplinary professionals, very few of whom are themselves expected to train disciplinary specialists or to do specialized disciplinary research (Solomon 1977,1978).[2]

Since World War II, improvements in the technologies of information processing and the rapid pace of scientific, technological, and social change combined to seal the fate of the individual academic scholar (just as the amateur scholar's fate was sealed with transition to the academic era). Early in the twentieth century, growth in the knowledge base was straining traditional academic means to produce even new social knowledge (Ben-David 1971:108ff). But after World War II, the information industry grew tremendously and the tasks of gathering information, interpreting it, and integrating it with existing knowledge were becoming increasingly complex. This provided a necessary condition for the effective replacement of the traditional academic scholar by "big machines" and/or his subordination to the will of those bureaucratic organizations in which big machines are "nursed." As was the case in engineering and the non-social sciences, the research tasks of social science and its branches are now too expensive to be pursued in a prewar manner.[3]

This is not to say that traditional institutional settings for academic sociology were immediately or necessarily weakened by this change in conditions under which social science research must be conducted. In many ways traditional institutions remain strong. When and where resources are available, universities and colleges are able to support the bureaucratic reorganization of research activity and purchase the very costly material necessary to conduct scholarly research today. During the 1960s, a decade which by the calendar would be considered postacademic, considerable growth occurred in sociology, the other social sciences, and all academic fields. But when the demographic, economic, and ideological trends — including those related to Sputnik —

that fostered the boom of the 1960s began to shift, academic institutions and research suffered a serious recession. The fields most seriously affected were those, like sociology (but even more so history, philosophy, and literature), which had not developed a strong practical component or a nonacademic research base. Today most academic fields in the United States are in a period of no-growth or decline in funding, personnel, students, new programs, and employment opportunities. The exceptions are in growth areas of the country, such as the West, Southwest, and Southeast, and in schools not considered traditional academic types: community colleges, institutes of technology, medical schools, public policy centers, etc.

Support of social science in conjunction with technical education and research is growing. This development is critical for the future of the postacademic orientation. The understanding that "to know Nature is to rule her" can be traced to the days of Francis Bacon and earlier; yet it remains a guiding principle at professional schools, institutes of technology, and in organizations where professional and technical fields are practiced. What is new is the rediscovery by these fields that human nature is a serious factor in our plans to "rule"; that technology is a thoroughly social force. Engineers, planners, and others have recently expressed their willingness to count scientific knowledge of human nature as a necessary component of their education: in professional engineering publications, in research training programs, and in curricular reforms at technical schools (Rossini et al. 1980; Johnston et al. forthcoming). The implications of such changes can be great for technicians, social scientists, and society at large. Social science, despite its limitations, has won an important measure of acceptance in the United States and elsewhere — perhaps most significantly in the Third World. The idea of placing social science on a more equal footing with other sciences as a knowledge base in engineering and planning is as old as sociology itself. But under present circumstances, the social sciences may at last be able to realize their potential as a most potent (and explosive) addition to technology's pool of resources.

Traditional academic institutions no longer have either the type of organization or money necessary to promote good social science. While the schools can pay at least part of a professor's salary, research and other activities must be supported in some other way. Increasingly, the pattern is for the best work of academics (by professional standards; see Zelditch 1979) to be subsidized in a nonacademic manner: either at a university but under the sponsorship of large public, quasi-public (private foundations), and private sector organizations, or within the confines of these organizations and professional schools on a permanent, temporary, or consultant basis. It is the prevalence of conditions such as these that marks the end of the academic era. Such conditions provide to the practicing sociologist a different set of options, needs, and opportunities than was the case when he or his teachers began to pursue sociological careers.

The Study of Science, Technology, and Society: A Field Is Born

This growth of social science beyond traditional academic boundaries is part of a general trend in the history of science that was forcefully brought to the attention of scholars and the public in Derek Price's *Little Science, Big Science* (1963). According to Price, and to the generation of sociologists, historians, and other students of scientific change influenced by him, extraordinary growth in all dimensions of professional science has occurred since World War II. It has been acccompanied by an apparently inexorable increase in the degree of specialization and interdependence in scientific activity. One product of this increasing division of labor among scientists is the proliferation of new academic interdisciplinary specialties.[4] These fields, ranging from biochemistry to bioethics, share the premise that many real-world problems are easier to solve through the application of perspectives that go beyond the limits of any single discipline. Among the most significant of these interdisciplinary (or cross-disciplinary) specialties is the one which Price himself is partly responsible for founding, the Study of Science, Technology, and Society (SSTS).

The history, strengths, and limitations of SSTS bear important lessons for the postacademic social scientist or for anyone who has seen the need to apply two or more scientific perspectives in trying to understand or change an increasingly complex world. Major works in the field, which appeared at about the same time as *Little Science, Big Science,* include the widely read writings on the "two cultures" by C.P. Snow (1964) and Alfred North Whitehead (1967), the historical studies of technological impact by Lynn White, Jr. (1962, 1968) and Samuel Lilley (1966), and the macroscopic analyses of Emmanuel Mesthene (1968,1970) and Robert Boguslaw (1965). More recent general studies that have become foundation works in SSTS include those of Talcott Parsons (1970), Fred Cottrell (1972), David Freeman (1974), Bernard Gendron (1977), and Noel de Nevers (1972). In 1977, indication that the field had come of age in its own right was provided by Price, his coeditor Ina Spiegel-Rösing, and sixteen other contributors who, under the aegis of the International Council for Science Policy Studies (ICSPS, of which Price was the first president in 1971), published a thorough survey of the "state of the art" and the future prospects of SSTS (Spiegel-Rösing and Price 1977).[5]

Elements of SSTS

According to this survey, SSTS can be divided into three major subfields: "normative and professional contexts," "social studies of science: disciplinary perspectives," and "science policy studies: policy perspective." Within each subfield much recent original work has been produced on criticism of science, the sociology of research communities, conditions of technological development, psychology of science, technology and public policy, military policy, and science and technology in developing countries.[6] Many current contributors to the SSTS perspective are acknowledged leaders in their own disciplinary areas. This fact, the variety of national and disciplinary backgrounds of participants, as well as its sponsorship by ICSPS, have already provided this young field with the stamp of academic legitimacy.

The subjects that constitute SSTS have inspired much useful and important research and policy formulation. Scores of interesting questions have been raised in this work and new directions for future research have been suggested. But in this strength also lies a key weakness of the field. As much as participants are sensitive to promising research directions, they are disconcertingly tentative about identifying the major issues and most fruitful questions that have arisen in the current relations between science, technology, and society. For this reason, much of this work is merely descriptive, bland, and even boring. Many of the works in this field tend to overwhelm one with information without providing a sense of proportion or a general program.

The self-conscious creation of a new interdisciplinary field that specializes in science, technology, and society is both audacious and paradoxical. The audacity of this commitment lies in the fact that the study of the interconnections among science, technology, and society is no less than the program of all the modern social sciences. As a program first expressed in the early days of the industrial revolution, there is nothing either new or special about it. Thus when SSTS is interesting, it ceases being a specialty at all. At the same time, it invites comparison with more traditional approaches not necessarily labeled "SSTS"; in this comparison, SSTS appears less than entirely successful.

The paradoxical aspect of the creation of SSTS is also related to the extensiveness of its subject matter, which contrasts sharply with the focus of most current intradisciplinary activity. As Big Science becomes more specialized, it also becomes more difficult for individual scientists to study anything but an incomprehensibly narrow portion of the world.

The public and those who formulate public policy are, or believe they are, increasingly dependent upon scientific counsel. Unlike the scientist, the public and policymakers tend to define problems in broad, emotive terms. Thus a structural communication gap has evolved in society's perceived dependency on science: scientists grow increasingly cautious and precise while society demands bold and sweeping answers. To close this gap, so that their science can fulfull its social responsibilities and so that they can take advantage of social opportunities, scientists who wish to communicate outside of an ever-contracting circle of like-minded professionals learn to "despecialize."

It is understandable, therefore, that interdisciplinary fields like SSTS are the ones most self-consciously focused on application. This is a source of both their strengths and weaknesses. The advantages of shaping a specialty to be policy-relevant are obvious and material. The business of funding the kinds of research that policymakers believe they need is big business (although even in the most capitalist countries, it is public-sector business). To focus on issues as formulated by government bureaucracies is one way to ensure the health and continued growth of a field. But as James Coleman (1972) and others (e.g., Horowitz 1971; Wax and Cassell 1979) have observed, allowing nonscientists to define scientific problems can also introduce intellectual and ethical problems.[7] One of these problems is a lack of depth or vision in the way in which disciplines are applied in interdisciplinary, policy-specific approaches. This lack of depth is evident in much of the work of SSTS, and should be considered a potentially serious pitfall for postacademic social science in general.

SSTS as Technical Virtuosity

Ina Spiegel-Rösing's otherwise excellent overview of SSTS is marred by an uncritical and basically favorable assessment of what has come to be called "scientometrics," the measurement of scientific activity (Spiegel-Rösing and Price 1977, ch. 1). Scientometrics, like any technology, is vacuous and its practice meaningless without theory. This was clear in the debate over a recent and very important type of scientometrics, citation, and co-citation analysis.[8] It is evident from this research that to "measure" scientific activity by means of counting citations without a detailed theoretical understanding of what it means, in all its social, cultural, and political manifestations, for A to cite B in publication P at time T under conditions C, etc., is worse than meaningless; it is a particularly insidious form of what Sorokin (1956) called "metromania." To suggest, as Spiegel-Rösing does, that scientometrics can perform important functions such as "mediat[ing] between the social studies of science and science policy studies" is to exchange scientific vision for accuracy.

The tendency to equate an interdisciplinary field (or discipline) with a methodology is familiar. As is the case in most social science disciplines, the appearance of this tendency in SSTS reflects the general tentativenesss characteristic of the field. The influence of Thomas Kuhn's theory of scientific revolutions (1970) in SSTS continues to be great. This is so not because his views on the rise and fall of different standards of truth, or "paradigms," are flawless; many of the points raised by Lakatos and Musgrave (1970), Frederick Suppe (1974), and other critics are sound. Rather, it is because Kuhn has expressed well a widely shared uneasiness. This is the uneasiness about our present age of uncertainty and about the implications of uncertainty for the growth of scientific knowledge. The tentativeness in SSTS is related to the fact that practitioners are especially aware that they themselves may have no paradigm, or that the paradigm to which they adhere can be (and probably will be) overthrown.

The SSTS approach to the social antecedents and consequences of technological change also illustrates these limitations. Writings in this field stress the growth of technology assessment, social impact analysis, and appropriate technology strategies, and they emphasize the importance to contemporary society of the Manhattan Project, the Pugwash Conference, Sputnik, resource politics, and NASA. But these concerns are not always moderated by historical, sociological, or philosophical perspectives. The sensitivity of SSTS is geared to that which is current and to the organization. Culture and personality play minor roles in the relationships among science, technology, and society for SSTS specialists. Generally lacking is the political and ideological depth which discussion of these issues reaches, for example, in the works of Mannheim, the Frankfurt School, or even Veblen. Though neither Mannheim nor Veblen would qualify as a practicing specialist in SSTS, their views on the impact of technology, the social sources of innovation, and technocracy have far more substance than is characteristic of most studies in SSTS.

Veblen, Ogburn, Mannheim, Durkheim, and dozens of other identifiable (though interdisciplinary) social scientists have related to the study of science, technology, and society in a critical manner largely absent in SSTS. This is partly a function of SSTS's relative youth and partly the consequence of the difference between the more theoretical social scientific tradition and the bureaucratically sensitive, elite focus upon which SSTS is based. Any comprehensive approach must also be selective. But a confessedly amateur bit of scientometrics indicates that much of the literature in SSTS is selective in a special way. A count of separate entries in the indices of the state-of-the-art survey cited reveals, for example, that Robert K. Merton, Harvey Brooks, Thomas Kuhn, and the Organization for Economic Cooperation and Development were mentioned on an average of twenty or more separate pages. On the other hand, Einstein has an average of two entries, James Coleman one, and Mannheim, Sorokin, and Copernicus are not cited at all. This suggests that SSTS is largely derivative, secondary, and in brief, dependent on Big Science, Big Government, and, as Spiegel-Rösing herself notes, "bigness" in general.

SSTS and Postacademic Social Science

Though it is a scholarly research and teaching-oriented field, SSTS is not a traditional academic discipline; it is a very special creature of our time. Its value to postacademic social science is therefore especially great. The collective product of this field is a valuable resource for those who seek to comprehend and alter the relationship between people and technology. But because SSTS practitioners are, foremost, academics, their work is prone to several weaknesses to which the postacademic must be alert. Not the least of these is an understanding on the part of SSTS "specialists" of their own roles and of what is relevant. Absent here is the concern for thousands of men and women currently working out the complexities of the society-technology relationship in research labs, policy centers, and professional and interdisciplinary teaching programs. It is by them, and not merely the science policy celebrities and international councils, that the history of postacademic social science is made. Exclusive focus on the world of Big Science, as Derek Price and his students are well aware, can contribute to the perpetuation of an inherent elitism that has often served to promote vested interests and state power rather than human development.

Irony and Technology in Application

In our discussion of Mannheim's program it was pointed out that group situations condition our perception of technological innovations, even for the technician, and that distorted accounts and significant omissions typically accompany the process. This view is shared by contemporary sociological students of Mannheim and by specialists in the relatively new interdisciplinary fields of technology assessment (TA), social impact analysis (SIA), and related technology policy approaches.

These approaches have made much use of the concepts of irony and detachment and of the social theories from which they derive, and they illustrate well how Mannheim's program is being applied today.

Irony and TA-SIA

TA and SIA were formed as distinct fields in 1967 when the U.S. Congress created the Office of Technology Assessment.[9] Science policy specialist Dorothy Nelkin (1977) describes the approach, first by quoting Congressman Emilio Q. Daddario. TA is a "a method of analysis that systematically appraises the nature, significance, status, and merit of a technological process" (428). Next, she elaborates:

> Technology assessments are intended to be comprehensive. One of the more elaborate approaches lays out seven steps to an assesment: defining a task, developing relevant technologies, defining the non-technological factors influencing applications of the technology, identifying areas that would be most influenced by application, tracing the process by which a technology influences a society, identifying alternate programs for obtaining maximum public advantage for the technology, and finally, analyzing the social consequences of each option (428).

Guidelines for social impact analysis were originally established as part of the TA framework. The need for a special focus on the social consequences of technology was outlined, as follows, by François Hetman (1973:269):

> The causal links which characterize the interactions between technology and society are of both a physical and a social nature. Therefore, it would be wrong to assert that such categories of side-effects as, for example, those on environment, can be studied through the utilization of the methods of the natural sciences themselves.[10]

Much of this work has focused on current technological innovations in developed countries. Perhaps because of the youth of the approach, it has been heavily oriented toward method rather than empirical studies or theory building. In part because technology's effects are so explicitly the outcome of conscious planning and so dramatic there, some of the most interesting recent examples of TA-SIA studies are concerned with the social effects of innovation in developing countries (Rossini and Bozeman 1977; Wolf and Peterson 1977). This research has evolved a strong anthropological component and draws extensively on previous research in rural sociology and agricultural economics (Sofranko, Fliegel, and Sharma 1977). Most significant from a public policy perspective, Third World SIA research is beginning to be incorporated into national and U.N.-sponsored work on developing appropriate technologies (Singer 1976; Hough 1979; UNESCO, various numbers).

Objectivity and Unintended Outcomes

Throughout TA-SIA research, in the Third World and in developed countries, run two themes that relate directly to Mannheim's views on the role of the ironic focus in understanding technology's effects: one is an interest in the unintended, unanticipated impacts of innovation; the other is the TA-SIA specialist's concern for the always elusive objectivity. Hetman (1973:268) summarizes the first of these themes by noting that "side effects manifest themselves not as global attributes of a given technology but as unexpected impacts on certain groups of population." Such impacts are referred to in TA-SIA literature as second-order consequences (after Bauer et al. 1969), and their relationship to the ironic focus has been partly explicated: "Bauer and his associates...occupied themselves particularly with 'second-order' consequences, and these may here be understood as having the same meaning as unintended effects. Not all unintended social effects of technology, by any means, are undesirable" (Schneider 1975:51). Nelkin (1977:429) has provided an extended discussion of the goal of objectivity, indicating that the nature of TA-SIA makes this goal a difficult, if not impossible, achievement. While TA-SIA "are based on the assumption that 'objective' identifications of the impact of technology will help yield rational solutions...assessment itself is a political process involving evaluation of the social desirability or undesirability of specific technologies" (see also Winner 1972).

The ironic focus has been stressed in Mannheim's and other sociological writing because of its advantages in providing an objective perspective on unanticipated consequences. Incorporation of the ironic technique into TA and SIA appears to be a viable amendment to the suggestion that follows from Nelkin's (1977) observation that the search for "the" account of technology's impacts should be declared futile and instead an adversary structure for TA-SIA be devised, involving opposition groups performing counterassessments — that the value-free approach should be abandoned altogether. While not denying the value of team research and multiple perspectives in TA-SIA, it might still be argued that they do not directly answer the need for a commonly accepted body of theory and method. From our review of the role the ironic focus on technology has played in sociology and related fields, it would appear that such a body of theory and method does exist. An examination of the social-scientific literature reveals that "there are endless ironies about technology. It is in many ways a most ambiguous boon to mankind... The sociologist inevitably has a special interest in the effects of technology within society itself" (Schneider 1975:50). Seen in this light, the ironic focus is an important element in the assessment of technology's social impact because it calls attention to a range of possible consequences of intentional social action suppressed (or in the extreme, inconceivable) by interest-bound actors.

The sociologist's reminder that not all technology's ironies are undersirable is important at a time when for many TA has come to mean "technology assassination." It is intended to serve as a caution to those who are too ready to stress a technology-versus-society mode and ignore the unlimited and liberating potential of technological innovation. Whether an outcome is desirable depends upon many variable aspects of human behavior to which the sociological perspective and the ironic focus sensitize the observer. Prevailing values can change in the course of extended action, perspectives of the various actors involved can differ substantially, and limits of the order of consequences beyond which the observer chooses not to hypothesize are always arbitrary. All these must be considered before the net effect, for good or ill, of a technology can be calculated.

Role of the Ironic Focus

Within these limitations, the ironic focus provides technology assessment with a social-scientific basis for evaluating the impact of innovation. The ironic focus guides the formulation of hypotheses stating that outcomes are dramatically different from intentions and/or anticipations. As a sociological technique, the ironic focus calls attention to the large number of possible coactors implicated in the web of social affiliations invariably associated with technological change. While the ironic focus cannot be used to decide the truth of hypotheses, its deep roots in both social science and the arts provide the technology assessor with access to an extensive body of accumulated examples and theoretical formulations. Schneider points to Ogburn's (1934a) discussion of the consequences of the automobile as an example of a very early technology assessment. One might also mention the observations of a substantial majority of social scientists, such as Durkheim's views on the relationship between technological progress and anomie, Freud's observations in *Civilization and Its Discontents,* and Marx's critique of capitalism.

In some respects, the course of Western social thought since the industrial revolution can be traced following the dialectic between technological change and the social criticism of technology's ironic consequences. Contemporary British critic Douglass Muecke, drawing on the literary work of Tolstoy and Kafka, makes a similar point:

In one obvious sense — that of technological advance — progress is a fact; it would be hard to persuade an astronomer with access to a computer or health authorities in a country which has eliminated smallpox or malaria that progress has not been made. They might, however, be open to the view that solving one problem is the surest means of discovering further unsuspected problems. Looking back, we can see how far we have come; but looking forward, "behold with strong surprise the new and distant scenes of endless science rise" (1970:31; also see Thompson 1948; Mann 1960).

The ironic focus provides TA-SIA approaches with a basis for taking account of social forces that impinge upon intentional action. It provides a rational basis for changing action patterns (e.g., changing plans for phasing in new technologies) and intentions or, at least, anticipations. It is part of an intervention strategy whose purpose is to avoid sudden reversals. It is a method for making a potential "victim" of technology aware of his plight in time to take effective counteraction (Arnstein and Christakis 1975:99ff).

In Mannheim's and other sociological uses of irony, involvement and thus interests are believed to narrow the focus of actors to a certain range of consequences. The use of irony in TA and SIA is based on a parallel assumption; that "outside" evaluation of the impact of technological change is needed because producers, consumers, and regulators all have a partial perspective because they are actors (many of these points are documented in a recent collection with the succinct title *Technology and Social Shock;* Lawless 1977). The ironic focus and the sociological way of looking at the world of which it is a key component are the types of "methods" outside of the natural sciences which Hetman feels are required in TA-SIA approaches. As Merton (1949:81) noted, they may be "indispensable elements in the theoretic repertoire of the social engineer. In this crucial sense, these concepts are not 'merely' theoretical. . .but are eminently practical. In the deliberate enactment of social change, they can be ignored only at the price of considerably heightening the risk of failure."

Role of the Sociologist in Transition

The academic era can be viewed as a century-long interlude in which most aspects of the agenda laid down by Adam Smith, Saint-Simon, Marx, and the other founders were set aside while other battles — including sociology's battle for legitimacy — were fought. Much academic social science, even cross-disciplinary fields like SSTS, has been motivated not by an interest in extending the vision of social knowledge for social development. Its sources lie in more secular interests: extending the power of certain disciplinary purists, ensuring that the fields appear sufficiently like physics to win public support for their practitioners, or the — ultimately conservative — interest in reproduction for its own sake.

This is not to say that such motivations somehow render the products of disciplinary academic social science valueless to those committed to the pursit of sociology for development. The postacademic social scientist is likely to be tested in so many and diverse ways that he cannot afford to avoid acquainting himself with any particular approach, regardless of how remote from applied concerns or from the social study of technology it may appear. But with the coming of the postacademic era, the style of sociology practiced by the majority since the delegitimation of Marx is no longer necessary or useful. The postacademic social scientist must be especially attentive to the work, styles, themes, and emphases of the minority — which includes, among others, the sociologists of technology discussed in part I.

These remarks should not be misunderstood as a prediction or an endorsement of the end of academic sociology. We are suggesting that its dominance is diminishing in comparison to other types of sociology and that its role is changing to one of service rather than mastery. Many sociologists and other social scientists believe that their discipline best serves the human interest when it is encouraged to "tend its own garden." In a formal sense this is entirely true, at least as true as is the application of the laissez faire model to the pursuit of any modern profession or occupation. One would never like to see the objectivity and independence of inquiry, for which sociologists and other scholars have struggled since the Enlightenment, subordinated to the will of particular social groups, political parties, or ideologies. Schneider's (1975:331) observations are well taken when he suggests that it is "sensible to have a certain separation of teacher, researcher, and practical-politician roles. . . [One] can only hope that such a separation will continue to be feasible in American society. . . such organizations as Federations of Social Scientists might merely become echoes of national goals, or instruments of official policy: or . . . they might come to do no more than afford bland, mild criticisms of 'things as they are.'"

It is wrong to suppose that the transition from academic to other organizational sources of support and work places for sociology is equivalent to a loss of objectivity or autonomy for its practitioners. Such a view assumes that academic sociology is free of extra, scientific pressures to conform. But as Veblen, Louis Wirth, Mills, the Frankfurt School, Mannheim, and others have argued (and as indicated in the preceding discussions of their work), academic sociology was shaped in many important ways by such pressures. The outcome was, more frequently than not, a distorted "command-suppressing" code that elevates accuracy and technical virtuosity to the status of goals, at the expense of the production of knowledge in the human interest. But sociology-in-application, in the formulation of social policy and in the technological innovation process, is not exempt from the distorting effects of external and internal social forces. Ben-David, James Coleman, Schneider, Murray Wax, and many others have warned of the threats which nonacademic milieux can pose to the craft of science (Ben-David 1971, ch. 2). The real change is not from a state of freedom to one of subordination, but rather from one potential source of subordination to another.

Postacademic social science cannot satisfy its own needs or those of the public with the same types of work that were once satisfactory. The real test of the merit of the discipline in postacademic contexts is the extent to which it can humanize technology and policy and thus contribute to development. Similarly, the principal test of the moral strength of the postacademic social scientist is not the pressure to accept academic canons of acceptability but the pressure to accept the "domesticated" equation that development is the accumulation of capital, capital-enriching technologies, and power for political-industrial elites. During the academic era social science did contribute to this domesticated type of development. The most common "solution" to the problem of its role in the development process was for it to remain factionalized into several pure disciplines and specialties within these disciplines and remain "above" politics. This ensured that it would be unable to confront fundamental issues in the study of technology in all their complex and transdisciplinary manifestations. Thus it has tended to accept and/or justify the current order. Such a passive solution on the part of industrial society to the problem of social science's potentially destabilizing effect on policy and the innovation process is no longer appropriate. Managers, engineers, planners, and the public are beginning to realize that they need social knowledge to operate effectively — so that they and we might survive.

The winning of a secure place for the social sciences within the academic establishment did not assure the unbridled pursuit of social knowledge; more often the cost was a "bridling" of this pursuit. Similarly, the fact that social science has an increasingly secure place within the nonacademic establishment does not guarantee that it will now be able to return, with no further difficulty, to the commitments of its founders. Despite the greater acceptability of social science perspectives, there still remains the same lack of sympathy toward the sociological way of looking at the world among engineers, planners, and policymakers expressed earlier in this century (Layton 1971) and which Horowitz (1969:549) summarizes:

> A consensualist vision of social life, a discounting of conflict models of behavior. This is directly traceable to the ideology of efficiency — and, of course, consensus is a more efficient style of social action than a conflict model, or one which recognizes the existence of sociological interests which are divergent at their sources, and hence cannot be removed by computerized accounting systems or abstract, community-imposed standards of consensus.

Such a consensualist vision articulates well with the development = accumulation formula. It represents a potent source of resistance, one with which the postacademic social scientist must contend, in his nonsociological colleagues and in himself, if he is to be effective.

Notes

1. Mannheim (1950, pt. 1; 1954, pts. 2, 3) discusses the decline of academic styles in connection with his general diagnosis. See also Price (1963) and Ben-David (1971) for details concerning resource shifts. In some respects, a "postacademic" engineering model was inherent in the social sciences from their very origins: "With remarkable consistency, early pioneers of sociology believed that engineering as a practical art and an applied technique offered the best supportive model for any meaningful science of society" (Horowitz 1972:420).
2. As we discuss in chapter 10, interdisciplinarity has become the focus of especially widespread attention among sociologists, philosophers, and historians of knowledge. Some idea about the general scope of this work can be found in the reviews of Walsh, Smith, and Landon (1975) and, especially with respect to team-based research, in Barth and Steck (1979), Taylor (1975), and Rossini and Porter (1978). Recent methodological treatments include Mitroff and Killman (1978) and Kash (1977). Several general social science journals now feature papers that discuss and illustrate problems and prospects of interdisciplinary studies, including *Knowledge, The American Sociologist, Social Studies of Science,* and substantive quarterlies such as *The Journal of Politics* and *Studies in Comparative International Development.* This impressive and growing body of literature strongly suggests that interdisciplinarity has come of age. The growth of interdisciplinary studies program in schools and universities and related research and funding trends suggests that even the academy has begun to emerge into the present.
3. Data from 1978-80 indicate that: "Colleges and universities perform about 30 percent of all federally funded research in the social and behavioral sciences, including about 47 percent of the basic research. . . . This situation indicates that there should be a nonacademic market for social and behavioral scientists" (Rhoades 1980b:5).
4. Noteworthy is the growing interest in the management of interdisciplinary, technology-focused team research. See Chubin et al. (1979), Rossini, Porter

and Zucker (1977), Walsh, Smith, and Landon (1975), Gillespie and Birnbaum (1977), and Newell, Saxburg, and Birnbaum (1975).
5. This volume was intended to serve as a standard reference work, and there is little reason to doubt that it will quickly become one. Other collections that provide surveys of the field include those of Burke (1972) and more recently, Burke and Eakin (1979).
6. Each of the fifteen chapters of Spiegel-Rösing and Price (1977) contains an excellent bibliography of these and related areas. See also Sharma and Quereshi (1978), Burke and Eakin (1979), and Gendron (1977).
7. The major political-ethical dilemma is related to the classic question of the scientist's autonomy in determining the ends to which his work is to be put. Mills, Mannheim, and other sociologists of technology have approached this dilemma as an aspect of the general crisis of publics in democratic society. Based on this perspective, the scientist's rights and power derive from the state of craftsmanship, the degree of democratic access to elites in society, and his own ability to attend to the formal (as well as the functional) rationality of his work. See also Bunge (1966), Layton (1971), and Skolimowski (1966).
8. See e.g. Small (1977, 1978) for a review of the "art" of citation analysis and comments on the meaning of citation.
9. Basic references in the TA literature include Hetman (1973), Arnstein and Christakis (1975), and *The Trend* (1974). There is a very large and growing body of literature in books, journals, and articles written by specialists in a variety of fields. The collection by Joel Tarr (1977) includes analyses of the historical impact of transportation, energy, and industrial innovations in the United States. Undergraduate and graduate courses and degree programs in technology assessment are now being offered at the University of Washington, Cal Tech, Washington University, MIT, Georgia Tech, and many other colleges and universities in the United States and abroad.
10. Opinions differ about whether the ironic focus, the dialectic, and related ideas are more than "mere" techniques of discovery. The exchange between Sorokin (1964) and Schneider (1964) centers on such an issue. The later comments of Schneider (1971, 1975) are especially useful. While he holds that these are techniques for formulating or expressing hypotheses, they are not simply aids or cosmetics. They are key components of a more general "sociological way of looking at the world."

8.

Feeling Helpless: The Idea of Autonomous Technology in Social Science

This and the following chapter return to the sociology of knowledge approach to the study of technology. These discussions apply the perspective in an examination of theories that presumably explain technology's impact on society. Here the focus is on the well-known pessimistic views of French philosopher and theologian Jacques Ellul. The next chapter assesses features of the equally popular optimistic views of the quality-of-life social indicators school. Both cases illustrate how the very complexity of the relationship between technological innovation and social change can encourage the production of selective, interest-serving theories. The purpose of these discussions is to use such accounts as negative and cautionary examples for the postacademic social scientist, whose goal is neither to lament nor to celebrate technological innovation but rather to contribute to its humanization.

The work of Ellul has been much debated in academic circles and in popular literature since the 1954 publication of *La Technique ou l'enjeu du siècle* (first English translation, 1964).[1] From this debate and related commentary — some recent, such as that of Mumford (1963, 1966, 1976), Marcuse (1968), Faunce (1968), Calder 1969, and Peter Berger et al. (1973), and some ancient — an important and widely discussed thesis has emerged. This thesis, called "the rule of technique" (Ellul 1964), the "Prometheus" theme (Landes 1969), "Frankenstein's Monster" theme, and most recently, "autonomous technology" (Winner 1977), focuses on the widespread loss of control by man over his inventions. According to this thesis, as industrial society has evolved, our social and political lives have increasingly been influenced by the need to service and sustain an ever-accelerating replacement of men by machines in work, leisure, and in all cultural pursuits. In the process, it is argued, human creativity and self-determination have virtually disappeared.

Ellul and other proponents of this thesis draw on diverse writings, both classic and current, in showing that this characterization of technology has a distinguished place in the history of social thought, and that the idea of autonomous technology is an integral part of our culture. But proponents attempt more, falling prey to political pitfalls inherent in the idea of autonomous technology. In attempting to show that the idea is not only a good one, but that it is also both frightening and correct, Ellul, Mumford, and the others have produced an ideology that contributes little to our understanding of the relationship between society and technology.

The Idea and Its Roots

The idea of "technics-out-of-control" has fascinated people in all cultures: In Hebrew, Greek, Hindu, and other mythological histories, we find scenarios in which deities punish mortals for eternities through the unintended side-effects that follow the application of forbidden know-how. This theme was central among the concerns of the earliest social scientists at the very beginning of the Industrial Revolution, and it forms a major part of the work of Marx and Freud.[2] Several more recent trends in social science and policy analysis also share an interest in autonomous technology. These include general theoretical treatments, like those of the Frankfurt School, John K. Galbraith (1968), and Don K. Price (1965); work that focuses on the ironies of technology, such as that discussed in chapter 7, and the new policy fields of technology assessment and impact analysis. Despite important differences between (and within) these approaches, they all stress the negative consequences of technological change that are not anticipated nor intended by human actors; what Robert K. Merton referred to some years ago as "latent dysfunctions."

The universality of the theme of technics-out-of-control and its importance in current social research show that questions like those raised by Ellul and his current leading disciple in the United States, Langdon Winner, are rooted in a long and respected intellectual tradition. Often only brief mention, or none at all, is made by these authors of some key sociological sources discussed here, such as the theories of Mannheim (whose *Freedom, Power, and Democratic Planning* relates in detail to Ellul and Winner's theses[3]), Mills, and Ogburn — thus proponents understate the importance of the theme. But to demonstrate that the theme is familiar, or even that it is becoming increasingly credible to scholars and to the public, is not to prove that technology is or is becoming autonomous. When such proof is attempted, as in Winner's (1977) retelling of the poignant story of how the snowmobile changed the culture of a Lapp reindeer-following community (Pelto 1973), evidence tends to be anecdotal; i.e., it is evocative but difficult to generalize. Perhaps, as Ravetz (1978:643) has suggested, the very project of attempting to show scientifically that technology is autonomous is ill-conceived: "The detailed evidence [Winner] adduces for his sweeping assertions is fragmentary and unconvincing. He cites a few cases (and some of them dubious) from war, where "small is beautiful" never did apply anyway. But it may be that theses so general as this are not capable of strict testing."

The question of whether technology is (becoming) autonomous is not merely general; it is audacious.[4] Winner, Mumford, Ellul, and especially Marcuse are aware of this. In response, they focus on some very evident ways in which technology may be out of human control. Among the most basic are that modern technology creates needs which can only be satisfied by more technology; and that technology transforms people — their beliefs, habits, and organization — to make them better adapted to technology's requirements. The broad question of autonomous technology is turned into an investigation of these two processes, "the technological imperative" and "reverse adaptation" respectively, as they relate to other aspects of modern society. These other aspects include the influence of technology on political systems, loss of human energy as a consequence of increasing complexity, and technical virtuosity.

The concept of technical virtuosity is an extension of Ellul's view of "the rule of technique" — closely related to Weber and Mannheim's concepts of *Zweck* (or purely functional, technical) rationality. As a tendency in modern society, it illustrates well the operation of the technological imperative and reverse adaptation. Technical virtuosity is mastery and the display of mastery of techniques, with no particular concern for the subject or object of their application. It is familiar in popular literature (e.g., *The Joy of Sex*) and as "methodology-ism" it is currently the scourge of the academic social sciences. It is a response to an imperative of technological innovation by virtue of the fact that more technology (ceteris paribus) requires more technical skills. With the increasing diffusion of technology, individuals and groups adjust their behavior, technical virtuoso performance with these techniques becomes a canon of evaluation, while the reality to which they are applied becomes a matter of indifference.

Concepts like technical virtuosity are fascinating. Yet despite the sense of fascination which their application to the study of social change can convey, they also contain a disappointing lack of political analysis. This lack becomes apparent in examining the autonomous technology thesis as a whole, and it is especially serious in view of the essential political component of the technological innovation process.

Validity of the Idea

While Ellul, Winner, Faunce, et al. are careful to observe that "techniques are not neutral," they generally fail to discuss the technical virtuoso (e.g., the methodologist) as a political actor. According to their analyses, the power of the technical elite — over technology and over others — is increasingly insignificant as the autonomy of technology increases. But this is a very partial view, as Mannheim and the sociology of technology have pointed out. In addition to the increase in numbers and prestige — in all fields — of the technical virtuoso, the diffusion of technological innovation has also brought increases in the power of the virtuoso "classes": power over recruitment and training of other elites and power over the paradigms of acceptable thought and practice. The possibility that technical virtuosity derives its effectiveness from the control that certain groups can exercise contradicts the autonomous technology thesis. For this reason, it is also a possibility not seriously considered by proponents.

Partiality of the Thesis

The main argument of the autonomous technology approach is that we are increasingly under the influence of tendencies like the technological imperative and reverse adaptation and, in these ways, technology is gaining the upper hand: it is coming to have unmediated control over people. The main weakness in this argument is not that it is impossible to show that these tendencies are real or that their effects are increasing with the diffusion of modern technology. The argument fails to convince because the very opposites of the specific tendencies and of the general process of technology "taking over" are equally real and just as surely increasing. The technological imperative and reverse adaptation represent two among a limitless number of ways in which technology and society interact. While it is true that technological innovation creates needs, it is equally true that it eliminates needs. While it is true that some of these effects are unanticipated by planners, it is also true that planning takes place among elites filling key positions which automatically makes for minority rule and increasing awareness of formerly unanticipated outcomes. Second- and higher-order consequences of technological innovation tend to contradict one another. Technology makes some people powerful in some ways, other people powerless in other ways.

Even if we accept the view that, on balance, tendencies such as the technological imperative and reverse adaptation have increased since the industrial revolution, it still does not follow that humanity as a whole is becoming increasingly helpless in the face of technology. Every technological innovation is subject to human control at each of several stages in the innovation process. Such control is always imperfect, and it is rarely identical to the conscious will of one or a few individuals, but control is exercised nevertheless. Proponents of the autonomous technology thesis are right to point out that technology is too easily dismissed as merely a tool whose effects derive solely from the intentions of the user. "Mere" tools do indeed shape the user — the "discipline of the machine," as Marx called it. But the tool user, or those who direct the tool user, can in turn reshape the tool. With the knowledge of the technological imperative increasingly available, technicians and planners incorporate this knowledge into their plans, thus controlling the manner in which technology controls people — the dialectic, as Marx called it.

Frankenstein's Monster

Langdon Winner's discussion of the "Frankenstein's Monster" theme underscores a characteristic attachment to the horror story side of technological innovation. This side has often been revealed, and it is one to which all responsible students of technology should be sensitive. But Winner (1977:307) tries to disclaim the shock effect of *Frankenstein* by drawing invidious (and downright snobbish) distinctions between Mary Shelley's treatment of the Frankenstein theme and subsequent ones:

> The fact of the matter is that the film scenarios have virtually nothing to do with *Frankenstein* the novel. In the original [sic] there is no crazed assistant, no criminal brain mistakenly transplanted. . . In place of such trash, the book contains a story offering an interesting treatment of the themes of creation, responsibility. . .[5]

"The fact of the matter" is that all depictions of the Frankenstein-Prometheus theme share a single purpose: frightening people by telling of the dangers of trying to imitate (or improve upon) nature with technology. Perhaps Winner is overly defensive because he (like every other American boy) had seen dozens of "trash" Frankenstein movies and comic books before discovering the genuine article. If so, he should reflect on what frightening movies and comics they were, and that they were as effective in communicating the Prometheus myth as Mary Shelley's horror story. What is really being communicated is not an analysis but a fear of technology. Insightful as they may be, such observations on *Frankenstein* and stress on the dangers of runaway technology tell only one part of a far more complex moral tale. For "not all unintended social effects of technology, by any means, are undesirable."[6] Many things must be considered before the net effect of technological innovation can be calculated.

The partiality to the frightening unintended consequences of technological change, combined with the failure to see the relationship between technology and society as dialectical and open, suggest that proponents of the autonomous technology thesis are more partisans than students of the dialectic of ideology and technology. The complexity of the question of autonomous technology and the way in which proponents bring true but partial evidence to bear suggest that the technics-out-of-control thesis is a distorting account. As a diagnosis, the autonomous technology theme, like other modern ideologies, provides an evocative, selective analysis of the "inevitable" social effects of a very complex, multidimensional modern institution. This analysis justifies some courses of political action while making others appear unreasonable or futile.

Considerable attention is given to showing that existing solutions to the problem of technics out of control, e.g., those of Lenin, Galbraith, and Daniel Bell, are all inadequate. As proponents understand the situation, none of the usual radical, liberal, or conservative strategies takes adequate account of the fact that "technology is now a conduit such that no matter what aims or purpose one decides to put in, a particular kind of product inevitably comes out. This state of affairs is not suited to political theory in any traditional sense."[7] As a therapy, a proposed alternative is offered in the movement that has been labeled "intellectual Luddism." Descriptions of this movement lack detail, but it appears to involve technology critique based on Ellul's main thesis. The possibility raised here that the autonomous technology theme is an ideology is one that proponents do not explore, perhaps because it requires a painfully high degree of reflection; it asks that proponents subject their own views to sociological analysis. Yet it is a possibility that would appear to be of interest to them, particularly to political scientists such as Winner, who explicitly trace their roots to the University of California, Berkeley, and to the Free Speech Movement (FSM) and the counterculture of the 1960s.

Political Context of the Idea

In the preface to *Autonomous Technology,* Winner quotes from

Mario Savio's Sproul Hall address (of December 1964): "There is a time when the operation of the machine becomes so odious, makes you so sick at heart, that you can't take part." In subsequent sections of his book, he employs the 1964-65 FSM position as a (sometimes hidden) agenda, elaborating on Savio's report and extending Savio's command to "put your bodies upon the gears and upon the wheels, upon the levers and upon all the apparatus and you've got to make it stop." Thus, a direct connection is drawn between Savio's solution, "intellectual Luddism," and the philosophy of Ellul. The relationship between diagnosis and therapy in the FSM analysis, in that of Ellul, or in Winner's has not been explored. Nor have proponents provided a systematic examination of why doctrines of autonomous technology and intellectual Luddism came to be formulated (or reformulated) by young social scientists and philosophers in the 1950s and 1960s. Such omissions are neither random nor minor. They point to a lack of self-conscious sociological insight that renders the autonomous technology thesis weaker than it might be, amounting to a fatal flaw in aspects of the FSM position the thesis appears to endorse.

The philosophers and social scientists who came to support the autonomous technology thesis after the mid-1950s are members of the group that Theodore Roszak (1969) labeled "technocracy's childern." These intellectuals are important actors in a recent staging of a very old battle: between classicist and romanticist, Philistine and bohemian, and more recently, establishment and counterculture. As a political struggle, it is motivated by competing claims for control: control of technology and especially control of the means of production and distribution of knowledge — the new consciousness industry. As proponents of the autonomous technology thesis have observed, this is not the struggle between proletariat and bourgeoisie. This is a struggle among intellectuals, just as most successful (and many not so successful) revolutions have been.[8] Other classes (or more characteristic of Ellul, "the society") are implicated in the autonomous technology thesis — just as other classes are implicated in the antithesis that credits technology with benign effects. But the promulgators and chief actors in this struggle for control are people like Jacques Ellul, Langdon Winner, Mario Savio, and other of technocracy's "children" and, of course, their "parents," members of technical and industrial classes.

Technocracy Versus "Technocracy's Children"

In seeing the children of technocrats as a cultural revolutionary class of intellectuals, and in pointing to the autonomous technology thesis and the program of intellectual Luddism as their new truth, Roszak provided the beginnings of the sociological analysis absent in other works on autonomous technology. But he ignored the significant fact that the countercultural critique is not a product of the 1960s (for it is a recurrent theme), and failed to see lack of correspondence between children, children of technocrats, and countercultural revolutionaries. Like Ellul, Calder, Winner, and Savio, Roszak substitutes partisanship for analysis.[9]

In the work on autonomous technology, one is given only vague and misleading hints about the social context of the autonomous technology thesis. From more sociological works, such as those of Mannheim, Schneider (1975, ch. 6), Gouldner (1976), and even Roszak, a clearer sense of the thesis as ideology emerges. Seen as one position in a political struggle for control, the idea of autonomous technology reveals telling ironies. Proponents argue that there is no human group or class from which control is to be wrested; no enemy other than technology itself. At the same time, an "enemy" position does indeed exist and has its proponents. This is the establishment view of progressive technology to which technocrats and members of the other groups subscribe. It is the largely unacknowledged position against which the autonomous technology thesis has been formulated, in both recent and classic versions.

In the struggle between technocracy and technocracy's children, this position complements and shapes the theme of technics-out-of-control in several ways. Where Ellul and Winner stress Frankenstein's problem, their technocratic counterparts, as the following chapter illustrates, continue to delight the corporate and individual consumer with life-prolonging, life-enchancing, and even life-imitating innovations. Where technology-out-of-control implies Luddite tactics to win back "our freedom," progressive technology implies that technology has already (or potentially) made us freer than people have ever been. Where autonomous technology speaks, ideologically, of there being no enemy, progressive technology — the ideology of "the enemy" — speaks of the end of ideology. Both positions express wishes as if they were facts. Each understates the degree to which ideology, class, and technology interact, thus also underestimating the cogency of the other.

The autonomous technology thesis is not the only position from which progressive technology and technocrats have been attacked.[10] Currently, however, it is a major — and evocative — antiestablishment critique. Unlike other critiques, such as classic Marxism, it goes so far as to deny the very existence of an effective establishment: "no matter what aims or purposes are put in, a particular kind of product inevitably comes out." For those who accept the thesis, this may be a very comforting revelation, since everyone should therefore be willing to accept the basic critique. If this is true, the movement to change the system through intellectual Luddism promises to be a universally popular one. But if one does not accept the premise that it makes no difference whose "aims or purposes are put in," the conclusion that no one can (rationally) be opposed to the autonomous technology thesis provides little comfort.

Technology and Political Action

To the extent that proponents provide a sense of the dangerous potential in technological innovation, they are to be credited with contributing an important weapon in the battle against dehumanization and a valuable corrective to the progressive technology thesis. But to the extent that they are suggesting that effective social change can no longer be accomplished through conscious restructuring of social relationships (but that it must, instead, be directed at technology itself), they have joined forces with "the enemy." Failure to consider Mannheim's views and other sociological work leads to a failure to see some major consequences of the extensive distribution, scale, and effectiveness of modern technology.

Perhaps in some senses technology is becoming increasingly autonomous. But consolidation of resources and the general systemic growth of modern technology have also made it increasingly easy for people to control people. Some have access to these key positions of control, others do not. And while planners and managers may remain unaware of the consequences of their acts, it is clear that these consequences can be more extensive and effective — for good or ill — than ever before. Technicians have their concepts of technological fix, technology assessment, and impact analysis. With these they anticipate and correct formerly unanticipated consequences. Thus it matters very much to those within the establishment *who* (e.g., old-line bureaucrats or technicians) controls "autonomous" technology. It also matters to technocracy's publics, to the people who — perhaps naively — would rather have a "friend" in planning, management, and administrative postions than a "foe." Perhaps it even matters to Ellul and Winner, at least insofar as they would like to have a say in who controls the technology necessary to carry off a successful campaign of intellectual Luddism!

It seems perverse for proponents of the autonomous technology thesis to eschew political action in the usual sense. To note that technology shapes our needs is not to deny the importance of gaining democratic access to the processes of design, innovation, and evaluation of technology. On the contrary, the awesome power of autonomous technology would make it imperative for there to be more democratic access to positions of influence than in the past. Yet Luddism is Luddism, however qualified, and it is a solution that is classically apolitical, fundamentally impractical, and highly unpopular (with good reason).

In its FSM variant, the autonomous technology thesis was a part (but an important part) of a generally naive analysis and strategy. Whatever else may have happened to the New Left between December 1964 and today, it did not succeed in slowing technological innovation. Perhaps other goals replaced this one, and perhaps technological innovation has been slowed. But a certain type of victory, of which Mario Savio spoke in 1964, has not become apparent. And while the New Left spoke of halting the machine and dismantling technology, the establishment as a whole continued to consolidate its control, and within the establishment technicians made gains in their struggle for positions of influence.

As Marxist and non-Marxist critics of the New Left have observed, the failures of the movement (and this is not to deny that it had successes) were partly due to the fact that its analysis distorted, rather than clarified, the nature of political control in modern society. It is a movement which, historically at least, is based on the acceptance of fundamental powerlessness. This is the kind of powerlessness one feels when lacking an identifiable human cause for one's troubles, the kind of powerlessness that Victor Frankenstein felt when his creation was completed. More prosaically, it is the powerlessness we feel just before we kick a machine for taking our money without giving us coffee. But people in all classes find it hard to accept that this kind of powerlessness comes from technology per se, when the evidence of their senses tells them that technology gives them power. It appears that the New Left did find it hard to get effective support for its version of the autonomous technology thesis. Increasingly in the public mind, the unflattering image of helpless young people kicking at machines was associated with the counterculture (see Luce 1972, for a particularly vitriolic attack).

Autonomous Technology and Ideology

The autonomous technology thesis shares the failings of New Left theory. While arguing often brilliantly that technics out-of-control is a credible theme, proponents are too easily convinced by their own partial analysis. Instead of asking further why the theme is credible, they simply use it to describe and prescribe. The result is a distorted diagnosis and a questionable strategy. A diagnosis that is more the helpless cry of intellectual outsiders than an assessment of the facts; and a strategy that is more an angry expression of impotence than a program for change. As followers (and critics) of the New Left have learned in various ways, feelings of helpessness combined with an attack on inanimate objects makes for a highly ineffective reconciliation of theory and praxis.

The idea of autonomous technology appears to be inherent (if not always explicit) in reflective social thought. It is a persuasive idea and thus a powerful tool for critics of the established technocratic order. Jacques Ellul, Lewis Mumford, Herbert Marcuse, William Faunce, Peter Berger, Langdon Winner, and others have produced a valuable analysis of the idea, and for this reason alone their work deserves careful study. But the analysis is also incomplete in a serious way. In failing to reflect on the political context of the autonomous technology thesis, proponents become (perhaps unwittingly) partisans of a hopeless cause. Just as the analysis of technical virtuosity falls short of examining the virtuoso as a planner or administrator who exercises control over (albeit autonomous) technology, the general analysis of the autonomous technology thesis falls short of examining the intellectual Luddite as a countercultural intellectual who seeks to seize control from the virtuoso. Were the autonomous technology thesis carried to this depth, proponents might see its chief weaknesses: the diagnosis distorts by understatement the degree to which control is exercised in technological society, and the therapy prescribes a strategy difficult for anyone (except perhaps a few marginal intellectuals) to support. The autonomous technology thesis is a significant contribution to the store of ideas about technology. But we must look to other perspectives for an adequate social scientific account of technology's "autonomy," as a theme in political life and as one aspect of a highly complex dialectic in modern technological society.

Notes

1. Ellul (1964). Especially noteworthy among these related works are Heilbroner (1967), Landes (1969), and various writings of the Frankfurt School, including Horkheimer (1974), and Horkheimer and Adorno (1973). See also the brief but thoughtful critique of Ellul in Schneider (1975, ch. 6).
2. Ellul and Winner discuss Saint-Simon at some length, but also relevant are the generations of French social theorists between Saint-Simon's time and Ellul's, including Comte, Durkheim, and Sartre. For other classic statements on the power of technology to transform social relations, see Schneider (1967) and Durkheim (1962). Major literary treatments include: Sypher (1971) and Leo Marx (1964).
3. Mannheim (1950). This book expresses profound sociological insight into technology's effects on society. It also provides a program for political action considerably more sophisticated than intellectual Luddism.
4. Perhaps, as Stan Carpenter (1978) points out, "autonomous" and "out-of-control" are not synonymous. Yet both states are characterized by a lack of human control, and that is the important thing. In either case, technology is

assumed to have the capacity to "do things" to people without people being able to control it. Whether technology in this situation is better characterized as "running itself" or as "out-of-control" seems an irresolvable semantic issue, rooted in the inappropriateness of predicating states like autonomy (or its lack) to technology in the first place.
5. This tone is not uncharacteristic of a defensiveness expressed in other parts of the book, a book that began as a Ph.D. dissertation.
6. Schneider (1975:51). Schneider's discussion points to a tradition, in which W.F. Ogburn figures prominently, of seeing the society-technology interaction as a process of mutual adjustment. In this sense, there is nothing exceptional about the manner in which technology constrains human behavior, for humans also act to alter technology.
7. Winner (1977:278). This statement, complete with a word like *inevitably,* summarizes well the technics-out-of-control diagnosis and partial therapy: traditional political theory is not suited to the situation. Had the author reflected on this statement from a sociology of knowledge perspective, he might have been less willing to defend it.
8. Roszak (1969) remains a valuable source on the relationship between the New Left and the autonomous technology thesis, despite the fact that the author's perspective is more that of cheerleader than analyst. Like other commentators on the New Left (e.g., Charles Reich 1975), Roszak seems to have let countercultural "smoke (!) get in his eyes." See Gouldner (1976:279).
9. See the critique of Roszak in Schneider (1975, ch. 6).
10. On the limitations of technology assessment, see also Winner (1972) and Tarr (1977).

9.

Technology, Social Science, and Quality of Life

Whether we like it or not, the fallacious portions of the negativistic theories have tangibly contributed to the present degradation of man and of all the great values, from the supreme value called God (or some other name) to the values of truth, love, beauty, creative genius, sainthood, and finally to those of fatherhood, motherhood, the family, duty, sacrifice, and decency in the treatment of man by man.

Pitirim A. Sorokin

Our discussion of distorted accounts of technology continues with an analysis of an approach in some respects diametrically opposed to the autonomous technology thesis. This is a variant of the work on measuring quality of life with social indicators, the City Index approach. It is based on the research of Raymond Bauer (1966) and others, and more recently, on the influential study of Campbell, Converse, and Rodgers (1976).[1] Its proponents stress the continuity between technological progress and attainment of "the good life." While the autonomous technology thesis points to the ways in which technology can turn on and harm its creators, the City Index approach argues, often with a convincing array of quantitative data, that increased investment in certain types of technological innovation invariably benefits society.

The City Index and Development: Measuring the Quality of Life

A series of studies done in the mid-1970s by a research group at the University of Nebraska (Todd 1977, 1978) provides an interesting and controversial illustration of this approach. The Nebraska study produced a list of one hundred U.S. cities, shown in Table 1, ranked in order of "attractiveness" according to their City Index. This index was derived by the researchers through combining the scores of each city on the variables listed in Table 2. Shortly after the publication of these findings, residents and supporters of the cities that rank "too low" on the list (Atlanta and Honolulu, for example) responded in professional journals, on editorial pages, and in letter columns of their local papers (see SDI 1978). Questions were raised about every aspect of the study, including the specific rankings and validity of the research itself.

This discussion concentrates on what are, for the postacademic social scientist, the major shortcomings of the quality of life and City Index literature. These issues are methodological rather than substantive. The point is not that Atlanta "deserves" a better score on Table 1, but that this approach to social research typifies "the fallacious portions of the negativistic theories" to which Sorokin refers in the quotation opening this chapter. The meaning of the phrase "quality of life" will be discussed only indirectly, though if forced to give a concise definition, one could do worse than begin with an examination of Sorokin's "great values." Most people who have grappled with the problem are aware that a satisfactory definition of "quality of life" is not easy to find or create. The phrase is intriguing as much for what it does not (or should not) mean as for what it does (or should) mean. In this book, we have largely avoided using "quality of life." Instead, the focus is on social development, though this might be viewed as the achievement of a "better life."

Development has been defined in terms of fairly standard ideals that emerged as part of the social programs of the Enlightenment: the attainment for all people of better health, longer life, greater freedom from drudgery and for creativeness, and greater degrees of political democracy. In a more subjective vein, we might go further and say that "the good life" (in the city or countryside) can be achieved only if people can (1) survive physically and (2) have a strong measure of "intangibles" such as hope, control, security, and legacy (Weinstein 1978). A technological innovation or a program of social change can be judged prodevelopmental, as contributing to quality of life, to the extent that it promotes material achievements such as health and democracy and that it preserves or improves the degree of hope, control, etc., that people have on a day-to-day basis. To discuss the validity of the findings of the City Index group, it would be necessary to know whether they have this definition of development in mind when they discuss "the good life." This is not clear from their comments. However, it is not a definition or specific findings that are at issue; it is the way in which certain crucial decisions are made in the City Index type of research.

Technology, Social Science, and Quality of Life   155

TABLE 1
Ranking of 100 Cities on Individual Factors

CITY	COMPOSITE RANKS All 80 Factors	Y_1	Y_2	CITY	COMPOSITE RANKS All 80 Factors	Y_1	Y_2
Lincoln	1	1	2	Tuscon	51	41	59
Madison	2	2	23	Albuquerque	52	79	28
Des Moines	3	3	15	Columbus, Ga.	53	64	41
Omaha	4	4	8	Tacoma	54	69	53
Greensboro	5	7	6	Toledo	55	53	58
Indianapolis	6	8	9	Mobile	56	62	49
Honolulu	7	5	34	Cincinnati	57	36	79
Tulsa	8	28	3	Long Beach	58	67	29
Wichita	9	12	10	Akron	59	39	68
Virginia Beach	10	9	5	Los Vegas	60	82	1
Jackson	11	6	35	Louisville	61	62	67
Spokane	12	18	22	Fresno	62	59	44
Fort Wayne	13	10	37	Worcester	63	25	86
Lexington	14	11	25	Pittsburgh	64	53	82
Salt Lake City	15	19	17	Kansas City	65	76	55
Lubbock	16	20	13	San Francisco	66	72	63
Nashville	17	23	18	St. Petersburg	67	74	48
Seattle	18	38	16	Rochester	68	29	89
Charlotte	19	31	12	Riverside	69	56	61
Knoxville	20	13	51	Chattanooga	70	70	50
Montgomery	21	39	14	New Orelans	71	59	70
Oklahoma City	22	42	7	El Paso	72	75	56
Houston	23	34	4	Santa Ana	73	64	47
Sacramento	24	35	19	Buffalo	74	46	88
San Diego	25	21	32	Washington, D.C.	75	78	78
Austin	26	17	54	Hartford	76	42	97
Shreveport	27	26	33	Miami	77	80	69
San Jose	28	22	27	Los Angeles	78	87	46
Little Rock	29	63	11	Bridgeport	79	61	85
St. Paul	30	16	74	Birmingham	80	82	72
Syracuse	31	15	73	San Antonio	81	70	75
Fort Lauderdale	32	49	24	Tampa	82	87	56
Milwaukee	33	14	71	Atlanta	83	86	66
Baton Rouge	34	45	21	Oakland	84	92	60
Minneapolis	35	29	64	Boston	85	81	92
Colorado Springs	36	58	31	Springfield, MA.	86	67	98
Portland	37	44	43	Norfolk	87	94	80
Anaheim	38	33	36	Dayton	88	89	91
Denver	39	50	45	Philadelphia	89	77	95
Rockford	40	51	38	Chicago	90	96	81
Dallas	41	72	20	New York	91	84	90
Fort Worth	42	51	30	Flint	92	97	75
Corpus Christi	43	31	30	Baltimore	93	85	95
Memphis	44	55	39	Paterson	94	95	87
Grand Rapids	45	26	64	St. Louis	95	91	94
Phoenix	46	58	26	Cleveland	96	90	99
Richmond	47	46	52	Jersey City	97	93	93
Columbus, OH.	48	36	62	Detroit	98	99	83
Jacksonville	49	48	40	Gary	99	100	84
Providence	50	23	77	Newark	100	98	100

TABLE 2
Individual Factors (Xs) for City Index

Economic Factors
1. 1973 per capita income.
2. 1973 per capita income CC/OCC: central city as percent of outside central city.
3. 1975 housing construction costs per square foot as percent of per capita income.
4. 1975 hospital room costs as percent of per capita income.
5. 1973 automobile and truck registrations per 1,000 population.
6. 1972 percent employees covered by bus transit (1/2 mile band).
7. 1975 electric (residential rates) 1000 KWH as percent of per capita income.
8. 1975 electric (commercial rates in dollars) monthly/1500 KWH.
9. 1975 electric (industrial rates in dollars) per 60,000 KWH.
10. 1975 natural gas rates for 1,000 cubic feet as percent of per capita income.
11. 1963-1972 percent change in retail sales in the central city.
12. 1963-1972 percentage point change in manufacturing employment.
13. 1972 value added in manufacturing/wages paid.
14. 1972 value added in manufacturing/employees.
15. 1970 median house value (CC/OCC) in percent.
16. 1975 building permits per 10,000 population.
17. 1975 building permits (CC/OCC) in percent.
18. 1975 per capita bank deposits in thousand dollars.
19. 1975 average annual unemployment rate.
20. 1974 telephones per 100 population.
21. 1974 per capita city debt as percent of per capita income.
22. 1975 per capita nonschool taxes as a percent of per capita income.
23. 1960-1972 annual percent chnge in accrued value of real property.
24. 1975 finance and general control government employees per 10,000 population.
25. 1976 municipal bond rating.
26. 1975 AFDC recipients per 10,000 population.
27. 1975 city fire rating.
28. 1976 cost of eating out as percent of per capita daily income.
29. 1975 per capita subsidized amount of food stamps in dollars.
30. 1974 commercial banking offices per 100,000 population.
31. 1976 percent of households with effective buying income under $15,000.

Demographic/Environmental Factors
32. 1973 population density (persons/acre).
33. 1970-1976 net population migration in percent.
34. 1950-1970 percentage point change in ratio of white population to total population.
35. 1974 voters per 10,000 voting population.
36. 1972 divorces per 100,000 population.
37. 1975 deaths per 100,000 population.
38. 1975 deaths from influenza and pneumonia per 100,000 population.
39. 1973 suicides per 100,000 population.
40. 1975 infant deaths per 100,000 population.
41. 1973 motor vehicle deaths per 100,000 population.
42. 1973 all other accidental deaths per 100,000 population.
43. 1975 hospitals per 100,000 population.
44. 1975 hospital beds per 100,000 population.
45. 1974 physicians per 100,000 population.
46. 1975 registered nurses per 100,000 population.
47. 1975 nursing home beds per 1,000 population 65 and over.
48. 1975 percent sunshine.
49. 1974 particulates mmg/m3.
50. 1974 CO mg/m3.
51. 1975 heating degree days.
52. 1975 average wind speed.

Crime Factors
53. 1975 robberies per 100,000 population.
54. 1975 negligent manslaughter per 100,000 population.
55. 1975 non-negligent manslaughter per 100,000 population.
56. 1975 rapes per 100,000 population.

TABLE 2 (continued)

57. 1975 burglaries per 100,000 population.
58. 1975 assaults per 100,000 population.
60. 1975 larceny per 100,000 population.

Recreation, Educational, and Other Factors
61. 1974 scholars per 100,000 population.
62. 1975 library volumes per 100 population.
63. 1975 TV stations per 100,000 population.
64. 1975 radio stations per 100,000 population.
65. 1975 hotel and motel rooms per 10,000 population.
66. 1972-1973 student/teacher ratio in public schools.
67. 1972 recreation and amusement establishment per 100,000 population.
68. 1972 eating and drinking establishments per 100,000 population.
69. 1974-1975 enrollments in higher education per 1,000 population.
70. 1972 percent of population covered by bus transit (1/2 mile band).
71. 1974 air passengers per 1,000 population.
72. 1976 circulation of daily newspapers as percent of households.
73. 1976 lawyers per 100,000 population.
74. 1976 contributions to United Way as percent of effective buying income.
75. 1976 per capita firm gifts for United Way in dollars.
76. 1976 museums per 100,000 population.
77. 1974 religious organizations per 100,000 population.
78. 1975 park acreage per 10,000 population.
79. 1975 golf courses per 100,000 population.
80. 1975 swimming pools per 100,000 population.

Examining the Approach

The City Index literature illustrates a particular ideological perspective on the study of the impact of technological change on the quality of life in urban society, one that has been called "positivist-technocratic."[2] This is not the only ideology that has been employed in attempts to understand and improve the quality of life in cities. The inventor of the City Index, Ralph Todd (1978), also calls attention to the "bad press" that the city has received from critics of technological society. Nevertheless, the positivist-technocratic perspective is dominant and well-established. The most striking positivist feature of the City Index approach is its use of quantitative indicators to measure quality. While the researchers are careful to stress the need for subjective as well as objective variables — especially the types used by Campbell, Converse, and Rodgers (1976) — what they and most other positivist-technocrats mean by "subjective" indicators is additional quantitative variables based upon public opinion surveys. These are also to be used to measure quality. But this point is moot since, as can be seen in Table 2, the City Index group does without any variables that tap opinions about quality of life or anything else.

A second characteristic this research shares with other positivist-technocrats is the attempt to be "value-free" in dealing with the measurement of quality of life and the relationships among individual and composite quality-of-life indicators. The City Index group employs the eighty variables (which they call "factors") for one hundred cities, singly and in various linear combinations. All these variables are expressed as rates or ratios, that is, as frequencies or amounts of something adjusted by the size of a population (of individuals, firms, etc.) or by a more general frequency or amount (e.g., per capita income). This method of operationalization is, in some ways, unbiased and complete. The variables chosen seem intuitively valid and their conversion to rates and ratios appears to eliminate the effect of city size.

Upon examination, however, it is obvious that crucial decisions have been made in the absence of explicit theoretical criteria. Even if we grant the highly unlikely possibility that these eighty variables were chosen over all other possible ones on the basis of an equiprobability selection (e.g., a random sample), so that their selection represents no particular predisposition, many of the decisions about whether a high or low ratio (or rate) points to high (or low) quality of life appear groundless. Perhaps some of these decisions are self-evident, for instance, that high per capita income indicates high "per capita well-being"; although even in this case, the possibility of an optimal income below some maximum or the possibility of a poorly distributed high per capita income are ignored.

How does one decide if a city with high "per capita subsidized amounts of food stamps" or a large "change in the ratio of White population to total population," has, all else equal, a "high" or, as the City Index group believes, a "low" quality of life? Whose life? The hard questions about the impact of the objects which social indicators indicate on quality of life, rather than about the impact of the indicators on one another, are glossed or suppressed by the City Index group and by the positivist-technocratic perspective in general. Not only does the approach lack an explicit basis for deciding about the meaning of scores on individual "factors," but the characteristic regression analysis employed takes us even further from knowledge about the quality of life. Two key regression equations produced in the Omaha study do, for example, provide some information about the statistical relationships among the "factors" but, in the absence of a theory, little else.

The dependent variables in these equations, the composite indices Y_1 and Y_2, are constructed through the process of giving equal weights to each individual "factor." Each city is assigned a Y_1 and a Y_2 value which

Equation 1:

$$Y_1 = 1789.35 + 233.41 X_4 + 1.50 X_{55} + 618.98 X_7 + 0.42 X_{58} - 21.05 X_{16}$$

t value: (5.500)**   (6.954)**   (6.556)**   (3.612)** (3.262)**
$R^2 = 0.81$
S.E. = 192.72

**Statistically significant at the .01 level
*Statistically significant at the .05 level

Equation 2:

$$Y_2 = 45.34 + 0.29 X_{55} + 0.11 X_{58} + 23.39 X_4 + 37.69 X_7 - 2.09 X_{16}$$

t value: (6.790)** (4.785)** (2.789)** (2.020)*   (1.639)
$R^2 = 0.70$
S.E. = 38.09

**Statistically significant at the .01 level
*Statistically significant at the .05 level

is the arithmetic mean of its scores on a set of the "factors." These composite variables are presumed to reflect quality of life in a more general sense than their individual components. But in this construction, the researchers merely combine the intuition that each "factor" contributes equally to the quality of life with the previous eighty intuitions about the relationship between high and low score and high versus low quality of life. The chances that equal weighting is valid, that, for example, one unit of per capita income is equal to one unit of carbon monoxide, in this or in most related exercising in indexing, is extremely remote.

The independent variables on the right of the equal sign in the two equations are the selected individual "factors" most highly correlated with Y_1 and Y_2 respectively. Equation 1 tells us that 81 percent of the time the value of a city on the Y_1 index can be predicted within ±192.7 points by adding 1789.35 to 233.41 times the city's hospital room costs (as a percent of per capita income), adding this to 150 times the city's number of nonnegligent manslaughters (per 100,000 population), adding this to 618.98 times the city's residential electrical rates (as a percent of per capita income), adding this to 0.42 times the city's number of assaults (per 100,000 population), and finally subtracting from this 21.05 times the city's number of building permits (per 100,000 population). The preceding is true only when we employ the data sources used in this specific study for the 100 cities in this sample. One cannot be blamed for failing to see any clear relationship between this equation and the question of quality of life in U.S. cities. The second equation is equally uninformative.

A third characteristic feature of this research is the manner in which it is expressly addressed to elites, to "industrial investors who demand lower costs, larger and more diversified labor pools" (Todd 1978:18). This is "policy-oriented" research intended to appeal to the values of those who can make or influence policy. From this perspective, "quality of life" means attractiveness to a particular social class, as opposed to people in general. This is a variant of a "scientific" conclusion which just happens to serve local and class interests. The findings illustrate the way in which conclusions and advice may be based neither on science nor on technological principles but rather on the opportunity structures that affect the investigator. We can see how interest-serving value judgments are smuggled into an argument via all the critical intuitions governing the meaning of variables, and done so in the name of scientific and technical credibility.

The City Index literature is treated here as typical of positivistic-technocratic approaches because it is not these findings that are subject to dispute but the predisposition of the approach. Some may see it as strangely coincidental that a group at the University of Nebraska "discovers" that its state capital and its home town rate first and fourth, respectively, in quality of life out of one hundred U.S. cities. But this is merely a caricature of all similar attempts to measure quality of life through regression techniques to achieve such goals as helping "industrial investors find the best conditions." The City Index researchers are more forthright than their fellow positivist-technocrats in "proving" that the best policy is the one that most directly benefits them. But this merely underscores the deeply ideological character of all supposedly value-free, more accurately theory-free, "scientific" approaches to technological impact. In the end, one has not learned very much from this research about the quality of life in Lincoln, Nebraska, but one does know that the town scores high on an intuitively derived and constructed potpourri of rates from official (and not always accurate) sources. One should be willing to concede this and see the more important point that this research is a naive application of a popular but very distorted perspective.

The positivist-technocrat attempts to solve problems of understanding and proper action which are virtually insoluble. In this vain attempt, he bootlegs into his analyses a set of predispositons disguised as operationalizations and a set of preferences in the name of rational policy. Whether a high versus low rate of anything that presumably is measured by a set of standard indicators leads to the good life is subject to endless debate. Similarly, the claim that policy to maximize in the manner prescribed by regression equations will reproduce in Atlanta or New York the reported good life in, e.g., Lincoln, Nebraska, remains unsupported. The approach employs the trappings of scientific method and the sensibility of equating progress with a shift in economic, demographic, recreational, and other indicators. But in doing so, it avoids important questions by answering unimportant ones.

Partiality of the Aproach

Although not posed in these terms, questions on the quality of life in U.S. cities raised by the City Index group are questions about the impact of certain technologies on society. The methods employed to answer them, like those shared with other positivist-technocrats, distort, largely by exclusion, the events they presume to understand and help to control. For more than two centuries, social scientists and ideologues have provided various accounts of the chain of effects linking technological progress, the raising of economic, demographic, recreational, educational, and environmental levels, as in Table 2, and the quality of life. Some, like the City Index group, have assumed that quality of life is correlated in a significant and positive way with innovations in industry, health, recreation, shelter, etc. Like the City Index group, many have stressed that the way to achieve "higher" quality of life is to have more of something, per capita, or per dollar. It is not unusual to find arguments made in the recent or distant past to the effect that "cities are today what they have always been — the centers of economic, social, and cultural opportunity.... All of these contribute in varying degrees to the good life" (Todd 1978:20).

The positivist-technocratic perspective fragments reality to attain the appearance of scientific accuracy and objective policy advice. As in the City Index analysis, but usually expressed more subtly, the conclusions follow directly from the nature of the questions addressed. The city in which quality of life is "highest" is that with the most "favorable" (this is not defined systematically) constellation of aggregate indicators of industrial and technological progress. To improve the quality of life in any city, according to this approach, policymakers must, through industrial and technological innovations, improve per capita rates or ratios of income, health facilities, TV stations, parks, and decrease the percentage of ADC recipients, etc. Neither the distribution of income nor the quality of health facilities, parks, utilties, and telecast programs, to which per capita access is improved, are ever actually examined. The meaning of such changes to potential ADC recipients is not considered. By focusing on a set of "independent" cases, such conclusions also ignore the fact that a city is not an undifferentiated island in a sea of nonurban settlements. Urban life is not isolated from nonurban life. A city is a node in a complex, multilevel system. It is nothing without its relationships with villages, towns, and other cities in local and worldwide networks. The "good life" in Lincoln, Nebraska, is the outcome of life (good, bad, or indifferent) in the Nebraska coutryside, in Omaha, in the other ninety-eight cities in the City Index sample, and even in Mexico City, Bombay, and Ho Chi Minh City.

Every one of the conventional quality of life indicators appears ambiguous when we consider the internal heterogeneity of even Lincoln, Madison, Des Moines, Omaha, or Greensboro. A low divorce rate may be good for some families — though it could mean that a lot of people who do not like each other stay married anyway — but it is also bad for some lawyers. Other indicators mean a good life for some groups but not for others. Some, like the City Index group's key variables X_{35} (non-negligent manslaughter) and X_{58} (assaults), can indicate the exact opposite of what they might seem. It could be that a town with a low X_{58} rating is a safe place. But since such a rate is taken from the Uniform Crime Reports, a low score can also mean that large numbers of victims do not report assaults; it can even mean that there are many assaults and the reporting system is imaginative.

Further analysis of Table 1 shows that there is an approximate, but significant, inverse correlation between quality of life in the cities in the City Index sample and percentage of Black, blue collar, and unskilled workers living in those cities. There is also a clear tendency for cities with low scores on the indicators to have a large suburban component — entirely omitted by the strange decision to count as "city" only the incorporated city proper rather than the Standard Metropolitan Statistical Area (SMSA) that the U.S. Census employs. This suggests something of a further intuitive predisposition in the analysis; that some cities are "good" because "bad" people do not live in them (poor people, Black people, people on welfare, etc.), and some cities are "bad" because "good" people live outside the city limits.

When this research is viewed as a variant of the positivist-technocratic approach to the study of the social impact of technology, it is easier to see it as partial and opportunistic. The basic flaw in the approach is its inability to account for the complexity and extensiveness of urban social life, to depict the variable impact that innovations intended to raise the value of social indicators actually have. In presuming to know how a particular rate that reflects technological advance relates to quality of life, the positivist-technocrat betrays a serious (and perhaps convenient) lack of sensitivity to the ambiguities and the unintended and often ironic outcomes of "progress."

Understanding Technology's Effects

The view that research such as that of the City Index group is inadequate is widely held. Both technocracy and positivism have been criticized since the early days of the urban-industrial revolution for portraying false descriptions and misleading prescriptions. As illustrated in the preceding chapter, it is both significant and unfortunate that critics usually offer only a different but not less distorted account. If positivism has "dehumanized" quality of life through the use of objective indicators and regression analysis, and if technocracy has mistakenly equated "better" with "more" (per capita or per dollar), the antipositivists and antitechnocrats have done harm in their own ways (Nelkin 1979).

Antitechnocratic and Subjectivist Critiques

The path between intent and outcome is rarely simple or direct. The centrality of the ironic focus in art and sociology, of functional analysis, and of dialectical approaches in social science attest to this. Insight into the nature of man is often closely tied to the discovery of "dissimulated outcomes," consequences that approach contradiction with intentions. We can see this principle clearly in the case of creative insight into the effects of technology on the quality of life. Discovery of the unintended consequences of progress has been a major concern in sociology since its mid-eighteenth-century origins. Perhaps the best-known example comes from Marx: that the capitalist classes, in order to survive, need to create, sustain, and exploit a working class — that will eventually destroy capitalism. But scores of other social theorists, including Adam Ferguson, Adam Smith, Max Weber, and Sigmund Freud, were equally struck by the tendency for capitalism, Protestantism, industrialism, urbanism, modernism, civilization, or of humanity itself to "sow the seeds" of self-destruction through innovative acts intended to "improve" things.

Current observations on the unintended outcomes of progress, some of which were discussed in the preceding chapter, are equally familiar.

The popular works of Jacques Ellul, D.H. Meadows et al. (1972), Alvin Toffler (1970), and Garrett Hardin (and Baden 1977), are examples of a large and growing number of books published during the last ten or fifteen years that stress the darker side of technological innovation. The antitechnocratic perspective is well established in these and related works. This counterperspective is based on the view that technological progress has inevitable destructive consequences. Instead of seeing technological advance in industry, commerce, and personal consumption as contributing to the quality of life, antitechnocrats and intellectual Luddites see it as a threat to life itself. According to these critics, urban society, insofar as it represents the concentration of technological solutions to human problems, is not a likely setting for "the good life"; at least not for long.

The positivist approach, too, has been under critical attack for some time, principally from hermeneuticists, subjectivists, and the *Verstehen* school (see also chapter 17). These critics would deny the very meaningfulness of objective indicators (that is, quantitative), including public opinion data, for understanding something so deeply a matter of experience as quality of life. For example, Richard Brown (1977) has recently provided an elaborate critique of positivism and a defense of subjectivism, as did Peter Winch (1958) in an earlier and more precise formulation. Similar arguments by the phenomenological sociologists (e.g. Schutz 1970) and ethnomethodologists (e.g. Garfinkel 1967) on understanding social life are well known to most contemporary social scientists. From these perspectives, social reality is seen as constructed not according to positivist principles but according to rules known (or at least followed) only by actors themselves. To attempt to explain social reality without knowing these "rules of the game" and without being aware of what it is like to be an actor can only be misleading. When such aspects are taken into account, the use of single or aggregate indicators, like those used by the City Index group to quantify the quality of life, thus violate the complex and deeply personal character of social experience.

The subjectivist and antitechnocratic perspectives are as cogent, but also as partial, as the positivist-technocratic. The approaches simultaneously criticize and complement one another. But at the very least, subjectivism's stress on experience and the antitechnocrats's stress on technology's undesirable side-effects show how the City Index type of research leaves an important set of questions unasked. The experience of living and coping in a modern city offers countless examples of how an innovation designed to improve the quality of life can make things worse.

We are all familiar with the "technological fix" syndrome. This is that the side-effects of a technological solution tend to require their own technological solutions which in turn have unintended side-effects, and so on. A vast new industry has grown around the need to correct environmental effects of industrial growth. One high technology is called upon to clean, recycle, and conserve the "by-products" of another. What will be the unintended consequences of the growth of the cleaning, recycling, conserving, and the other "fix" industries? According to the technological fix syndrome, they, too, will be "solved" with new technologies. We sense this syndrome; it may even trouble us. It affects the quality of life, but as the subjectivists point out, positivist social science has systematically ignored this more subtly "felt" side of progress.

The Sociology of Knowledge Perspective

The sociology of knowledge is of special interest in this light because of its analysis and critique of positivism and technocracy. Mannheim and more recent sociologists of knowledge have argued that the principal modern symbol systems, science, ideology, and even language itself, have characteristically been shaped by being used in attempts to understand the effects of technological change on quality of life. The effects of technological change are, in comparison to the intended consequences, so vast, complex, and far-reaching that they can never be fully comprehended. No accounts provided either by positivism, technocracy, or their critics have yet been able to portray the effects of innovation without distorting reality. Because science, ideology, and language must be selective (they must both exclude and include relevant data), none can say in any definitive sense whether the consequences of technological progress constitute a blessing or a curse. Yet people, including the City Index group, presume to know what the ultimate consequences will be. They act, counsel action, or promote policy based on such a presumption. Social scientists, engineers, and planners all use theories and ideologies to understand technological change; and we all use such understanding to guide action. But if, as the sociologists of knowledge suggest, current scientific and nonscientific understanding distorts the world, we may be making some grave mistakes by acting as if we know more than we do.

At issue in social theories about the effects of science and technology on quality of life and in our daily experience is a fundamental question about social change. People would like to know, and many social scientific and ideological perspectives suggest that we can know, whether progress will occur through investment in specific technological innovations.

This type of question asks for an account of the direction of social change, an "eschatology" (Horowitz 1972, sec.2) or, in the more bureaucratically sensitive language of technology assessment, it calls for reliable "scenario construction" (Arnstein and Christakis 1975:213ff). The complexity and extensiveness of technological impact suggest that there are many possible eschatologies, that several reasonable scenarios can be constructed around the theme of the effects of any innovation on quality of life. But which is correct? Theories in the sociology of knowledge suggest that neither technocracy, positivism, nor any other ideology can by itself provide an answer. Proponents of the various perspectives usually claim correctness for their own analyses and falsity for all others; yet they all provide systematically distorted accounts.

In criticizing the excesses of positivism, scientism, or opportunism, subjectivists argue that the scientific method is at fault. Some claim that there is no real value in making distinctions between art and science, explaining and moralizing. But in making such claims, they weaken the power of science, art, and ethics to make their contributions to quality of life. The scientific method is more often violated than demonstrated in positivistic research like that of the City Index group. Subjectivists are wrong to call for dispensing with social science when so little has been done and so much that is presented as social science is ideology made to appear objective. The social construction of reality, the subjectivity of actors, our feelings about progress, and all the related phenomena subjectivists argue must be incorporated into the study of social life can be (and on rare occasion are) incorporated in a scientific manner. Science is a way of knowing; so too are art and morals. If the label of "science" has been appropriated by a certain type of opportunistic ideologue (who typically proclaims the end of ideology), it is the opportunism and the ideology, rather than science, that deserve criticism. In criticizing the ahistorical, atheoretical optimism of technocracy, the antitechnocrat frequently adopts an equally ungrounded pessimism. While it is often enlightening to emphasize that specific technological innovations have had negative consequences, it does not follow that technological innovation in itself is ultimately destructive. While technocracy can be criticized for equating high rates and ratios with quality of life, it is no better to equate them with inevitable deterioration of life.

Third Way in Technology Studies

The most disconcerting implication of the City Index approach is not its own shortcomings but the apparent lack of less biased alternatives.

Neither positivist-technocracy nor its critics seem able to offer an adequate analysis of the effects of material progress. Neither the hopes of positivist-technocracy nor the cynicism and fears of the critics can give an undistorted account of what technology does to or for the quality of people's lives. The "new and radical fact of human history," as C. Wright Mills called it, is that neither the University of Nebraska Group nor other, more sophisticated positivist-technocrats, nor their ideological critics have any special methods or disinterested perspectives with which to understand or control quality of life. "We are," indeed, "on our own" (Mills 1959:185). The City Index research is a naive application of the positivist-technocratic perspective. More significant weaknesses are in the approach itself rather than in one group's use of it. At issue is the inability of the approach and of counterpositions to deal with questions addressed. The City Index group is not "guilty" of anything except being social animals and thus trapped, like many professionals, in the prevailing methodological miasma that passes for understanding.

In other chapters of this book the inadequacy of prevailing scientific and ideological approaches to understanding the effects of technological innovation on quality of life is explored further. If valid, the implications of this thesis for planned social change are profound. If none of the prevailing sources for understanding the relationship between progress and quality of life can provide an undistorted account, other sources must be sought, invented, and/or applied. But do such sources exist, in theory or in fact? As we have argued, in Mannheim's third way, they do exist in a germinal form, at least. With this program and with the broad set of concerns that constitute the postacademic style, we have the beginnings of an alternative to positivist-technocratic, subjectivist-antitechnocratic, and other accounts of technological innovation. It is possible that, as a result of searching for such an alternative, we may even find an antidote to, in Sorokin's words, "the present degradation of man."

Notes

1. The social indicators movement has branched out into several substantive specialties including development indicators, science indicators, and, as discussed in this chapter, quality of life indicators. See also Sheldon and Moore (1968) and Andrews and Withey (1975).
2. This somewhat cumbersome term is based on the discussion in Gouldner (1976). The "positivist" part refers to an approach in social science which uncritically applies techniques that, it is presumed, are valid for answering certain types of questions about physical and biological behavior. The "technocrat"

part refers to an approach to social change that uncritically conceives of solutions to social problems in terms of technological innovation. Since technocracy assumes or proclaims the centrality of technocratic "classes" in planned social change, it can be, like all ideologies, an interest-serving account. Positivism and technocracy are not identical, although, as in the case of the City Index literature, they often complement one another.

10.

Sociology: Pure, Applied, and Interdisciplinary

This chapter concludes the first part of our discussion of the ideas and institutional changes that mark the transition to postacademic social science. Its focus is on the distinction between pure and applied sociology, which has affected the evolution of the academic discipline, and current trends toward interdisciplinary activity. The main point is that the pure/applied distinction is false and misleading, and that related differences between disciplinary and interdisciplinary orientations are of far more consequence. In the second instance, the choice to pursue social research, teaching, and other activities with one or the other orientation will affect the way in which such work is conducted, the nature of its products, and the uses to which they are put.

Disciplinary approaches act as an impediment in the types of work postacademic social scientists are likely to do. The recent growth of nonacademic activity for sociologists, anthropologists, etc., has revealed the limitations of academic orientations in the study of technology and in general. This has led to a renewed and growing interest in approaches that transcend disciplinary boundaries. During the past few decades, the method and substance of interdisciplinarity has become a favorite topic of social research and a favorite strategy for social researchers, teachers, and sociotechnical professionals. These developments suggest that, in yet another area, social science is finding its way back to its founding commitment: to study the social world in all its complexity and with all its impurities, not so that social science may prosper, but in order that man might be better served by his inventions.

Pure and Applied Approaches in the Non-Social and Social Sciences

The attempt to classify science as pure or applied is a part of our culture which extends back to the industrial revolution. In concept, it can be traced to the distinction in Greek philosophy between pure *(episteme, noesis)* and practical *(techne)* "ways of knowing"(Kranzberg 1979). It has had a bearing on the lives and work of scientists and other scholars and on the evolution of science, technology, and modern knowledge generally. It has acted as a self-confirming label that varies in its referents from time to time, between nations, and among particular sciences. The distinction has served invidiously and/or to mask other, often ideological, differences between schools of thought. In academic social science the pure/applied distinction has served both as a defense for certain styles of work and as an explanation for why the disciplines have developed as they have. Regardless of how we choose to evaluate the effects of the distinction, viewed from the 1980s, it appears indefinite, unproductive, and archaic.[1]

The Pure/Applied Distinction as Invidious

It is difficult to conceive of the pure/applied dimension in historical context without thinking of it hierarchically. From the beginning of the era of academic science in England, France, and especially Germany, a line was drawn between the more noble, more important production of scientific knowledge and its necessary but less glorious application (Ben-David 1971:126-27). This distinction was applied in several ways. One of these was the decision to include the knowledge-producing fields — physics, chemistry, biology — within the framework of natural philosophy. From their institutional beginnings, the highest academic degree in these fields, as well as in philosophy, was the Ph.D. In Germany and elsewhere, the prestige of philosophy and the older humanities was extended to what came to be known as the pure sciences. In exchange, these sciences adapted their distinctive scientific method to the classic styles of the humanities. Their research came increasingly to be evaluated through peer review procedures and to be published in professional journals. Their pedagogy and their norms of student-teacher relations increasingly resembled the scholarly orientations practiced by their colleagues in philosophy, history, and literature.

Fields that specialized in practical invention and technique, though based, ideally, in pure science, were relegated to a marginal status in the universities or were consigned to lower status types of organizations: professional schools, labs and workshops, polytechnic schools, and institutes of technology. Initially these fields did not offer Ph.D.s or even Master's degrees. And though graduate education in technical fields has grown considerably since World War II, the ordinary degree held by engineers and technicians is still a B.S.; for many, it is still only an

Associate (junior college) degree (Pettit and Gere 1963; Horowitz 1972:431). Part of the explanation for this separate and unequal treatment of knowledge production and application lies in our cultural (and ideological) valuation of "head" over "hand" work — with exact parallels in Brahmanism, Plato's *Republic,* etc.[2] The personalities of individual scientists and the evolution of national academic institutions also contributed to this situation. Within many of the academic sciences, the distinction came under attack and disciplinary and institutional differences based on it began to break down soon after they were established (Ben-David 1971:126). Yet in principle, the distinction has persisted through the academic era and to this day.

The distinction between applied and pure science seems to rest on whether scientific activity is pursued in relation to definite practical problems or for its own sake. This appearance has proved to be difficult to verify. More certain is the fact that, until recently at least, pure science in Europe and the United States has enjoyed greater social prestige, has served as a professional model within disciplinary organizations, and has been more highly rewarded in other ways. Even this needs to be qualified, since the dependencies between theory and practice were evident from the very beginning — to experimental scientists especially. But as long as research was sufficiently cheap so that universities could compete with other organizations for its support, academic science could be kept relatively pure, insulated from technical applications. At least the flow of knowledge "from" the scientific academy "down to" the technician could be maintained.

As the academic era proceeded, the hierarchical connotations of the pure/applied distinction in the non-social sciences took on greater importance than any substantive ones. It became increasingly obvious that the distinction is, at most, pragmatic. Knowledge pursued in relation to solving a practical problem, such as telegraphy or the construction of an atomic bomb, can readily lead to the most highly abstract theoretical formulations, of the most "basic" types, in the pure scientist's repertoire. Such research can even have this effect and fail to achieve its intended goal. Consider for example the numerous insights into the behavior of subatomic particles that emerged from research conducted, but never actually incorporated, in the invention of television. Similarly, knowledge pursued with no other purpose than to test or somehow alter basic knowledge formulations, such as the early work in genetics which led to the identification of DNA, or the more recent extensions of this that led to the invention of new life forms, can, sometimes with surprising speed, be turned into the most useful, banal, or even dangerous technologies.[3]

What is pure science in one laboratory may be applied in another, the "basic" knowledge for one generation of pure scientists becomes working principles for applied scientists of another. Who is to judge whether the work of one scientist is purer or his concerns more basic than another's?

This last question is not rhetorical. Because of the arbitrariness of the pure/applied distinction it becomes very important, if the distinction is to be maintained, to understand who renders such judgments and through what criteria. This is an empirical question in the sociology and politics of science, one whose answer must entail a decision made by scientists, funding agencies, or others, and not an unambiguously identifiable characteristic of scientific activity itself or its product. The academic era ended earliest for the most highly technology-dependent sciences such as physics. Today, nonacademic sources of support are the norm even for academic scientists by way of organizationally sponsored research. With these changes, even the invidious aspects of the pure/applied distinction have lost their force: the value of putting pure research "above" applied has been obviated by the great leveler of academic distinctions — material resources. Now, as most scientists will acknowledge, pure science needs applied science for its material survival, just as once applied science was, or was believed to be, dependent on pure science.

Purity and Application in Sociology

These observations are meant to apply to science in general, but most of the evidence that underlies them relates most clearly to the non-social sciences. The case of sociology, except in broad outlines, is somewhat different. This is due to peculiarities of its subject matter — human relationships — and also the late date at which the academic social sciences became institutionalized. During the era of amateur social science, the program of sociology was formulated with an explicit applied orientation. The models upon which the new sciences of Smith, Saint-Simon, Comte, and Marx were based were not those of the academic classics and humanities nor those of the natural sciences. To distinguish their work from that of conventional social philosophers, the preacademic self-consciously identified with the problem-selection criteria and methods of the industrial arts and politics (Horowitz 1972:418ff). They sought to know the world so as to change it. As Durkheim observed, they were not "quietist scholars" but religiously inspired men of action, very much creatures of their revolutionary times.

In the wake of the splitting and institutionalization of academic sociology and, to varying degrees, economics, psychology, and political science, the drive to share the prestige and authority of the humanities became an early strategy in the battle for legitimacy (Ben-David 1971:127ff). The perceived need on the part of the early economists (Anderson and Buck 1980) and academic sociologists, of whom Weber and Durkheim are chief examples, to disassociate social science from ideology further reinforced the discipline's official stand in favor of purity. The early academics classed the founders as ideologues; Marx in Weber's and Herbert Spencer's case, Saint-Simon in Durkheim's. From the late 1800s on, the serious business of proving that the job of the discipline is to understand the world, not to change it, began it occupy a good share of the energies of academic sociology. The sites at which social change is made — political movements, religious and welfare organizations, labor unions, R&D labs, factories, elite circles, and other "natural laboratories" of social science — were either avoided by or closed to early academic social scientists. Even the limited and asymmetric interaction between academic and "less-than-academic" non-social science was substantial in comparison to the interaction between early academic sociology and the "sociology of the streets," offices, factories, etc.

Even these distancing factors were not sufficient to "purify" sociology entirely. Spencer, Weber, Durkheim, and their lesser-known contemporaries and students did insist that sociology and the endorsement of any particular course of social action are two separate matters. But despite their sincerity and goodwill, it was impossible for them, as it is for us, to deal with the questions they confronted (decades before abstracted empiricism refined the art of posing "value-free" questions) without addressing applied concerns. As "scientific" and hyperbolic as was the language of the early academics, their diagnoses and therapies were thinly disguised. This does not invalidate their observations. The engagement, moral awareness, and the same desire to change the world through social science that drove the ideologues who inspired these observations also gives the observations of the early academics special validity. The differences between early academic sociology and what we call today "professional consulting" or "policy science" are all differences of style rather than substance.

The practical application of social knowledge received new impetus with the establishment of academic sociology in the United States in the 1890s in a society immersed in the progressive era. This was reinforced when U.S. sociology began to assume a dominant international role with the emergence of the Chicago School. Academic economics, the discipline whence most early U.S. sociology departments arose, had by 1900 already proved to be a profitably applied enterprise, in policy and technical contexts. The journalistic and social-problems orientation of the Chicago School, its pragmatist philosophical foundations, and, as Gurvitch has noted, the harshness of life in U.S. and European cities after 1929, gave credence to an applied sociology. With Ogburn, this even led to a full-blown sociology of technology.

The decline of the Chicago School's influence in U.S. sociology in the years just before and through World War II coincided with a new, and this time better organized and more successful movement of purists. The commitment to description rather than action was revived. This commitment was strengthened by the example of the successful presumably pure sciences,[4] the incorporation of highly quantitative research techniques (with their bureaucratic research ethos), the continuing quest for legitimacy in the face of more abundant resources for academic activities, and by more prosperous times in general. As a result of this most recent circulation of elites in academic sociology, the pure sciences orientation has come to rule the profession; applied sociology is at best tolerated and at worst it is considered a lower-status calling.[5]

Ironically, the pure/applied distinction has all but lost its meaning in the non-social sciences that academic sociology seeks to emulate while academic sociology has "purified" itself to a point far beyond the dreams of Weber and the other pioneers. Earlier academics struggled in vain to disguise their moral commitments behind what they believed was scientific language. Today the cultural if not the numerical majority of academic sociologists struggle in vain to hide their lack of moral commitment — not to be confused with commitment to personal goals, career aggrandizement, etc. — behind evocative words and phrases like "deviancy," "religion," "social stratification," "family," "community," "political beliefs," and "group conflict." They are capable of engaging as sociologists not with deviancy, religion, etc., but only with indicators, variables, specifications, data sets, regression models, variance reduction, nonrecursiveness, index construction, and equally "pressing" methodological issues. For example the author of a recent paper on transition to adulthood in contemporary United States published in the *American Sociological Review* observes, in the classic prose of abstracted empiricism, that

> only one previous attempt has been made to estimate a simultaneous-equation model in which age at marriage and educational attainment are hypothesized to reciprocally influence each other. . . .That analysis produced the anomalous result that age at first marriage had a negative effect on educational attainment. It seems likely that such a finding arose from the fact that, with the exception of one variable — age at graduation from high school — the exogenous variables used to estimate the model via two-stage least square had little effect on age at first marriage. During the

estimation process, age at graduation from high school, therefore, in effect, was substituted for age at first marriage. Since this variable was related negatively to educational attainment, the effect of age at first marriage was also found to be negative (Marini 1978:487).

There can be little doubt about the purity of this type of sociology. The only question that remains unanswered is "What does this have to do with the social scientific study of human beings?"

Critics of Pure Social Science

After making one's way through not just a few sentences but pages in bimonthly and quarterly journals, books, monographs, and week-long annual conventions filled with misplaced modifiers, split infinitives, agentless causes, and "findings" that simply "arose," it is not difficult to sympathize with the ideological and subjectivist critics of positivism. It *is* disturbing that such strange and alienated thinking about social relations can pass for understanding. But it is a mistake to place the blame for this albeit dominant and highly rewarded type of sociology on the scientific method or on a specific "bourgeois mentality," as ideologues and subjectivists tend to do. The blame, if blame is to be placed at all, lies with individual academic sociologists, particularly the academic elites, their professional culture, and the manner in which they interpret and apply scientific method.

Implicit in the methodological creed of abstracted empiricism is the view that the sociological enterprise is equivalent to application of the scientific method to "problems" and data pertinent to human relationships, and the consequent partitioning of results into portions labeled "literature review," "theory," "model specification," "discussion of variables," "methods of analysis," "findings," "relevance of findings to theory," and the inevitable "need for further research." By acting out — one wonders how many of them believe in what they do — this stilted and overinflated view of the role of scientific method, abstracted empiricists obscure the moral and political bearing of their work. In defense against the arguments of subjectivists and ideologues, the empiricist responds that he is merely being a pure scientist. But the programs of these academic critics to replace pure scientific method with pure poetics or polemics too easily dismiss the very important though partial role scientific method can play. They also miss the main point that by its very nature social science — perhaps all science — is a very impure business.

Inspired in part by the momentum of their own traditions, but also in reaction to the excesses of abstracted empiricism, the critics of pure social science attempt to construct alternative sociologies as partial as those they criticize. The debates rage: alternative journals, counterorganizations, radical caucuses, and departmental purges arise, become institutionalized, and die out; competing views are represented in major symposia on theory and method, and students are recruited to one or another camp in the battle for defining what shall be legitimate pure sociology. When resolution occurs, it is invariably through political, as opposed to intellectual, processes. In the absence of consensus on theory and method, the only criteria universally applicable are those related to technical virtuosity. For the dominant sociologists, it is virtuosity with multiple regression techniques like causal path modeling; for their critics, it is virtuosity with the techniques of the aesthetic philosopher or the political rhetorician.

It seems obvious that the calling of sociology requires that one be a path modeler, an aesthetician, a rhetorician, and more; that one know the place of each orientation. Yet this is subject to serious dispute. For one thing, such holism appears to ignore the growing need to specialize. But specializing in the study of a limited range of subjects or problems — required of any scholar if he is to be effective — is not the same as specializing in the exclusive use of statistical, ethical, or ideological modes of analysis. Even then, the abstracted empiricist argues, the purpose of social science is to advance knowledge, not to relate that knowledge to extrascientific concerns; and knowledge creation is sufficiently challenging to make it necessary to leave these other matters to those outside the profession.

Neither of these arguments, especially the latter, is easy to refute. But viewed in context, their limitations are obvious. Claims made for purity by abstracted empiricists and the exclusivism implicit in both the claims and counterproposals have played important roles in shaping the discipline and the profession. They serve to justify certain types of intellectual activity and disqualify others as authentic academic sociology. The abstracted empiricists, in particular, have chosen to follow a half-baked and outdated model of successful non-social science. But they have also been thorough and politically astute in promoting their program. As a result, they have rendered the opposition relatively weak and defensive. The program of pure positivist sociology is so well articulated that, as long as practices such as peer evaluation and criteria such as "command" of relevant literature, competence with "the" methods, and due caution about one's findings prevail as the only means whereby sociologists are certified, rewarded, and legitimized, it will be difficult for all but a few unusually powerful marginal individuals to practice another type of sociology and survive with the academy. Fortunately, however, outside of and beyond the academy, things are different.

The Bearing of Purity on Postacademic Concerns

The purpose of these remarks is neither to refute nor reform academic sociology, in its abstraced empiricist or alternative versions. Not only would such intentions be virtually impossible to realize, but they are also misplaced. One need only reflect on those instances (such as China up to 1979) in which sociology has been declared illegal to realize that even its most odious, most dehumanized variants are better than no sociology at all. The discipline has contributed in ways great and small to the store of man's knowledge about himself, and there is every reason to believe that it will continue to be a necessary and vital part of our culture and academies.

Today academic sociology is neither the only type of sociology practiced nor can it be judged a better or higher form. It has a place in the discipline, even in the education and certification of nonacademic sociologists. Yet it is a limited place, one that can by no means encompass all the activity properly called legitimate sociology.

The idea that to be purer is to be better or more authentic and the very distinction on which the ranking is based have no factual basis in the history of the non-social sciences — which is not to say that the idea has not had self-confirming effects. The claim that the "purest" sociology is the pursuit of questions about only those aspects of human relations which can be encompassed by abstracted empiricist or competing styles and rules of inquiry is not a description at all. It is a definition, one that excludes from consideration things essential to postacademic social science. If a commitment to this type of purity is a defining feature of sociology, the postacademic social scientist is not a sociologist. But one does not have to believe that such purity defines sociology, though it may define academic sociology. Nor must one believe that the less-than-pure sociologist is necessarily interested soley in the application of knowledge in a narrow sense.

This book is about an orientation that combines social scientific research about technology, its social antecedents, and its consequences with the application of the results of such research to problems of technological design, innovation, and diffusion. Within this broad framework, one can identify a range of approaches, from what might be called pure research (the production of highly general and theoretical knowledge) to applied, "nuts-and-bolts" activity. To stress, as we have, that postacademic social science is problem-centered and service-oriented is to say nothing about how soon or how directly a particular research project or line of inquiry can be expected to "solve" a problem at hand. As every non-social scientist and technician is aware, often very basic theoretical issues need to be resolved before a highly specific innovation can be achieved.

The social scientist, too, should be clear about the distinction between sociological problems that are academic — even if they are limited and specific (which they usually are) — and sociological problems that are general, theoretical, or deal with broad principles. Academic problems arise from academic debate, unsettled questions, and unresolved issues that appear to require solutions from the point of view of academic elites. When debate centers, for example, on positivist versus subjectivist claims, when unsettled questions are about the proper "inclusion level for dummy indicators," and when unresolved issues relate to the "anomalous result that age at first marriage had a negative effect on educational attainment," these concerns may become the object of academic attention. Theoretical or basic questions can arise in these areas, but they can just as readily arise (and some pragmatist philosophers would argue that they are more likely to) in practical contexts, through serendipity, and in many other ways.

Of special interest are those instances in which basic research is inspired by the need to design and innovate technologies to promote development, e.g., when attempts to design better irrigation systems lead to fundamental discoveries in hydraulics. Here pure sociology (in one sense of the word) is not necessarily academic. Certainly, both academic and nonacademic orientations can be applied in dealing with practical issues; but in academic settings, such issues get on the research agenda because they have been authorized by the gatekeepers and, at least manifestly, have measured up to academic standards. For the postacademic such authorization, while always worth considering, is neither a necessary nor a sufficient condition for his interest in an issue. The end to which knowledge is to be put, not the level of specificity or generality, purity or applicability, professional acceptability or irrelevancy, is what distinguishes academic from postacademic social science.

For these and other reasons we have used the label "postacademic" rather than "applied sociologist" (or "social problems specialist"). The distinction between pure and applied sociology shares connotations with the general distinction between pure and applied science, and has special connotations related to the discipline's internal politics and its battle for legitimacy. All these factors make it very difficult to sort out the descriptive and evaluative levels of meaning of such labels as "applied" or even "nonacademic sociology." As ill-defined as the category "pure sociology" is, its opposite, applied sociology, by virtue of the fact that it is a residual category, has an even murkier referent. If it must be presumed that pure sociology excludes the subjective concerns of the people it studies, moral purpose on the part of the investigator, and aesthetics and social criticism as intergral parts of the sociological enterprise, then postacademic social science must be applied sociology. But much that is labled "applied sociology" might be of little or no direct interest to the postacademic, and much that the postacademic social scientist might do as an interdisciplinary, technology-centered, developmentalist teacher, researcher, or practitioner would be labeled "pure science."

In the type of work in which the postacademic social scientist engages — teaching sociology or other disciplines to professional and technical students, participating in social impact research, consulting with technicians and professionals on problems of design, performing policy analyses and/or formulation — it is essential that there be a sensitivity to the intimate and dialectical relationship between social theory and social action. This relationship must not be obscured by compartmentalizing theory and practice into separate and ranked categories such as pure and applied. Regardless of the advantages that this distinction (and its invidious connotations) may have in academic milieux, it can be upheld only at great cost outside of academe. For in separating the theoretical and practical aspects of sociology, or worse, in arguing that because practice has prescriptive and aesthetic dimensions not encompassed by the scientific method, sociology cannot deal with it, one denies sociology the moral authority it might have to help change society. This is the authority which social-scientific knowledge richly deserves, not to be unequivocally accepted as final truth, but to be heard and taken seriously because it matters to man's lot.

Transition to Interdisciplinary Social Science

In the review, in part I, of the critiques and programs of the amateur and academic sociologists of technology, we were careful to stress the shared commitment among these scholars to multi- and interdisciplinary social science. It was also noted that the programs explicated, or that follow closely from the critiques, characteristically lacked a social agency for their extensive application, let alone their institutionalization. While the programs may have been sound, there was no practical way to carry them out. This applies particularly to interdisciplinarity. A key factor in the main drift of academic sociology toward purity and virtuosity is the lack of money and other material resources to support or promote interdisciplinary activity.

The current increase in interdisciplinary activity, which includes but extends far beyond the creation of fields such as SSTS, provides further evidence that the era of academic social science has all but passed. The hidden agenda in these disciplines, evident in the works of writers from Mandeville to Mills, is now being made explicit.

Disciplinary Commitment

As other non-social and social sciences staked out their territories (in part under an economically and politically determined territorial imperative), academic sociology was left with an "expertise" best defined negatively: sociology "is" neither economics, statistics, psychology, nor anthropology. This has also meant that sociology is eminently dispensable in the presence of any and all of these other fields. From outside of the discipline, the inclusion of sociology in scientific work, even social-scientific work, appears superfluous. As long as sociology is guided by those whose interest in technology as subject and object is at best passive, it is superfluous. There are internal reasons too. In reaction to the role in which it was cast as the study of "leftovers" and through the application of an elaborate system of internal controls, such as "quality control," sociology was defined out of certain types of work felt to be the property of other disciplines. Or, if this is put in absolutes, then a sociologist, a group, or a school of thought could (and still can) prosper within academic sociology without having to make more than a token acknowledgement of similar and perhaps essential work in other disciplines.

This tendency toward strict disciplinarity has been further amplified by sociology's "second front," its battle for legitimacy. The social character of academic settings, in First World countries especially, made it increasingly important for professional sociology after World War I to give every appearance that its criteria of recruitment, certification, and rewards were as rigorous and cognizant of scientific standards as those of any other field. To achieve this appearance in a problem area where rigor and the common application of the scientific method, though necessary, are only minor aspects of mastery, it became necessary to abandon or downplay the more important (but more messy and intuitive) aspects of the sociological way. A particular type of mastery had to be deemphasized so that academic sociology would appear legitimate and sufficiently scientific: mastery of the classics in social thought (whether or not they are properly sociological), of developmentalism, of value-explicitness, and of the subtlety, intricacy, and historicity of social relations. Yet it is this type of mastery that leads most conclusively and directly to the need for interdisciplinarity. Thus the academic sociologist, at the risk of survival of his own career, has been constrained to deal with a "predisciplinary" reality [6] for which his disciplinary orientation is no match. The goal of explaining a social event or phenomenon using only sociological, as opposed to economic, psychological, anthropological, etc., variables is doomed from the outset to partiality, triviality, or worse, inapplicability. In the language of the abstracted empiricist: "the attempt to reduce the variance in our models" in which all relevant variables are "specified with sociological operationalizations" is futile. Eventually, it is going to be necessary to "bring into our equations 'exogenous' factors" that cause us to look outside of academic sociology.

To look beyond one's discipline in a serious and concerted manner is also to place one's work and career beyond the exclusive scrutiny of disciplinary judges and authorities. The cost of such a commitment may be great. For the profession it may be costly because it opens practitioners to criteria that are at best different and at worst too difficult to satisfy. In the case of sociology, it threatens to undermine the discipline's hard-won legitimacy. For the individual sociologist it may be costly because professional forums for evaluation within the field are threatened by that which they cannot control; disloyalty and incompetence are suspected, or something more personally or politically threatening is suspected for which labels of disloyalty and incompetence serve as socially acceptable proxies. As a result, both the profession and the professional sociologist are constrained to play a safe, narrow, and hopeless disciplinary game.

A further tendency in sociology is that while the organized profession and the academic discipline have won their legitimate place, in the process they have been virtually fused into a single doctrine, structure, and population; a closed and unhealthy situation. This has occured at a time when the academic discipline is losing sources of material and social support and personnel and when it must now compete with the emerging interdisciplinary postacademic field. In these circumstances, there are two sides to the immediate prognosis for interdisciplinary sociology. On one side is the momentum of external and internal forces that are carrying the field increasingly toward interdisciplinariy: interdisciplinary team research, teaching, and consultancy opportunities continue to grow in number and improve in intellectual quality.[7] On the other side is the powerful inertia of the organized profession and the academic discipline that it so tightly controls. The existence of these forces and the influence they continue to exert on recruitment, certification, and reward systems, are relevant to the institutionalization and survival of postacademic social science.

Growth of Interdisciplinarity

The conscious pursuit of interdisciplinary social research on technology or technological applications can take many forms (Porter and Rossini 1978; see also ch. 7, n.2 this book), even excluding the many variants of paralled but "unrelated" research strategies that shade into consciously interdisciplinary work. Most generally, they can be divided between approaches oriented to use by individuals and approaches meant to be used in group research, teaching, and practice, that is, work in which various individual specialists in particular disciplines participate. These approaches are not mutually exclusive. Within sociology, the two approaches have been divided between specialties: the sociology of knowledge in the case of individual interdisciplinary orientations and group dynamics, and, to a limited degree, the sociology and history of science, in the case of team approaches. The first approach has been discussed in connection with the concept of *Freishwebende* and the work of Mannheim. The suggestion is that the aim of the interdisciplinary individual is to achieve a "classless" state of mind in relation not only to secular interests, of the usual sort but also to those of his academic "class," i.e., his discipline (Mannheim 1968, pt.3). The second approach is of more direct interest in the present context. Themes that are central in group research, such as division of labor, collective outcomes, conflict, and consensus, are virtual first principles in sociology. The theory and practice of this crucial group dynamics setting, interdisciplinary team teaching, research, and practice, are aspects of social science in which the application of scientific principles to one's own activity is especially pertinent.

Whether of the individual or group variant, interdisciplinary science, social science especially, is problem-focused. The common ground upon which the interests of the various disciplines or individuals are based in such work is the situation, the problem, or the phenomenon. It is assumed that these can be meaningfully studied and shaped, if appropriate, only from two or more perspectives. Like the practice of interdisciplinarity in general, this is a matter of degree. In some respects, there are no disciplines that exclude everything encompassed by some other discipline. At the same time, some topics better lend themselves to interdisciplinary approaches than others. Whether a particular problem is approached in a self-consciously interdisciplinary manner depends on several factors, including economic ones. But the ideal-typical site for interdisciplinary work is one in which the main focus is on a problematic situation. The focus of such work is on something external; it does not "arise" from "the literature" or from theoretical developments within a body of thought.

The distinction between problem-centered, interdisciplinary work, and that inspired by and better approached within the confines of a single academic discipline, is not identical to the distinction between pure and applied orientations. In the sense that it is problem-motivated and problem-focused, interdisciplinary work is "applied" to something other than the growth of disciplinary knowledge itself. But such problems and their solutions need not have immediate practical bearing to be especially susceptible to interdisciplinary approaches. Fields that illustrate these differences well include linguistics and systems theory, two eminently interdisciplinary but not necessarily "applied" foci of research and teaching.

Interdisciplinarity and Problem Centeredness

By placing "the problem" at the center of scientific activity and teaching, interdisciplinary approaches do not lend themselves to exclusive reliance upon disciplinary criteria and authorities for such matters as validation of findings, certification of methods, or recognition of discovery. Whether a problem is formulated in "correct" sociological (or historical, or psychological, etc.) terms, whether "proper" sociological techniques are employed, or even whether the "true" sociological authorities accept the solutions become secondary concerns in interdisciplinary work. Similarly, the point of interdisciplinary work is not to satisfy prescribed canons of language, technique, and discovery in any particular discipline, or even to achieve a degree of agreement between two or more disciplines that, for example, one's language is proper, that techniques are appropriate, etc. Though these may be desirable results, disciplinary canons serve a different, referential function in interdisciplinary work. The language, techniques, and modes of certification in such work must first be judged pragmatically: Has the problem been "solved" in an adequate way (postponing, for the moment, discussion of the criteria by which adequacy is decided)? If so, the language and techniques are "correct" and the solution authentic. In seeking this end, the practitioner draws on his discipline or disciplines, but principally as means to the definite goal that characterizes interdisciplinary work. The language, techniques, and authority of the discipline(s) are not final arbiters but are themselves tested in relation to their contribution to problem solving and in comparison to other disciplines.

The applicability of the principle of problem centeredness depends crucially on how it is decided that a solution is correct or adequate. These criteria include rules for making such a judgment and for determining who is to make it. They are highly complex and vary from one problem area to another and even between individual problems within an area. Despite these complexities, in relation to the broad set of phenomena and situations the postacademic social scientist is likely to encounter, one can speak of the general character of criteria for deciding whether a problem has been solved.

Postacademic social science is concerned with the incorporation of social knowledge into the study and practice of technological design, diffusion, and innovation, and it is committed to the promotion of developmental ends. The adequacy of the solution to a problem in such work depends upon several things. First, such a solution must remain faithful to the principles of the specific technologies under investigation. To take an extreme example, an effective explanation of the social components of a technology — whether it is mechanical, medical, industrial, or social — cannot "prove" that such a technology cannot exist when in fact it does, or that it cannot function in the way it actually does function. Less trivially, a design or innovation strategy in which social knowledge is incorporated must "work": in the more obvious sense of a machine that still runs even after the sociologist lays hands on it, as well as in the more elusive sense of a machine whose social impact is as intended — development-encouraging. However these types of testing standards as to whether the technology "works" are determined and applied, it is they, and not academic standards which must serve as criteria for solution to a sociotechnical problem.

Such a test also raises the issues of monitoring "solutions," since whether a sociologically explained, designed, or innovated technology functions correctly depends upon longer-term and secondary factors. This is especially true when what "works" is interpreted as "promoting development." Solving a sociotechnical problem requires adequate scope, framework, testability, etc., for monitoring performance of the technology during a significant portion of its working life, and for assessing its social impact in actual or realistic settings. A solution must be distinct, technically correct, and must also be subject to monitoring, by way of technology assessment and social impact analysis as well as engineering evaluation. It must be subject to reformulation in the face of new evidence; it cannot be a closed solution.

If a sociotechnical solution is achieved in accord with disciplinary procedure and stands that test of legitimation so much the better. If the discipline has something to offer toward the adequate solution of a soicotechnical problem, its competent application by an interdisciplinary practitioner ought to be evident in the fact that the solution satisfies the criteria of distinctness, technical correctness, and openness. If a discipline can lead systematically to solutions of such problems, it has something to offer the practitioner. But disciplinary correctness is not sufficient or even necessary to ensure correctness in the approach of the interdisciplinary social scientist or in that of any other interdisciplinary work.

Interdisciplinarity and Multiple Realities

An obvious assumption upon which interdisciplinary approaches are premised, especially in the study of technological innovation, is that referred to as "multiple realities." In problem-centered work where distinctive criteria apply, reality must be encountered in a relatively unrefined and context-laden form. In this type of work it is inappropriate to simplify beyond a certain point, though greater simplification may be entirely acceptable by academic disciplinary standards. For disciplinary reasons, it may be desirable to assume that a particular group or other social formation exists without a technology of interest or that a particular technology exists outside of society. But in postacademic activities, interdependence between social and technical factors must eventually be faced and resulting problems solved. Because the reality of the postacademic social scientist is so raw and complex, it cannot be comprehended or effectively altered from only one point of view. The multiple reality of technological design and innovation demands multiple perspectives.

> Technology is like a great mountain peak. It looks different according to the side from which one views it. From one vantage point only a small part may be seen, from another, the outlook is clouded. Yet we may get a clear view from another side. Few of us see it from all its sides; so each of us is likely to have a limited conception of its nature. It is desirable, then, to look at technology from various points view; for in this way we get a less narrow picture (Ogburn 1957:3)

Interdisciplinary researchers, teachers, and practitioners must be prepared to practice exchange or, as Peter Berger (1964) refers to it, "alternation" between the various perspectives from which the "mountain peak" is viewed. Like all scientists, the interdisciplinary researcher seeks the certitude of intersubjectivity. His purpose in taking alternate viewpoints is not to produce a proliferation of different answers to a single question in some perversity of scientific method. It is to seek convergence among viewpoints in providing more holistic and thus realistic accounts. For him it is not sufficient that others within a particular discipline agree that what he discovers is true or that the policies he proposes are appropriate. As partisan observers of an uncertain world, it gives subjective satisfaction that our observations are borne out by others who know the field. But in the problematic and often life-affecting concerns to which interdisciplinary approaches are increasingly brought to bear, it is perhaps more important that others who do not know "the" field, but know another field, also agree. The certitude that arises from interdisciplinary agreement is of a different magnitude than that arising from disciplinary, intersubjective agreement.

This illustrates the elementary sociological (and commonsense) principle of division of labor: a difficult task that can be divided among several people will be performed more efficiently if so divided. To take advantage of this principle the work must be organized and coordinated. To realize its potential advantages over disciplinary activity, one must distinguish between the right and wrong ways to pursue interdisciplinary teaching, research, and application, and put this distinction into practice. This adds another item to the list of criteria for determining adequacy of interdisciplinary work: it must include rules for relating the individual contributions of each discipline; a management and coordination plan. Without such a plan, even the best intentions of the most competent practitioners would be difficult, if not impossible, to realize.

Managing Interdisciplinary Work: Some Principles

The study of the management of interdisciplinary activity is a young but rapidly growing field of interest among social scientists, philosophers, and others (Gillespie and Birnbaum 1977; Newell et al. 1975). Because organization and coordination are essential in this type of work and these goals are not easy to achieve, research and commentary in this area have revealed serious vulnerabilities in interdisciplinary work. Such work can be as ineffective as poorly done disciplinary work, if not less effective because of the confusion emerging from the "sound of poorly orchestrated individual tunes."

Such problems and the more promising prospects for well-coordinated interdisciplinary activity have been considered in the literature on "research on research."[8] Especially noteworthy and relevant to the concerns of postacademic social science is growing interest in the management of interdisciplinary, technology-focused, team research (Chubin et al. 1979; Rossini et al. 1977; Porter et al. 1980; Walsh et al. 1975). Emerging from specific studies in this area and from programmatic commentary on technology-focused research (Solomon 1977, 1978) is a general set of guidelines for the pursuit of interdisciplinary problem-centered activity. In order to give a sense of what is currently viewed as the best way to conduct interdisciplinary research, these guidelines may be summarized in three general points.

1. A stress on the role of democratic process and leardership (Rossini et al. 1978). A commitment to democratic procedure is often a time-consuming way to coordinate diverse, specialized contributions to problem-solving, but it is generally time well spent. This is because of direct benefits in comparison to more authoritarian and narrow approaches and because democratic pro-

cedure helps establish a rapport and trust that facilitates effective participation of individuals. The principle is familiar: group structure and communication must be so arranged that each participant has adequate opportunity to affect procedure and results, and no participant has an advantage in having his views dominate. Leadership is best exercised through the role of a facilitator or organizer. A leader in interdisciplinary group activity, in making his views known, should be prepared to set his authority aside and participate as an ordinary group member. Tendencies exist to think of one discipline as inferior or superior to another, and these cannot be eliminated at a stroke. But an essentially democratic process in interdisciplinary research, teaching, or practical activity can dampen the distorting effects of such perceptions. Emphasis on problem-solving and a leader's willingness to keep the group task-oriented further help maintain an environment in which democratic procedure is both applicable and effective.
2. A stress on pluralism. Once an interdisciplinary team has been constituted and the often difficult questions of size and disciplinary composition have been resolved (Taylor 1975; Newell et al. 1975), it is important that team members be encouraged to present and consider all possible viewpoints. While time constraints cannot be entirely ignored, participants should be alert to the liabilities of premature closure of such complex issues as the social effects of a particular technological innovation. This includes the tendency known as "group think" (Janis 1972). Within the bounds of time and the parameters of the problem at hand, conflict between opposing interpretations is to be encouraged. Its opposite, "tyranny of certainty," presents special dangers to interdisciplinary work, in part because of psychological predispositions of scientists.[9] The level of individual expertise required for such work, the character of democratic process, and the nature of the task can easily combine to convince participants that consensus equals unassailable truth. Like democratic participation, pluralism can be unwieldy. But it can be managed effectively through reiteration of main points until a group decision or its approximation is achieved.
3. Emphasis on conflict resolution. The question of which individual or disciplinary viewpoint is correct when differing opinions arise should be handled in a "science-advocacy" setting (Nelkin 1977). Appeals to authority, argument, and didactic have important roles to play in such instances. But ultimately participants must refer difficult conflicts to experimental and practical test procedures. When scientifically informed viewpoints are in conflict, pragmatic criteria must rule: conflicts should be expressed in terms of testable hypotheses or predictions. In applying this guideline, one must still face all the issues associated with the use of scientific method generally. But a well-established ideal for conflict resolution, presumably shared by members of interdisciplinary teams, does exist.

These ground rules are very simple and abstract, and uncertainties are bound to emerge even in their most faithful application. Yet a degree of closure can be realistically expected because of the nature of the task. In sociotechnical research, teaching, and practice, the goal, however difficult, is clear: to comprehend how a particular technology "works" in the fullest possible meaning of the word and/or to make it work better.

This goal can act as the "master" discipline to which all other disciplines must be addressed.

There is much more involved in effective interdisciplinary work than mere adherence to ideals such as democratic procedures, pluralism, pragmatism, and problem-centeredness. An entire "science" of interdisciplinarity has evolved, in part as a response to the growth in demand and opportunities for this type of activity. Yet even these simple ground rules do not qualify as the only or even preferred modus operandi for the academic social sciences today. The academic pure science model, against which the critiques and philosophies undelying the interdisciplinary approach were formulated, is quite different. While most academic social scientists, especially in First World countries, subscribe to democracy and pluralism as general moral or political values, they are usually unprepared to employ such principles professionally. Instead, a relatively dogmatic, theoretical, and exclusivist model of individual expertise and an elitist theory of scientific knowledge (Dahrendorf 1967, ch. 5) have predominated. The fact that the interdisciplinary orientation is at least being articulated and debated is itself a significant aspect in the transition to postacademic social science.

Purity, Disciplinarity, and the Postacademic Calling

The distinction between pure and applied social science and commitment to disciplinary orientations are part of the heritage of the academic era. While granting that they served effectively in the legitimacy battle in which sociology and related fields were involved during that era, such distinctions and commitments have outlived their value. In nonacademic settings and increasingly within the academy, material conditions, such as availability of funds for research, and theoretical trends have strengthened the case for interdisciplinarity and problem-centeredness. Development of the academic side of social science will remain important, especially as opportunities for sociotechnical work increase. But it will depend more than ever on interchange between theoretical and practical approaches and on collaboration with researchers, teachers, etc., outside of one's immediate discipline.

These trends are significant because they are occurring in conjunction with other currents of social change within professional communities and in society in general. Breakdown of the pure/applied distinction and the movement to interdisciplinary activity articulate with several other trends: the reformulation of Mannheim's program discussed in chapter 4, creation of SSTS, convergence of TA/SIA and sociological perspectives on irony and technology, critical examination of ideological and utopian accounts of technology's impact, and definite material and resource shifts in scientific and educational institutions. Their point of articulation is in a range of activities and interests that go by many names: sociotechnical research, sociology of technology, applied sociology, service teaching, etc. However diverse, this range constitutes a specific postacademic calling to which thousands have already responded.

The transcendence of the pure/applied distinction and of the boundaries of disciplinary propriety is being achieved or attempted every day in postacademic settings: team-oriented technical research projects, technology policy consultations, and classrooms in schools that provide interdisciplinary and professional curricula. This act of "transcendence" is not a metaphysical feat but a practical response to real situations fast becoming the norm for professional social scientists and technicians. Taken together, these many individual acts of boundary crossing constitute a new (but old) style of social science with an important role in helping people live with their technologies. With this role in mind, we have identified features of the transition to the postacademic era which might provide landmarks for the practitioner. The value of problem-centered interdisciplinary activity is apparent in all of these; this type of activity is fundamental to the postacademic orientation. To the extent that it is pursued and improved upon, the application of social science in postacademic settings and its development within the academy will be made more effective, and technological society will benefit accordingly.

Notes

1. To some extent, social scientists have been forced to uphold this distinction to qualify for certain classes of funding. Non-social scientists are also affected by this situation. The U.S. National Science Foundation, the Social Science Research Council, and many other funding agencies continue to distinguish between pure and applied research (though not always under these names). Part of the reason is that, especially when money ultimately comes from the govenment, it is easier to justify applied research. But to do this means that a distinction between applied and pure is made and certain types of research are definitively placed in one or the other category.
2. The theory that abstract knowledge is and ought to be "above" practical knowledge can be found in the philosophies of many cultures. There is no logical or psychological basis for this theory although it may have a sociological explanation: such philosophies are formulated by those who themselves specialize in abstract knowledge and are often in the direct service of ruling elites. The view that practical knowledge is inferior justifies the inferior position of the masses and rationalizes the philosophers' identification with the elites. In the works of Jean-Jacques Rousseau and Marx, we find important early-modern challenges to this doctrine. Philosophical pragmatism provides not only a challenge but an alternative view of the status of practical knowledge.

3. The experiences of Ananda M. Chakrabarty, inventor of "oil eating" bacteria, illustrate the arbitrariness of the pure/applied distinction. When Chakrabarty was doing pure research at the University of Illinois Department of Biochemistry between 1965 and 1971, he was not concerned with application: "He realized the genes involved in the degradation of hydrocarbons are not in the chromosomes, but in small, auxiliary parcels of genes called plasmids. Because he was working in a university environment, and trained to think in terms of basic research, Chakrabarty said he did not recognize the concept's potential. Chakrabarty got the chance to give the concept a practical application when General Electric Company wooed him to Schenectady, New York, in 1971 to work on a biological solution for the problem of oil spills" (Landman 1980:1). On June 16, 1980, the U.S. Supreme Court overruled the U.S. Patent Office's earlier rejection of G.E. and Chakrabarty's application for a patent.
4. The idea that in the non-social sciences the pursuit of knowledge was intentionally kept separate from practical concerns is partly mythical. At most, early physicists, chemists, etc., had a naive theory of the relationship between science and politics and did not believe it necessary to discuss the close connection between their own political ideals and their work as scientists. Two applied social scientists have recently summarized the situation: "The assumption about the nature of scientific knowledge incorporated in such disciplines as economics follows from a highly selective reading of the history of modern science. The essential features of the reading are familiar. They include claims that the natural sciences owe their power to their successful separation of the pursuit of knowledge from normative considerations; that abstraction and quantification are the soundest procedure for gaining true and useful knowledge; and that knowledge is ultimately sustained by being grounded in empirical reality" (Anderson and Buck 1980:224).
5. During 1979-80, debate on this issue was carried to the Letters to the Editor column of *Footnotes,* the professional newsletter of the American Sociological Association. The ASA and other national and regional social science associations now include panel discussions on applied sociology at their annual conventions and in their professional publications. See *Sociological Practice* a new independent journal edited at the University of Maryland, Baltimore.
6. The concept of "predisciplinarity" was recently used in connection with solar energy research by the sociologist-engineer from Benin, Emile Paraiso (Bota, Weinstein, and Walton 1979:15).
7. Another measure of this is the growth of interdisciplinary applied journals. To note just one recent example: volume 1, no. 1 of a journal entitled *Environmental Impact Assessment Review,* edited at Massachusetts Institute of Technology, appeared in March 1980. Its contributors and editorial board include professors, consultants, researchers, and policy specialists from several different universities, Argonne and Oak Ridge National Laboratories, several private research organizations, and a variety of government agencies.
8. This literature now includes several journals dedicated to the study of research as social activity such as *Research on Research* (Elsevier Publications, Amsterdam), *Knowledge* (edited at Princeton University), and *Social Studies of Science* (Sage Publications, London).

Sociology: Pure, Applied, and Interdisciplinary 191

9. For an interesting case study of the effects of psychological predispositions in bringing about premature closure in the interdisciplinary work done in connection with the U.S. Space Program, see Mitroff (1974).

PART III

Science and Technology for Development

Introduction

This part takes up in some detail a theme mentioned on several occasions in earlier chapters — the commitment to development. Like the ironic focus, interdisciplinarity, and the sociology of knowledge perspective, developmentalism was an important component of the doctrines of the founders of social science, the work of Karl Mannheim, and the other sociologists of technology. Like these other themes, it is in the process of being revitalized after a period of relative neglect during the academic era. This renewed focus on development has many sources within and external to the social and technical fields. Perhaps most significant is the recent rise of Third World nations. Successful movements for political independence in Asia, the Middle East, and Africa begun shortly after World War II once again brought the complex issues associated with development and underdevelopment to the attention of professionals and publics throughout the world. Subsequent struggles for social, economic, and cultural independence in these areas and in Latin America have raised questions about the role of technology in society with an urgency (and uncertainty) not felt since the early days of Europe's industrial revolution. As the postwar period progressed, it became increasingly apparent that the future of man and the science of man was going to be profoundly influenced by the course of development in the Third World and in general.

At a time when the impact of the Third World on development issues was just beginning to be felt, Karl Mannheim provided a framework for understanding this impact as part of his general program. Since then Mannheim's observations on development and the Third World, and those of many other social scientists, technicians, and policy specialists in several fields, have been explored, debated, and more recently, applied (for an early survey see *Scientific American* 1963; also Horowitz 1972). This work has been responsible for the creation of several interdisciplinary forums, organizations, journals, research specialties, and practical opportunities in "the development game"; and it has also helped reshape traditional academic disciplines.

Though extremely diverse, this new developmentalism is premised on a reinterpretation of the relationship between technology and development inspired by the rise of the Third World: Technological innovation of the type that characterized development in European and other First and Second World countries can no longer be expected to have the largely positive effects it was once believed to have. The promotion of improved life chances and lifestyles "for all" can no longer be viewed as the automatic outcome of the spread of labor-saving, energy-, and capital-intensive technologies through normal economic and political channels (Richardson 1979). Historical research on Europe's development and sociological, economic, and anthropological work in Third World countries have shown how deeply interdependent the underdeveloped countries and today's developed countries have been since the industrial revolution. This has cast serious doubt on the equation between R&D investment and European development in all its dimensions; and it has brought into question the view that such investment will necessarily bring progress to today's Third World.

At one time it could be believed with relative assurance that the development of society — European society, in any case — and technological innovation were one: "So long as modern capitalist society was an expanding system with underdeveloped countries to absorb men, capital, and energies, there was always an outlet [overseas] for the misuse of power and extreme forms of exploitation." But with the rise of the Third World, "imperialism, the cause of recurrent international friction an economic upheaval, seems to have reached a point of saturation. The world is divided; there are no more open spaces with free homesteads for immigrants, and the backward peoples have been awakened through communication with and education by their rulers or trading partners" (Mannheim 1950:9).

These concerns and the degree to which they have influenced current international relations are perhaps nowhere better expressed than in the recent series of calls for special meetings and designated "years" issued by international organizations such as the United Nations Economic and Social Council and the UN Development Program. In August 1976, member nations began to formulate a long-term program relating closely to the central interests of Mannheim, sociologists of technology, and postacademic social science. This is to "examine systematically how science and technology can best be employed to accelerate human development and promote human welfare" and "to explore how best to harness science and technology to the process of socioeconomic development."[1]

Since well before this announcement was issued, much related activity has taken place among government and academic international development specialists and other professionals. A large and wide-ranging body of new literature on the classic theme of science and technology for development and on the variety of issues which the theme entails has been produced (Lund Conference 1977; Mattill 1977; Wionczek 1977). UN documents and responses issued by national governments, nongovernmental organizations, and scholars echo many of the questions raised during the past two centuries. Recently posed questions concerning "fundamental issues of social control over science and technology, and the uses of science and technology for meaningful development" (Lund Letter no. 1:8) extend far beyond specific meetings and designated years and to the very roots of the social and technical disciplines.

In the following chapters, the theme of science and technology for development is examined in light of current programs and more traditional social-scientific concerns. As is the case with themes discussed earlier, we find a simultaneous growth of interest in a sociotechnical orientation and growth of opportunities for applying this orientation in other than traditional academic contexts. Inherent in this convergence is appreciation of the complex, ambiguous, and often ironic character of the development process. These intellectual and material trends are being shaped in response to recognition that attaining higher levels of development through new (especially imported) technologies is a necessary but also extremely uncertain pursuit. A chief cause of this uncertainty is the ineluctable human factor, often overlooked during the academic era but now acknowledged to affect the application of even the most narrowly technical development strategies.

Note

1. Statement by Panel of Administration Witnesses, hearing before Subcommittee on Science, Technology, and Space, U.S. Senate Committee on Commerce, Science, and Transportation, 15 December 1977. Current information available from the U.N. Office of Science and Technology, United Nations, New York 10017.

11.

Development and Postacademic Social Science

The idea of development as the rational pursuit of social progress gave meaning to the birth of modern technology and science, including social science. For the founders of this new outlook, the overt purpose both of technological innovation and of the study of the relationship between innovation and society was development. Between the split and the beginnings of the postacademic era, this commitment to development and to the social dimensions of technology in general was characteristically of minor concern. This was because social progress came to be viewed as the automatic outcome of technological innovation, thus requiring no special knowledge beyond that needed to achieve technical breakthroughs; because the study of the relationship between technology and society was cast as inherently ideological; and because the academic social sciences had other battles to fight which exacted commitment to disciplinary rather than developmental ends.

The changes we have discussed as aspects of the transition to postacademic social science, including the ideas of Mills, the Frankfurt School, and Mannheim, the growth of interdisciplinarity, and increased support for social science outside of academic contexts, have brought the development concept to the forefront once again. The social scientist is once more being asked to produce and apply knowledge with a purpose other than mere accumulation. But this time, the sense of purpose arises not as a philosophical response to industrial revolution but as a practical response to requests of the technical community. In technology assessment work, policy analysis, social feasibility studies, and a variety of interdisciplinary research and teaching contexts, social science is now fulfilling a role first envisioned more than two centuries ago as part of the scientific basis for development.

This chapter assesses the reemergence of developmentalism and the uses to which the development concept has been put.[1] The ideals and realities of development have been important parts of social history since the industrial revolution. One consequence of this is that such ideals and realities are rife with normative and ideological implications. It is therefore necessary that the postacademic social scientist be especially clear about what a commitment to development means and how it relates to the goals of other professionals and of society at large. In this review of definitional issues, development models, and ideological dimensions, development is a notoriously elusive goal for social science to pursue. Nevertheless, for both the epistemological and moral reasons stressed by Mannheim, Mills, and others such a pursuit is an essential anchor for the applied interdisciplinary orientation of postacademic social science.

Earlier we referred to a distinction between development per se and development as "domesticated." Our point was to show how the concept of development that emerged during the three births of sociology (the "per se") differs significantly from the course of social change promoted, achieved, and labeled as development in contemporary technological societies. The basis for this incongruity lay in many social and intellectual forces of the industrial era, including the rise to power of certain elites such as the "captains of industry" and the "dictators of the proletariat," the splitting of science and ideology, and the relative ignorance, affected and real, by early industrializing countries of the extent to which their wealth depended on the impoverishment and often elimination of other nations, especially those in today's Third World. This disjunction between development as an Enlightenment ideal and as the accumulation of technologies — in an R&D society — which best serve the perpetuation of certain types of elites is decisive in the history of our societies and social sciences.

Development Ideals and Realities

The distinction between development as an Enlightenment ideal and as it is practiced and theorized about in post-Enlightenment societies is common, both in explicit treatments and, by implication, in the literature of social philosophy, economics, and other social sciences. Gunnar Myrdal (1970), Talcott Parsons (1951), Paul Baran (1957), Mannheim, and Joseph Schumpeter (1950) are among those who have given their attention to this distinction and related issues, from quite different politico-ideological perspectives. Others who have attempted to grapple with such issues include Robert Nisbet (1969), Everett Hagen (1962), and Szymon Chodak (1973). Of most direct relevance to the problem at hand is the comprehensive work on development begun by Mannheim and extended by Mills and Irving Louis Horowitz.

"Three Worlds of Development": Ideals and Applications

In this approach, a focus on the history of the concept of development is combined with a sociological analysis of development as historically operationalized (Horowitz 1972, sec. 1). The conclusion drawn from this comparison is that not one but two broad models exist upon which applications have been based in a wide variety of industrialized nations. They are (1) the capitalist, consumerist, free enterprise model of Western European democracies and (2) the command-state, production-oriented, controlled-enterprise model, especially that applied since World War II in the nations of Eastern Europe. These two constitute the First and Second Worlds of development, respectively, though no pure case or pure application of either development strategy now exists. While the two models remain distinct, the nations and societies of the two worlds interact more freely and as intellectuals and planners in both worlds reformulate their strategies, the models themselves shape one another.

These complexities are amplified in the face of another, Third World of development. This is the set of development models and activities of nations which by a formal definition based on NATO (First World) and Warsaw Pact (Second World) membership lay on either side of the "Iron Curtain that descended on Europe" after World War II. Third World countries are those which have been colonies of First or, less frequently, Second World nations. The Third World approach to development is less well integrated and more poorly defined than that of the other two worlds. The Third World has, rather than a model, a very large set of theoretical combinations of elements from First and Second World models. Despite the mixed appearance of development strategies in even distinctly First or Second World countries like the United States and the Soviet Union, those of the Third World are phenomenally eclectic and in part purposely so.

Another factor complicating the course of contemporary development is the variable way in which these ideals are put into practice. There are several steps in this process. The formulation of First and Second World models and Third World combinations as development strategies constitutes the first step toward operationalization of the Enlightenment concept. The second step is the translation of such strategies into tactics of state policy. Application of, say, First World principles, even to First World societies, varies significantly in time and among societies. The third step is the execution of these policies, including how they are carried out, which agencies have particular responsibilities, how "soft" is the state (see chapter 12), and what influences other than state-authorized affect the manner in which policies are implemented and monitored. Such details will vary even among societies with "identical" policy orientations. A fourth step is the evaluation of policy outcomes. Since development is a policy-saturated process, even if the policy is based on classic laissez faire principles, it matters much how decision makers judge whether development has occurred. Such procedures and guidelines vary significantly from society to society. Operationalization of development in all three worlds occurs in three or four linked but hardly perfectly coordinated stages: model building (and revising), policy formulation, policy implementation, and evaluation of results. Taken together, these separate the ideal from the reality of development. As the stages proceed, development as practiced is ever more likely to differ in significant and ultimately ironically opposite ways from the now-remote historical ideal.

Role of the State in Development

Underlying current operationalizations of development is the prominent role played by the state and elites. This role was not taken seriously enough by Enlightenment intellectuals, up to and including Saint-Simon and Marx, who first explored the implications of the development ideal. For some time, neither intellectuals nor their publics were prepared to accept the vulnerability of science and technology to traditional authority. As a result, the most liberating discovery of human history, that with the revolution in industry, polity, and society of the post-1750 period, man's destiny was now under his own control, became a mere partially-fulfilled or never-fulfilled promise, a blind behind which all manner of state-authorized policies and practices could be perpetuated.

Adam Smith, Saint-Simon, Karl Marx, and their preacademic contemporaries were impressed by the power of technology to enervate, and were taken with the magnitude of the development ideal. They were so impressed that they could not account for the ability of arbitrary power to withstand the "withering" tendencies of rationality, of a scientific world view, or of scientific socialism. But in the wake of fission, development was reoriented to the purposes of the state and authorized elites rather than the state being rendered benign or superfluous by development. In a classic illustration of the "iron law of oligarchy," those in power have, since the split, used the development ideal to maintain power.

Under certain circumstances, the maintenance of state power has not been the exclusive or even the most explicit outcome of the application of First, Second, or Third World development strategies and tactics: social and economic progress that measures up to Enlightenment ideals does occur, in good times for many segments of society and in bad times for a few. But it is also frequently the case that partial or uneven development must be promoted, in violation of the holistic character attributed to it by the preacademics: some must be impoverished for the enrichment of others, certain political rights must be curtailed for the sake of economic growth, the aspirations of some social groups must remain temporarily or permanently unrealized for the sake of political solidarity.

A final irony occurs when even the taking of lives in a systematic and authorized manner can be counted as an investment for development. This occurs when, as one variant of this strange logic goes, the "order" is charged with promoting development and the elimination of some individuals and groups is viewed as necessary for the stability of the order, or — in another variant — certain classes, e.g., peasants, are viewed as so stubbornly counterdevelopmental that their continued existence is, if it comes to that, too costly to be endured by a developing order (Horowitz 1980, esp. chs. 2,5,6).

The domestication of development in all three worlds has been expressed in a creed and program assuring that social change will take the course state authorities determine, in order to satisfy their ends, including perpetuation of state power. At times, economic growth, political stability, and the welfare of social groups do serve as ends in themselves; at times, they serve only as rhetorical proxies for the hegemony of social and political elites. But the development toward which postacademic social science must be oriented is none of its better-known domesticated varieties — although it shares with them remote roots in the programs of Smith, Saint-Simon, and Marx. Answers to the question as to what constitutes development, provided in the realities of capitalist, socialist, mixed, or Third World societies, are morally unsatisfactory for a critical social-scientific approach. Such answers obviate the need for a postacademic approach. In none of them is a decisive role reserved for the incorporation of independent, critical, and morally explicit social-scientific insight into the design and creation of technology.

In all three worlds, technology and development have been shaped to nourish one another. This is reasonable and necessary. Technology as knowledge to transform the relationship between man and nature is a sine qua non for any type of purposeful improvement in social and economic conditions. But technology is not entirely autonomous, though like development programs of which it forms an integral part, it is rife with unanticipated outcomes. What it "does" is decisively affected by who controls its design, innovation, diffusion, and R&D. In its domesticated varieties, development as a state prerogative has been pursued with technologies whose design is also a state prerogative.

The manner in which technology and development relate in practice is determined neither by technological nor developmental imperatives alone, but always with the firm guidance of political, economic, and social elites. Unless the technical professional is willing to be a servant and a piece of equipment himself for such elites, he has little to offer the domesticated innovation process. While it may not seriously affect the quality of the work of a technician or a non-social scientist to act in the interests of the state, such affiliation usually distorts the work of the social scientist beyond the point of usefulness. Just after World War II, George Orwell (1968b:64,69) put this case concisely in a series of articles on the effects of state control on "exact" (i.e., non-social) science and the technical fields, on one hand, and social science, history, and literature, on the other.

> A totalitarian society which succeeded in perpetuating itself would probably set up a schizophrenic system of thought, in which the laws of common sense held good in everyday life and in certain exact sciences, but could be disregarded by the politician, the historian, and the sociologist. Already there are countless people who would think it scandalous to falsify a scientific textbook, but would see nothing wrong in falsifying an historical fact. It is at the point where literature and politics cross that totalitarianism exerts its greatest pressure on the intellectual. The exact sciences are not, at this date, menaced to anything like the same extent. This partly accounts for the fact that, in all countries, it is easier for the scientists than for the writers to line up behind their respective governments. . . [but] the destruction of intellectual liberty cripples the journalist, the sociological writer, historian, the critic, and the poet in that order.

For the postacademic social scientist, prevailing official definitions of development (or as Orwell points out, of anything else) and prevailing courses of technological design and innovation intended to achieve such development must be treated not as articles of faith but as data, as political and historical variables, and as merely part of the story.

Bourgeois, Marxist, and Idealist Development Models

A recent review of these issues that applies Mannheim's sociology of knowledge perspective (Gellar 1979) has shown how both the First and Second World concepts of development, which the reviewer calls "bourgeois" and "Marxist" models, are thoroughly ideological. Proponents of each, including both the elites whose survival depends upon these models and the academic social scientists who serve as their apologists, insist that theirs is scientific but that the other is ideological. This analysis (which shares much with those of Mills, Horowitz, Gouldner, Myrdal, and several others) details the ways in which proponents point to the laws of history, scientific data, and rational argument which putatively prove that the "Western way" or the "Soviet way" is the true way and how the "other way" is antidevelopment propaganda fodder. Recognizing the political uses to which development is put, Gellar concludes that the word has no exact referent, that it is inherently normative; it is inevitably part of an elaborated and distorted code whose diagnosis and therapy serve narrow group interests. None of the Third World revisions, syntheses, and refinements of the bourgeois and Marxist models are exempt from this charge.

In seeking an alternative, Gellar considers a third, moralist-idealist approach to development. This "third way" makes no claims to being purely scientific. Citing the work of Dennis Goulet (1977), Franz Fanon (1963), Paulo Freire, and even Mao-tse Tung, Gellar opts for a treatment that is forthrightly moralistic and openly idealistic, that appeals to the ethical impact of planned change and seeks criteria for development in the comparison between achievements and humanistic ideals, rather than between the "bottom line" of a double-entry accounts book and official goals such as production quotas. One can find reason to quibble with this characterization of a third way. There is a disconcerting incompleteness in such an alternative to bourgeois and Marxist approaches, and there is a strong element of arbitrariness in the decision that, for example, Franz Fanon or Mao-tse Tung are adequate spokemen for it. Granting these limitations, the analysis identifies an essential error of omission in First and Second World models and their variants: that the domesticated varieties have reserved no role for morality or idealism in the development process. For this reason, the alternative deserves careful consideration. The omission of morality and idealism underlies an essential identity between current First, Second, and Third World development theory and praxis, and underscores the difference between domesticated varieties of development and those of preacademic theorists.

Marxist, bourgeois, mixed, and Third World development strategies all call for increases in wealth, health, and wisdom, for resource shifts from less to more productive ends, for industrialization, urbanization (or more recently, deurbanization), self-determination, capitalization, proletarianization, land-use intensification, land redistribution, education, stabilization, and democratization. Everywhere, these and similar goals are targeted as the best possible outcomes of application of technical know-how and equipment. These are all goals whose solution is inherently technological. Different ends require different technology mixes, different innovation and management procedures, etc.; some of these technologies are industrial, others communication- or energy-related, others medical, and many are social. But intrinsically, domesticated development is the pursuit of these goals through technological means.

Absent in all these strategies is a theory of the morality of development, a set of ground rules stating the ethical limits beyond or in violation of which development is not "authentic." Like any other game, development must have some rules other than "do it." Yet nowhere do we find procedures for insuring that it be the ideals of development (whether Marxist, bourgeois, or other) and not the groups who claim to be the priests, prophets, and guardians of the ideals that rule the development game. Lacking in domesticated development and in preacademic theories is what from our perspective would appear to be a first principle of development modeling and practice: a willingness to take ideals seriously enough neither to idolize them nor to become entirely cynical about them. As a result of this neglect by practitioners and theorists, idolatry and cynicism rule the ways in which First, Second, and Third World societies proceed to develop. Omission of morality and idealism in domesticated development programs fits into the purposes of the state and elites. Without a public check on whether a nation "plays by the rules," or without some means of ensuring that the ideals of development — which are usually uncontroversial — are treated with the appropriate degree of respect, anything goes. At worst, horrendous crimes are committed and/or justified in the name of development and in complete violation of its core ideals.

The institutions charged with the conduct of development and application of technology to this end, the state and its agencies, may be inherently incapable of morality and idealism.[2] Because technology is elaborated technique, it is in itself amoral: it can be used, abused, destroyed, and perfected by people of any ethical persuasion. This is not to say that individuals, especially social scientists, cannot or should not be moralistic and idealistic about the uses of technology for development. In the absence of higher authority in civil society, it is imperative that social scientists be so oriented. While domesticated development serves its purposes for some, the fact that moralism and idealism have been "bred out" of its several varieties means that it cannot be accepted at face value as an operating principle for postacademic social science.[3]

Development and Universalism

One aspect of the development model formulated during sociology's three births relates closely to the role of ethics and ideals. This is the theme of universalism; in the words of Schiller, the *alle Menschen* for whom development is intended.[4] It may be true that when poets and philosophers of the Enlightenment spoke of the beneficiaries of the fruits of development as "everyone," they were speaking poetically or metaphorically rather than philosophically or scientifically. But it is equally likely that they meant it. Whether it was presented as the near vision of Smith or Saint-Simon or the far vision of Marx, there was in the doctrines of the amateur social scientists a focus on the time, way, or place, and at least a possibility, of long, happy, creative, fruitful lives for all people in the world. According to this vision, an event or invention was to be judged pro or antidevelopment to the extent that it appeared to lead to that possibility.

What even the most far-sighted of these visionaries could not conceive was the nearness of the time when a proposition such as "event X or invention Y affects everybody in the world in a certain way" could be tested. The technologies of development and for monitoring the impact of development are today worldwide in distribution and scope, and recognized as such. It has become possible to put the universality of these ideals to the test. On the basis of such test, judgments can now be made about the extent to which a change is authentic development in the original sense and independent of what it might be called according to a bourgeois, Marxist, or mixed (domesticated) lexicon. Now the Enlightenment ideals can be interpreted and operationalized, as pitifully approximate as these operationalizations must be at times, in terms of the development of every single person in this world. With this possibility, it is meaningful to consider freeing ourselves from the collectivist and nationalistic biases that necessarily characterize the type of development promoted by ruling elites.

In the ideals of the preacademics and in the reality that has intervened between their time and ours, postacademic social science has a basis for an independent model of development. This model, with its stress on universalism, can correct for the absence of morality and ideals in domesticated models. It provides guidance for some crucial decisions not addressed in First, Second, or Third World orientations. It can disallow as inauthentic or wanting, a program of social change in which some people and groups develop at the expense of others; it can, for example, rule out as a gross perversion any program that improves the health, wealth, wisdom, creativity, or happiness of some people by enslaving or murdering others.

Universalism is a very abstract — many would say naive — goal. But it is not the only theme in the preacademic program that provides substance for a postacademic alternative to the development models and activities of the academic era; and it is unusual in that it explicitly acknowledges the systemic character of the world in which the domesticated models are now being applied, formulated, and revised. In conjunction with the often abused but still essential emphasis of the preacademic program in which development is viewed as the realization of human potential, stress on "for all" may be the beginning of an answer to the question of who are to be the recipients of sociology and technology.[5]

Developmentalism: Ideology Conditioned by Interdisciplinarity

Postacademic social science, too, has distinct ideological connotations. While its role is to provide a general but identifiable sense of purpose for the pursuit and application of knowledge about the relationship between society and technology, this orientation also presents a challenge to proponents of domesticated development within and outside of the social sciences. Group and individual interests are invested in the perpetuation of domesticated development and, willingly or not, in the promotion of amoral, cynical, or fanatical development strategies. To promote a third, postacademic way to development is to come eventually into conflict with the groups and individuals whose welfare, as they see it, depends upon the promotion of some combination or variant of the other two ways.

Development as a moral ideal and technology in its broadest sense are closely connected: technology is the means whereby people may improve their lot. Development in its domesticated variants is dependent on the specific technologies the state authorizes, promotes, ignores, suppresses. In the First, Second, and Third Worlds, development "priorities" and the need for maintaining state autonomy guide the latter in R&D functions. As a result, technologies derive their meaning from, and in turn give meaning to, activities pursued in the name of Marxist, bourgeois, or mixed development models. One cannot escape ideological implications when dealing with problems of technology for development.

The distorting tendencies of developmentalism as ideology are limited by the fact that the study of technology and development, together and as separate specialties, is an inherently interdisciplinary undertaking. "To speak of interdisciplinary relations in international development is redundant because development, as a distinct field of study, is itself an amalgam of many professional and technical specialties" (Horowitz 1972:420). This has been known for a very long time, in both U.S. social science and in the classic work of its European predecessors. But with growing awareness by technicians and technology-policy specialists of the social impact of innovation, with increasing opportunity for interdisciplinary work of all types, in brief, with the coming of the postacademic era, we have witnessed a blossoming of interest and activity in interdisciplinary technology and development studies.

This recent work does more than merely echo the technocratic ideals of Saint-Simon and Comte or the holism and critical objectives of Marx concerning the role of technology in modern capitalist and postcapitalist society. Yet it takes as seriously as did the founders of social science the thorough interpenetration of social, economic, political, cultural, and scientific dimensions of technology and development. Current interest in these issues is not only infused with an acknowledged interdisciplinary bias, but, for the first time in history, it is equally informed by practical experience in technological design and innovation activity. And practice has shown that the world of technology and development is a "predisciplinary" world.

Interdisciplinary approaches to technology and development are both scientific and normative. One can readily appreciate the value of careful, disciplined hypothesis testing and formulation in the study and design of technology. But because technology is so deeply imbedded in social relations and development is so subject to ideological interpretation, scientific procedure alone is insufficient for guiding all the choices crucial in this type of work. Here, as in few other areas in social science, personal interests, secular values, and group norms have telling effects on problem formulation, measurement decisions, etc. Interdisciplinarity has a very important purpose to serve: it allows the mechanisms of pluralistic procedures to mitigate such effects, particularly the "secularizing" effects of the discipline.

The tendency to disguise value judgments as scientific can be minimized in other ways as well. One of these, discussed at greater length in chapter 16, is the promotion of value explicitness. This practice not only keeps the researcher or research team honest to themselves about what constitutes a "benefit," "progress," "development," "effective functioning," etc., but also permits value conflict and conflict management to proceed openly rather than as the hidden agenda of presumably technical dilemmas. Most important, value explicitness brings a crucial humanizing dimension into the discussion of technology and development.

Even if the theory and practice of innovation for development is merely an accounting process writ large, it remains essential that the accounting begin with items such as the value of the life and freedom of human beings. Without an explicit interdisciplinary awareness of his own values and those of his coworkers, the student of technology and development need never account seriously for such "items" as the length, productiveness, and quality of human lives. *With* such awareness, it may be possible to understand and design technologies in light of their contribution not only to growth in GNP, rates of urbanization, speed with which messages can be sent, and the degree to which technical functions are integrated, but also to the promotion of life and its quality — for all.

Notes

1. While approaches such as political development, nation building, and dependency theory have risen and declined, developmentalism continues to be pursued in several fields (Horowitz 1978; for an earlier comprehensive statement, see Horowitz 1972). Basic works of dependency and postdependency orientations include Wallerstein (1961, 1966, 1974) and Frank (1967, 1969). The works of Braudel (1966, 1973) and Rodney (1972) have also been influential in this area. Interdisciplinary development journals in which these various approaches are featured include *Economic Development and Cultural Change* (University of Chicago Press), *Studies in Comparative International Development* (Transaction Periodicals Consortium, Rutgers University) and of special interest to technology studies, *Technology and Culture* (University of Chicago Press).
2. See the concluding chapters of Horowitz (1980) for an extended discussion of this point.
3. Many groups and individual scholars are involved in an attempt to apply such a critical humanistic social science to development planning (for a bibliography see LRIS 1977-79). This has proved to be an intrinsically interdisciplinary effort. The main principles of this applied social science that are of recent interest include: (1) concern with the moral consequences of social knowledge and state power; (2) stress on the diversity of social-scientific perspectives; (3) appreciation of the differences in tempo of work in applied versus theoretical social sciences; and (4) an attempt to distinguish between elite and democratic uses of social facts in planning.
4. A fair-sized monograph could be written about who exactly Enlightenment intellectuals had in mind when they spoke of progress "for all," "everybody," "mankind," etc.
5. Perhaps it is best to leave these remarks at this general level. To seek a more detailed account of the type of development postacademic social science should address would be like trying to describe the "other" society in detail, i.e., presumptuous and counterproductive.

12.

The Soft State and Development Administration: Toward an Appropriate Planning Technology for the Third World

The commonwealth, then, to maintain its independence, is bound to preserve the causes of fear and reverence, otherwise it ceases to be a commonwealth. For the person or persons that hold dominion can no more combine with the keeping up of majesty. . . the open violation or contempt of laws passed by themselves, than they can combine existence and nonexistence.

Baruch Spinoza

When modern technology comes to a Third World society (or to a less-developed sector of any society), both the technology and society are changed as they attempt to accomodate one another.[1] The thesis, which underlies several recent trends in academic and sociotechnical fields, such as the appropriate technology movement, expresses a relatively new perspective on the process of diffusion. Between the time of A.L. Kroeber's classic work on diffusion and recent studies of the Ohio State project (Brown 1981), a generally monological or "one-way" model dominated research and policy on the transfer of technologies (as well as commodities, values, and cultural items generally). According to this model, a diffused item, basically unaltered, will eventually find a place in a new cultural environment. Even if a certain amount of resistance on the part of the new hosts is encountered, it can be eliminated or rechanneled to allow the item to play its proper role. Since the early 1950s, much research and theoretical work on the diffusion process has shown this model to be a woefully inadequate depiction of reality. In light of this criticism, new "two-way" diffusion models are becoming the norm in scientific work and policy relating to technology transfer. With their uses, new dimensions of the innovation process have been revealed, raising serious questions about the role of technology in development programs. This chapter discusses some of the complexities associated with the use of transferred technology in development planning. The focus is on the soft state, an especially dramatic, unanticipated outcome of the supposed one-way diffusion of First and Second World know-how to the Third World. Based upon an analysis of the causes of this key irony in development studies, we consider some remedies suggested by the concept of appropriate technology transfer.

Softness, Mock Bureaucracy, and Corruption

Softness was first identified by Gunnar Myrdal (1968, 1970b) as a pathological condition of public administration, especially in the Third World countries. It refers to the widespread inability of government bureaucracies to comply with rules and procedures they are mandated to follow. According to Myrdal, this condition interacts with economic, environmental, demographic, and political factors to inhibit the attainment of development goals. Despite considerable theoretical, definitional, and methodological problems associated with the concept of soft state, it is an undeniable part of Third World society. Public bureaucracy in Third World countries shares the usual range of inefficiencies and informal elements found in bureaucracy everywhere; and softness is a characteristic of all states, regardless of geography, history, or economic conditions (Myrdal 1970a, ch. 8). But Third World bureaucracy has also a unique character that perhaps must be experienced to be appreciated. In these countries, public administration is often conducted under the assumption, usually shared by administrators and the public, that the rules governing organizational functioning are themselves subject to negotiation. Much innovative and unanticipated behavior, often labeled "corrupt," is not only tolerated but is virtually institutionalized. At the same time, formal rules are known and acknowledged to be legitimate — but not necessarily right.

This double standard was recognized in an industrial context by Alvin Gouldner (1954), who labeled it "mock bureaucracy." While this term, like "softness" itself, may be more evocative than descriptive, it captures the flavor of public organization in the Third World: patterned, tolerated behavior on the part of superiors, subordinates, and clients which knowingly deviates from formal rules. Widespread mock bureaucracy has been acknowledged as an impediment to development by social scientists such as Myrdal (1968, 1970a,b), Nicholas J. Demerath

(1976), William Chambliss, and Robert Seidman (Chambliss and Seidman 1971; Seidman 1979). Others, like Irving Louis Horowitz (1972), Fred Riggs (1964), and James Scott (1972), have discussed the phenomenon under different names appropriate to their theoretical perspectives. Despite this scholarly attention, the inclusion of softness as a factor in policy considerations in Third World countries is ordinarily not as a bureaucratic or administrative condition. Rather, under the labels "corruption," "bribery," and "nepotism," it is treated as an individual crime (Peters and Welsh 1978).

While softness and corruption are often closely related and mutually supportive, they are not the same thing; to confuse them can have serious implications in dealing with the condition as a problem to be solved or controlled. In the same spirit, it is likely that softness has social benefits and functions as well as dysfunctions (Merton 1949, ch. 1). To characterize it as a pathology is not to suggest that all aspects of Third World society are harmed by administrative noncompliance. Softness is not anarchy, and for those who know the operative rules and subtle ways in which they relate to formal rules, the behavior of public bureaucracy can be both predictable and satisfying. Yet the fact that softness is not identical to corruption, and the distinct possibility that softness has social benefits, do not contradict the need to formulate policy to contain it. But such policy cannot be purely a matter of criminal law and justice. If it is to be formulated, it must be done with due awareness that softness serves social needs and thus involves tradeoffs.

Soft Organizations and Corrupt People

Some characteristics and effects of softness and corruption are similar, but the concepts are distinct. Corruption and softness differ in two principal respects. First, softness is foremost an attribute of social collectivities, organizations. It is a patterned response of role occupants, for it exists in the structure of bureaucratic relations themselves. Corruption refers to the behavior of individuals, and to groups only by analogy or arithmetic. A group may be referred to as corrupt if its behavior is like that of a corrupt person, or if it contains a large number of such people. Corrupt behavior, in turn, is (according to *Webster's Unabridged Dictionary,* second edition) "depravity; wickedness; perversion or deterioration of moral principles; loss of purity or integrity." It implies activity that violates well-established social norms. Organizations can deal in "depraved," "wicked," or "perverse" pursuits and thus rightly be called corrupt. Similarly, a corrupt organization might be composed of people with "deteriorating moral principles." But the mock bureaucracies in Third World public administration are neither necessarily wicked in their operation nor their personnel especially depraved. It is possible that softness corrupts; and it may be that soft states recruit corrupt people. But corruption in an organization is at most symptomatic, while softness is a far-reaching aspect of its structure (Weinstein 1974). A second key difference between softness and corruption lies in the normative connotation of the label "corrupt." Corruption in public life refers to the violation of a public trust that is embodied in a shared understanding about what constitutes correct behavior. Softness results from conflicting sets of normative standards, each of which has authority in the minds of bureaucratic officials and the public. The "mock" character of bureaucracy in a soft state lies in the coexistence of (1) official rules acknowledged and (2) actual rules followed. Administrators and others who relate to bureaucracy in the Third World may be doing what is expected by the second type of rules while "observing" the first type when possible. Such behavior is not corrupt, although it may be manipulative. Under the circumstances in which public administration takes place in the Third World, it can be "virtuous" inasmuch as it abides by local standards. Ultimately, corruption and softness may both be related to anomie. But whereas corruption is a "normlessness" in relation to a clear standard, softness is a normlessness of too many options.

Softness and Corruption in Mutual Support

According to Myrdal (1970a:208) softness and coruption may both permit or provoke each other in circular causation and may have cumulative effects inhibiting the achievement of development goals. Corruption may limit a state's ability to achieve national political and economic development in several ways. The presence of corruption introduces an element of irrationality into all administration at each level of the decision-making structure. Corruption constitutes a deviation from planned, anticipated outcomes, introducing incalculability and a corresponding obfuscation of accountability. In order to combat the irrationality of corruption and thus speed up production, minimize costs, etc., increased controls over funding and distribution and utilization of other resources become necessary. These controls must be imposed at the cost of increasing power and invulnerability of the overseeing bureaucracy. As Riggs (1964) and Wittfogel (1957) have noted, this complicates aspects of national development by making bureaucracy even less accountable, efficient, or responsive to the needs and demands of other national interest groups. Finally, to the extent that corruption is omnipresent, it may preclude necessary reform, undermine governmental legitimacy, and generate civil strife, further limiting the ability of a nation to realize its developmental goals (Dobel 1978).

Indirectly, corruption and softness lead to loss of legitimacy due to government failure to realize its own economic projects. More directly, groups whose demands are not met or who are excluded from the distribution of undeserved goods (those who do not benefit from corruption) will also cease to consider the government in power as legitimate. With the institutionalization of corruption in an unchallenged bureaucracy (or in any other interest group), the bureaucracy tends to resist any changes in the status quo. The result may be the inability and unwillingness to formulate policies that would overcome softness itself. The attempts of those who seek to reform or replace the government in power may result in further political instability and, in the extreme, violence. Unfortunately for the soft states, softness creates conditions that encourage corruption and make its eradication more difficult.

Corruption, like softness, may also be functional for a political system. It may improve the acess of nonelites to political leaders, allowing them to influence decision-making processes. Bribes and other unofficial payments may be withheld from elites viewed as unresponsive to group demands. Hence corruption may be employed as a means of guaranteeing governmental accountability. To the extent that these arguments are true, corruption may be viewed as "a necessary and not harmful lubricant for a cumbersome administration" (quoted by Myrdal 1970a:208) and may ultimately serve as a means of ensuring political stability. Such an assessment ignores obvious moral problems of softness and corruption (Myrdal 1970a:228). Perhaps of greater importance, it does not address the serious effects they can have in the crucial context of development administration.

The Process of Development Planning

When we conceive of softness as a characteristic of social organizations reflecting the presence of two or more normative systems ("polynormativism"; Riggs 1964), its role in development administration and some directions for formulating policy to contain it become clearer. A critical locus of softness in Third World public administration is in the formal economic and social planning sector: national and regional planning boards, ministries, and related agencies. In this sector the inability of bureaucracy to comply with its formal normative framework, or "soft planning," is directly related to the society's inability to develop rationally. If a poor nation is to achieve economic and social well-being, it must have some means of guaranteeing that scarce human, monetary, and material resources are judiciously utilized. The pattern adopted by the overwhelming majority of Third World countries is through formulation and implementation of a development plan which explicitly identifies the measures to be taken by the public and private sectors to achieve stated goals. Such planning also provides criteria for formulating intermediate-range policies and identifying future goals. The development planning process is by no means simple. It includes:

1. A survey of current economic conditions, e.g., national income, productivity levels, industrial trends, foreign trade, etc..
2. A survey of the social situation, e.g., population changes, health, education, and housing capacity.
3. An evaluation of progress achieved under the preceding plan and/or regime.
4. A statement of the general objectives of social and economic policies.
5. Estimates of growth, or targets, for each major economic or social component during the period covered by the plan.
6. Suggestions of measures designed to raise the rate of economic growth, especially to stimulate saving and investment, increase productivity, and improve the institutional framework of economic activities like land reform or the reorganization of markets for commodities, labor, and capital.
7. A program of government expenditures (Lewis 1968).

The failure of development plans to meet their stated goals has been noted by many theorists and planners, both within and outside the Third World. Although reasons cited for these features vary widely between observers and countries, five basic conditions appear to be operative:

1. Demographic. Population increases and migration flows occur in ways unforeseen when development goals are formulated.
2. Economic. Costs of production, market structures, and related factors change in ways unanticipated in the plans; sheer scarcities are underestimated in targeted goals.
3. External political relations. Geopolitical systems change in ways unanticipated in plans.
4. Internal political relations. Political instability affects implementation and administration of plans.
5. Environmental. Natural disasters, weather, and other environmental factors inhibit the attainment of planning goals.

All countries are, to some degree, affected by a combination of these conditions. Planning under any circumstances is a precarious task, but it is rendered especially difficult in societies subject to the additional uncertainties associated with softness.

The Soft State and Development Planning

Despite obstacles inherent in planning for the development of Third World countries, few scholars and fewer Third World development administrators are prepared to face the near or far future without a sense of priorities and objectives for the nation as a whole. Whatever the shortfalls that seem inevitably to occur in the process of establishing and pursuing production quotas, targets, and other socially directed ends, the alternative, no public sector planning, appears far less adequate. If planning is going to be undertaken in the face of economic, demographic, and political uncertainties, softness and mock bureaucracy must be counted as factors which will affect planning outcomes (Delacroix and Ragin 1978).

Softness in development planning is reflected, critically, in the lack of achievement of formally stated and codified objectives. Demerath (1976) discussed softness as a key factor in family planning programs in India and elsewhere. He found that the targets established for these population control efforts were generally unrealistic in relation to the ability or willingness of family planning administrative organizations and their clients to operate by formal "rational" standards. Resource scarcities have not been a serious obstacle in the well-funded USAID programs that Demerath discusses. But scarcity of effective implementation and a lack of enforcement of administrative procedure have been critical. Seidman (1979) makes this point in his research on development planning in Africa. In Nigeria, Ghana, and other countries, planners, realistically enough, set goals in the awareness that material inputs, skilled labor, and capital are scarce. Yet as Seidman pointedly argues, they erroneously presume an unlimited supply of law. As a result, planners characteristically overestimate the extent to which administrators and public can or will comply with the formal pursuit of development quotas, targets, and innovations.

Softness in development administration arises, as it does in other public administrative contexts, from the presence of formal rules and techniques in a social system in which these rules have no traditional moral force. The outcome is a mock acknowledgement of formalities and the simultaneous pursuit of informal ends rooted in local traditional society and culture. The values meaningful to the people involved tend to be oriented to different purposes from those which development bureaucracies are intended to serve. Thus when administrators ignore development directives to deal in particularistic ways with each other and with clients, they are not behaving randomly nor, in the absence of corruption, in narrowly self-interested ways. Instead, "bending," "going around," and outright ignoring of formal directives is the result of conforming to well-established, accepted but informal, and even illegal, standards.

To the extent that softness helps maintain these standards and the traditional social solidarity they may imply, it can in general be functional. Bureaucracy may not behave as it should, but society is, in some ways, served nonetheless. In development planning such "benefits" of softness must be measured against the very serious sacrifices that a lack of compliance to administrative procedures entails. For this reason it is essential to consider the softness that occurs in the planning sector as a focus of public policy. Effective planning requires knowledge not only of demographic, economic, and resource parameters but also of the causes and consequences of softness itself. Explicit acknowledgement of the tendency for noncompliance is a first step in rendering the planning process more effective.

Development Planning As an Inappropriate Technology

Development planning is a technology originating in the transfer and diffusion of administrative strategies from the industrialized countries of Europe and North America. These planning strategies are today generally undertaken by Third World nationals in Third World capitals. Yet their roots in First and Second World values and normative assumptions are evident. Planners educated in industrialized countries or in their style plan within a state administrative structure that is a legacy of the colonial era. Like many other transferred technologies, the planning know-how incorporated into Third World development programs is designed to be effective in social circumstances different from those it now confronts. The key premises of most development plans misjudge Third World realities. One of these is the unquestioned efficacy of comprehensive planning. Plans of five years and more establish long-range goals and allocate scarce material and human resources to particular projects (this is true of both First World, e.g., Britain, and Second World models; see Berry 1975). This large-scale, explicit, and long-term commitment of resources and offical prestige to a particular set of goals incorrectly assumes that Third World development is a closed problem: that boundaries are fixed during problem solving; that the process is marked by a predictability of final solutions; that the process is conscious, controllable, and can be reconstructed; that solutions are provable and logically correct; and that procedures are known which directly aid problem solving (e.g., algorithms or heuristics; see Richards 1974). It is also assumed that the planning principals are able to agree on the goals of the plan and the means to their realization.

In planning situations characterized by these conditions, logical and computational means-end analyses can be pursued. However, Third World planners do not operate in such closed systems. They plan in a far more open context, characterized by changing boundaries, unanticipated events and consequences (hence unpredictability of results), uncontrollable influences such as foreign intervention or a drop in the price of a principal export commodity, etc. (Thompson 1971; Riggs 1964). In the absence of the closed environment necessary for comprehensive planning, many First World economists have argued for abandoning the approach in favor of incremental solutions and "muddling through" (Lindblom 1959; Hirschman 1958; Hirschman and Lindblom 1962; more recently, Lloyd 1978).[2] Lloyd, for example, suggests that the "analysis of very complex problems may be facilitated if the effort to formulate a thoroughgoing definition of the problem is delayed or even omitted from the analysis. . . [to avoid] an oversimplification or erroneous definition which, in turn, leads to an irrelevant solution." These observations best apply to the making of tactical decisions, particularly in industrially developed countries. In the Third World, planners must take account of such fundamental tasks that "thoroughgoing definitions of the problem" cannot, on moral, political, and technical grounds, be delayed or omitted. Thus arises the characteristic dilemma of Third World development: the need for extensive and coordinated innovation in an environment unsuited for the use of available means to innovate.

Approaches to social planning diffused to the Third World, like many industrial and agricultural technologies, have failed to take account of the significant cultural, social, and historical differences between the area in which the technology was created and that to which it is transferred. Technical knowledge designed to help overcome environmental and biological limitations fails to achieve its intended aims. Instead, new problems arise as society attempts to adapt to the requirements of unfamiliar "solutions" (Robinson 1979; Evans and Adler 1980). The effect is familiar: the transferred technology is poorly utilized, scarce resources are expended in servicing it, and expected outputs are not realized (Seidman 1979:3). As in other instances of inappropriate technology transfer, the diffusion of planning strategies has left the technology itself ineffective, incapable of doing what it is intended to do because of the combined influences of origin and destination. The ineffectivness of administrative procedure in development planning has its source in the apparent contradiction between commitment to new rational, technical planning principles, and the viability of traditional local norms and standards. In the exchange, formal acknowledgement of technical principles and actual behavior that violates them coexist, embodied in soft administration and realized in its product, "mock" development.[3]

Causes of Soft Planning

If traditional First and Second World assumptions about the strategies for pursuing development are inappropriate for the Third World and productive only of softness, one must ask why Third World planners persist in using them. The causes of adoption of inappropriate technologies and of First and Second World planning approaches represent a complex mixture of internal and external, historical and contemporary factors. In seeking to formulate policy that will contain the spread of soft planning or mitigate its effects, three general sets of causes require attention: those associated with colonialism, with neocolonialism, and those whose origins are indigenous to Third World countries.

Role of Colonialism and Neocolonialism

Colonial and neocolonial relations have contributed to softness in the Third World through introduction of an authoritative but foreign set of norms and values into an ongoing culture. This situation was first discussed in connection with the sources of mock bureaucracy in industrial organizations. In Gouldner's (1954, ch.5) study the fact that certain rules are systematically acknowledged but broken was illustrated by the practice of workers in a gypsum plant who observed the plant-wide nonsmoking rule in the mine but not in the front office. When pressed for an explanation, employees reported that in the mine smoking could trigger an explosion, thus the prohibition was viewed as integral to the production process and in the best interests of all workers. The smoking prohibition in the front office had simply been imposed by the fire marshall and was not viewed by the workers or the management as deserving of compliance, except during inspection by the authorities. The imposition of a set of norms by an outside agency resulted in institutionalized disobedience and selective compliance with regulations. This case is instructive because it can easily be related to patterns of disobedience and selective compliance in the Third World, resulting in part from rules imposed by the colonial experience.

Colonial Legacy

Myrdal (1970a) and others have discussed the direct and indirect colonial legacy of social indiscipline in many African and Asian nations that have become independent since the late 1940s, especially those that did not achieve independence easily. The difficult struggles for national liberation and independence resulted in the institutionalization of what had before been only informal patterns of noncompliance; that is, national civil disobedience to colonial authority, of which the most striking example is provided by Gandhi's movement in India. These attitudes were often transferred to authority in general, including the postcolonial state. At times, the anticolonial anarchic attitudes that proved to be so successful were eventually adopted and utilized by political opponents of postindependence indigenous governments. Consequently, as Horowitz (1966:339-42) and Myrdal (1970a:213-14) point out, new governments were unwilling and unable to place obligations upon citizens for fear of alienating them. They sought to avoid betraying revolutionary ideological commitments and alliances and generating behavior that would be beyond the scope of their ability to control. Popular dissatisfaction would threaten the legitimacy of the new states and might encourage or facilitate foreign penetration and new domination. Rather than demand the necessary sacrifices that accompany economic growth, the governments employed positive methods, including exhortation, subsidies, and other financial inducements to assure popular adoption and compliance with their programs. In addition to being a costly method of achieving economic growth, such government-sponsored financial inducements have proved problematic.

The attempt of Third World governments to move in the direction of what Victor Thompson (1974) calls "compassionate behavior," actions designed to mollify the population and enlist its support, is a prime contributing factor to both individual corruption and national social indiscipline. Compassion cannot be dispensed objectively. Its dispensation is affected by the different levels of influence and power of various groups competing for scarce resources: the bureaucrat's own political, familial, tribal, and/or religious loyalties and perceptions, and numerous other subjective factors. This arbitrariness contributes to softness since it prompts the politics of self-interest and abandonment of laws as prescriptions for behavior (Dobel 1978). In this atmosphere of self-interest and limited adherence to laws, a free-for-all environment is created in which bribery, personalism, and inefficiency are likely to flourish. As a result of this heavy reliance on politics of inducement and government control, huge sums of scarce capital are expended for short-term popular gratification rather than long-term capital investment, and bureaucracy becomes overtaxed and unable to discharge its responsibilitites. This increases opportunities for deviant behavior within bureaucracy (Scott 1972) and society at large (Dobel 1978). These factors create a corresponding decline of popular faith in the laws and in governmental legitimacy and an unwillingness and/or inability of ruling elites to support any changes in the status quo (Myrdal 1970a:233; Dobel 1978:969-70).

The incursion of bureaucratic control at all levels and in all areas of public policy has deleterious effects upon governmental accountability (Riggs 1964:26-27). One reason for this is that the bureaucracy's monopoly on resources and expertise (in addition to government protection) prevents other groups, such as opposition political parties, labor unions, and other voluntary organizations from accumulating enough power to influence policy or exercise any effective countervailing checks on the bureaucracy. With such a monopoly, the inefficiency of bureaucracy, which prevents the successful implementation of programs necessary for development, becomes a primary aspect of the general environment in which softness flourishes.

The inimical influence of competition between local and externally imposed values contributes to the inability of states to implement and enforce development programs. The polynormative societies of the Third World lack a single set of criteria upon which behavior can be evaluated. Many of them have strong extended family ties; thus nepotism is widely practiced and largely accepted. At the same time, adoption of rational planning principles encourages merit criteria for employment and advancement. Thus a Third World administrator can find justification for adhering to one or the other set of quite different value systems with relative impunity. This behavior, considered corrupt by some analysts, is viewed as "flexibility" by others. Regardless of the moral quality of this behavior, planning, the achievement of national integration, and the assessment of bureaucratic accountability become most problematic. The degree of predictability and calculability of programs and actions integral to the successful deployment and allocation of resources to long-term plans is quite low.

The paternalistic character of superior/subordinate colonial relations generates additional conditions that may contribute to mock bureaucracy and, in turn, to softness. The creation of paternal obligations that fall heavily upon the poor encourages individuals in every stratum of society to attempt to get away with as much self-serving behavior as possible. Given the extractive nature of the colonial relationship, withholding wealth from the colonizers constituted a serious legal transgression. Such disobedience was often dealt with by the introduction of military troops into an area to ensure compliance with production quotas and colonial directives. In reaction to such cruel treatment, the colonized learned to practice petty obstructionism such as foot-dragging and overzealous compliance with operating procedures to delay production, accompanied by low levels of discipline and morale. These patterns have persisted to this day. In the postindependence states, continued reliance of the government upon the military and upon institutionalized violence has assured leaders unparalleled power. These reactions to colonial domination have articulated with the negative impacts of unchallengeable bureaucracy upon national development planning organizations.

The Influence of Neocolonialism

The waves of independence movements that swept Asia and Africa after World War II ended the official colonial relationships between these nations and those in Europe. The achievement of national political and economic development was retarded by a set of neocolonial factors that perpetuated and further institutionalized preconditions for softness already plaguing these nations. Such factors may be functions of more developed nations' necessary quest for political security via economic strength or, from a Marxist perspective, they may be inherent contradictions of the capitalist economic system — the need to "dump" surplus capital (as argued by Baran and Sweezy 1966) or the need for oligopolistic businesses to expand continuously. It is difficult to dispute the general claim that political and economic decisions in many Third World nations remain under the influence of foreign corporate executives and political leaders of both the First and Second Worlds (Galtung 1971). One result of this neocolonial relationship is that development planning may be beyond the capabilities of Third World countries. Under foreign and multinational corporate influence, Third World countries often become involved in the production of luxury goods for export. Capital-intensive modes of production may be promoted in spite of the absence of requisite economic and cultural infrastructures, resource scarcity, and manpower considerations.

The influence of First World multinational corporations and Second World state bureaucracies in the internal affairs of Third World countries has been studied by a number of perceptive analysts (Sampson 1974; Barnet and Muller 1974; Chase-Dunn 1975; Horowitz 1974).[4] Hymer (1970) has shown that Third World nations do not control vital stages of production in their own economies, since many steps in the production process are performed elsewhere. In the wake of the widely acknowledged failures of import substitution, Third World nations depend upon foreign governments and multinational corporations for the transfer of technology, processing of goods, marketing, and/or capital. Multinationals and state agencies create clientele groups within the host nation whose interests, privileges, and status are directly linked to the foreign organization. These organizations are often protected explicitly (e.g., by the Hickenlooper Amendment) or implicitly by the trading and political policies of powerful nations in which they are headquartered (Sampson 1974; Lake 1976).

Wealthy nations may influence (by design or by accident) the development policies of poorer nations with a variety of mechanisms. Cohen (1973:198) notes that "rich nations generally escalate ('cascade') their tariff structures by stages of production where rates are low for raw materials and higher for finished products. This biases opportunities for development." The result of this policy is that trade in raw materials is encouraged while movement by Third World nations toward processing raw materials and manufacture is frustrated. Protectionism of less competitive industries in wealthy nations, military assistance packages, tying strings to foreign aid, outright political blackmail and, as in the recent case of Afghanistan, military invasion are other means by which influence is exercised in the Third World.

The behavior of indigenous leaders is circumscribed by parameters imposed by the political and corporate leaders of First and Second World nations, and foreign aid serves to enrich state officials and bureaucrats rather than improve the national quality of life. To the extent that multinational penetration results in uneven development in different sectors of the economy in a Third World nation, while generating a revolution of rising expectations and aggravating the maldistribution of wealth, levels of legitimacy and popular acceptance enjoyed by the government are likely to decline. This may stimulate the formation of government opposition and instability, the intrusion of international actors into the political situation, and the exodus of foreign capital. In the face of these problems, Third World people may seek recourse from individual patrons. This further undermines the drive toward national unification; it promotes national disintegration along tribal, regional, and ethnic lines. Thus colonialism and neocolonialism introduce a complex framework of "outside" rules and directives which, in combination with local norms, values, and systems of stratification, fuel the tendency toward mock bureaucracy and softness.

Local and Traditional Sources of Softness

Corruption and softness in many Third World countries are also related to nonmodern and illegitimate patterns of local behavior which may have existed long before the establishment of colonial or imperialist relations. Nepotism, favoritism, and "kickbacks" are well-documented examples of ancient and virtually universal innovative behavior. In many instances, of which Korea is a well-known case, political independence and modernization have left intact centuries-old practices such as bribery and clan loyalty, that subsequently became operational features of the modern state (Horowitz 1972:414-19). In these and related ways, tradition plays a significant role in a complex set of relationships affecting development efforts.

In most Third World countries, traditional institutions give rise to a series of informal — unofficial and/or illegal — "feudalistic" relationships. Lacking significant economic, communications, and transportation infrastructures, central governments have been unable to exercise effective control of or provide services for the far-flung citizenry. Without government to insure the most fundamental welfare needs of the citizens, traditional relationships are established and/or reinforced. Cases in point are the hacienda and latifundium systems in Central and South America, the *Jajmani* system in India (Klass 1978), and the plantation system in antebellum United States (Goldschmidt 1963).

The extensiveness and intensiveness of patron/client relationships in today's Third World vary greatly from country to country and between regions and groups. Yet to the extent that systems of mutual obligation are institutionalized, the development policies of the central government constitute a destabilizing intrusion into traditional authority and behavior patterns. To the extent that the central government attempts to displace the local patron as provider of welfare, recipient of popular loyalty, or the ultimate judicial arbiter, it can expect at best, noncompliance from patrons and at worst violent resistance. In either event, costs to the government are great. In the face of such institutionalized patterns, the effective exercise of government control over the implementation, enforcement, and realization of developmental policies is most tenuous.

The causality also runs in the other direction. Traditional systems of loyalty are also reinforced by softness and corruption. Softness encourages the self-interested behavior associated with patron/clientism which in turn undermines laws intended to promote social discipline. To the extent that people are serving and served by masters different from state authorities, the calculability necessary for effective planning or adminstration in general cannot be assumed. Instead, mock bureaucratic compliance with formal directives is the rule. Eventually, traditional systems contribute to and are sustained by the breakdown of the political legitimacy of central authorities, the inability of government to demand sacrifices from its citizens, and national disintegration. In these ways, patron/clientism and similar systems weigh as heavily as colonialism and neocolonialism against the Third World's quest for development.

Combining these observations with our earlier remarks, a framework for understanding and dealing with the soft state begins to emerge. Softness is a condition of administrative organizations in which mock compliance with formal rules is widespread. It is a condition characterized by the coexistence of official, modern, "external" codes of bureaucratic conduct and local, traditional codes of interpersonal behavior. Individuals, administrators and clients, typically experience dual commitments. In the behavior of a soft state, official codes are typically given second priority in the face of enduring traditionalism. One result is behavior that appears corrupt from the official perspective but which may also be serving legitimate local ends. Another result is a lack of effective functioning of administration itself. As has been emphasized, softness can be a detriment to development in many ways, but its most direct impact is in the administration of development programs themselves. In this realm, "legitimate corruption" and ineffective administration render the development process irrational and unlikely to succeed.

Formulating Antisoftness Policy

Social and economic development in today's Third World countries is a thoroughly policy-saturated type of social change. At the center of a typically complex network of national, regional, and local development-related activities is the planning sector. Here, policy is established which is intended to coordinate diverse outputs to achieve predetermined national goals. The planning sector has, in theory, a crucial role to play in Third World development. While it is true that it is constrained by material scarcities, a colonial legacy, various external factors, and difficult local political conditions, effective planning can still act to make the best of available resources. Development planning has not fulfilled its important role. One chief reason for this failure suggests the need for an additional type of development policy aimed at regulating softness itself (especially in development planning administration).

Taking Account of Softness

Softness is a general pathology of public administration organizations traceable to the coexistence of two or more sets of behavioral codes and expectations in a single administrative context. In the planning sector, one of these codes is the modern technology of coordinated planning transferred from First and Second World societies. The other is the (usually heterogenous) cultural context to which the planning technology must make some, seemingly impossible, adjustments. The direct effect of softness, failure to achieve stated organizational objectives, may weaken a society's ability to develop wherever and whenever softness occurs. Softness in development planning guarantees that regardless of the general performance of the economy and social service organizations, development as a coordinated and coherent activity will not take place.

It may be true, as Hirschman, Lindblom, and others have argued, that economic growth and social welfare have often been sacrificed for the sake of such rational development strategies. It also appears that softness can benefit society in providing security and stability in the face of the uncertainites of modernization. These points suggest that soft development planning may not be entirely harmful. Planning itself may cost more than it is worth; softness may be a survival adaptation. Even granting these points, as long as development planning is pursued in Third World countries (and there is every reason to believe that it will be pursued for a long time to come), it will be necessary to take account of the softness factor. The high cost of planning in relation to its benefits, which economists attribute to the large number of uncertainties and uncontrollable factors, is due in part to softness: to the central uncertainty of whether an organization will follow its own rules. To reduce softness may be to cut the cost of planning to a degree sufficient to justify continued planning.

The social benefits of softness, too, must be viewed in perspective. While softness is not corruption, it is often conducive to bribery, nepotism, and favoritism for narrow personal gains. Corruption and softness are different problems, but their coexistence cannot be denied. This relationship should be stressed in the face of the claim that the positive functions of softness justify its continuation. These ambiguities suggest that priorities and values must be made explicit in the formulation of softness policy. Parties at interest must decide how much social solidarity is required, the cost in lack of development (and in corruption). Alternate means to achieve the security provided by softness need to be explored. While these are the issues that arise in any attempt to render appropriate a traditionally inappropriate technology, they have not ordinarily been posed with specific reference to softness.[5]

A significant first step toward controlling softness in development planning is the recognition that it exists as an organizational malaise whose source is in the transfer of an inappropriate social technology. Based upon such recognition, policy can be devised to effect organizational change, revise planning approaches, and to diffuse a "hardening" of adminsitrative activity from the planning center to other sectors of Third World society, economy, and polity. Those who see effective planning as futile or softness as beneficial will have an interest in only limited controls on softness; those who continue to believe in planning efficacy, or who hold that the costs of softness outweigh its benefits, will seek more extensive controls. But if debate between these positions is to be conducted at all, softness must be identified as a meaningful target for development policy.

Planning and Elites

One curious paradox of planning administration in today's Third

World countries is that planning is an elite specialty explicitly pursued in the interest of the masses. As Mannheim observed, planning increasingly occurs within roles, "positions of influence" technically and politically far beyond the reach of the public. In both totalitarian and democratic industrialized countries, the effect of planning by inaccessible elites is, in Mannheim's view, a denial of human rights and a limitation on human potential. In Third World countries, regardless of political structure, the effect of the elite planning technology diffused from the industrialized countries is even more basic than this. Third World societies have traditional modes of conduct meant to govern relations between elites and masses. In some countries, patronage prevails, in others traditional modes reflect tribal, caste, or clan relations. But everywhere well understood, socially functional ways exist for superiors, subordinates, equals, and kinsmen to relate to each other. In the presence of these norms and values, elite planning strategies aimed at serving the nonelite by providing the fruits of development act instead to reaffirm these traditional elite/nonelite norms and obligations.

The paradox of elite planning for nonelite ends contributes to the mock character of development adminstrative organizations in several ways. In traditional elite systems, the well-being of the masses is of secondary, derivative, concern. Instead, the well-being of the "microsystem," the patron and client, elder and child, or high caste and low caste are seen to be the mutual and reciprocal interests of both parties. When modern development planning takes place in the myriad of daily exchanges among officials, administrators, clients, and public, mutual interests along traditional lines take precedence over the "rational altruism" inherent in formal, official procedure. The resulting tokenism toward official procedure, as well as behavior that appears corrupt and which can indeed breed corruption, require attention if softness is to be controlled and development to occur at all.

Development planning can be rendered less soft to the extent that it can be made to vitiate, rather than support, traditional elite systems. Traditional stratificiation systems can inhibit development in many ways, but in development administration their effects are crucial. To eliminate these effects of traditional elite/nonelite relations, it is necessary to encourage, even enforce, widespread participation in the planning process. The purpose of such participation is to guarantee the self-interest of those for whom the planning is pursued. Policy to control softness in development planning should be aimed at extending the technical and political base of the planning process. Development goals must be formulated in consultation with those whose cooperation is most essential in achieving the goals. The costs, allocational aspects, and targets associated with development planning must be debated in public forums. In today's Third World and in the industrialized nations as well, planning goals and means are not formulated in this way; they are decided on by a few specialized and modernized elites. The outcome, in the Third World at least, is not better planning but mock planning.

Planning and Community Standards: The Chinese Case

An appropriate planning technology must somehow incorporate local community standards into formal administration. Informally, community standards play a decisive role in how development (and other) administrative organizations function. But as long as they remain unrecognized, their influence will continue to be counterdevelopmental. Since Selznick's (1949) classic study of the TVA organization, theorists have known of the power of cooptation in modern administration. Yet in the central set of planning organizations in the Third World, clients, publics, and even administrators remain uncoopted. The entire process of formulating and implementing planning goals is managed as a remote, almost abstract, exercise. Means, goals, and the planning technology itself are not the products of local culture; few participants have a personal interest in compliance, but most have much to gain, as they see it, from abiding by traditional and neofeudal rules of deference and reciprocity. This contradiction between formal and local standards can be ameliorated to the extent that formal and local be made to coincide; this in turn would be made easier through promotion of mass participation in development planning.

As White (1973) has observed, the successes of China's development programs are partly due to what he calls "the Maoist critique of bureaucracy." This critique, which includes the famous *Xia-Fang* program of occupational exchanges between elite and mass, can be seen as antisoftness policy, at least within Chinese communist contexts (Halty-Carrere 1980:74-86). Mao's attack on organizational ineffectiveness was thorough and, according to the present government, even excessive. But regardless of their special and controversial character, these policies gave China a working development program in which economic productivity, urban-rural relations, and population growth were planned and effectively controlled. The Maoist critique refers to the bureaucracy and planning administration transferred from both First World countries and the Soviet Union. Mao was sharply critical of the Soviet approach to peasant nonparticipation, as Gandhi was critical of the British approach. Though China's example of appropriate planning, like other aspects of its society and revolution, cannot be literally applied to any other Third World country, the essence of China's approach, the spread of participation in planning, incorporation of local values and standards, and the undermining of traditional elite systems, bears a lesson for all: softness can be contained by policy, and its containment in planning administration can contribute to the attainment of development goals.[6]

Controlling Softness

The formulation and implementation of softness policy would appear to be an important but extremely difficult task. One source of this difficulty is the general lack of recognition of softness as a problem. Policy to control softness and particularly to "harden" development planning, must begin with a focus on the organization. The object of such policy, in the short run at least, is to eliminate mock bureaucracy and not, as the problem is now approached, to punish corrupt people. Even if softness is recognized as a distinct organizational problem, policy to control it remains subject to the very condition it is formulated to control. some states may be too soft to be able to implement effective antisoftness policies. This would seem especially likely in light of the fact that some parties (perhaps all to some extent) have an interest in keeping things soft. These factors, as well as continuing interdependence between Third World and industrialized countries, world resource politics, and other geopolitical relations, make it far easier to suggest the formulation of antisoftness policies than to follow such a suggestion.

Despite these obstacles, the attempt to understand and control softness should be undertaken by Third World development planners. As economists and other students of development have observed, much that influences the development process — economic imputs, environmental factors, political conditions — is beyond the control of Third World planners. At the perhaps narrow margin, however, Third World planners, administrators, and their publics do have potential control over the goals they establish and the means they choose to achieve them. By persisting in the use of inappropriate planning technologies, Third World planners are forfeiting control of even this component of development. Instead of a well-functioning planning process, the current situation is one in which plans, goals, and targets get bypassed in favor of other interests. Softness pervades administration. If this control is to be regained, an appropriate planning technology sensitive to local interests and standards and which overcomes traditional stratification principles must be devised.

Notes

1. In a recent study of the impact of a new bicycle factory in a village in the Ansansol region of West Bengal, India, Morton Klass (1978:254) summarizes this two-way effect neatly: "The village and the factory now live side by side, if uneasily, and they interact. The factory has changed the village and its community, and the factory has changed in response to the values of those who staff it. Neither, it would seem, can ever be the same again."
2. For a thorough review of the strategy of incrementalism and related approaches to development in contemporary academic political science, see Higgott (1980).
3. The implication here is that the soft state has the same basic source as soft development administration, which is a specific part of the state apparatus in Third World countries. Because planning, in theory at least, plays such a crucial role in the activities of the state, it is difficult to conceive of effective antisoftness policy that does not attack the problem here at its root in planning administration.
4. As Horowitz (1974) and others have noted, it is not simply a matter of U.S. or Western European "capitalist" domination. If we are going to speak of a worldwide economic system to which Third World countries are peripheral or subordinate, then Soviet and other command economy states are equally implicated as core or dominant powers.
5. Recent studies that presume to be comprehensive guides to effective development planning in the Third World, but which are marked by a profound disregard of the role of softness, are Halty-Carrere (1980) and Kidd (1980). For all of their sensitivity to political and economic "realities," neither of these works addresses the question of why the sound and well-known solutions they propose have not been implemented.
6. This is not meant as an endorsement of Chinese communism or of totalitarian rule. Our ignorance about the Chinese development experience remains abyssmal. If the sacrifice of life and liberty in China "for development" has been as great as many critics have indicated, this is something to be decried, not emulated. But the principle of China's "anitsoftness" policy is valuable and can be analyzed without engaging in the polemics often associated with the use of "the Chinese case."

13.

Appropriateness Versus Appropriation: Alternative Approaches to Technology Transfer

Technology is neither good nor bad, nor is it neutral.
                                                Melvin Kranzberg

From over three decades of experience in the transfer of technology to the Third World, a key trend in development theory has emerged. This is "the increasing awareness of the disruptive effects of technology which has led to a growing dissatisfaction with narrow assessments that do not take sufficient account of the social and human consequences of technology transfer, and to a growing recognition that these factors may prove as important as economic or employment factors" (Singer 1977:39). Those who have followed the recent growth of interest in the international application of social impact analysis, technology assessment, and appropriate technology strategies are familiar with this formulation.[1] Social scientists have demonstrated conclusively that the diffusion of modern science and technology to Third World countries does not necessarily, or even ordinarily, bring development. Social development theorists and policy specialists have begun the search for "appropriate," "intermediate," "alternate," or "indigenous" technologies that may be designed in or tranferred to Third World countries. National organizations such as the U.S. Agency for International Development have issued proposals for appropriate technology R&D. And, under the aegis of the International Labour Office, UNICEF, and other international bodies experimental appropriate technology centers have been established throughout the Third World.

Continued emphasis on technology as a politically neutral means to specific desired ends has allowed decision makers in the industrialized countries and their critics to evade serious problems related to control and supply. By ignoring questions raised in the Third World concerning who owns technology, proponents of appropriate technology strategies have failed to confront crucial issues in the formulation of science and technology for development.[2] Control of technology (appropriate and otherwise) stems in part from a well-organized and established, but seldom discussed, system of proprietary rights. These rights are institutionalized through national and international organizations that award patents. The patent system has made it possible for First World multinational corporations (MNCs) and Second World state enterprises to monopolize technology and use it for political and economic gain. This often occurs at the expense of social development, as in the recent conflict between IBM and the government of India, the dispute between ITT and Chile under Allende, and many other instances. Unless this system is carefully considered, the appropriate technology movement is likely to fall short of achieving its stated aims.

This chapter examines questions of ownership and control of technology through a review of literature on the appropriate technology movement and studies focusing on the role of the patent system in technology transfer. Our purpose is to explore the need and potential for Third World countries to appropriate technology for their own social development ends. While the relationship between ownership, control, and development has been considered central since Marx's time, it has received very little attention from proponents of appropriate technology strategies. While moves toward nationalization of corporations and industries have been made in most Third World countries, the question of appropriating technologies owned by corporations has not, until very recently, been seriously considered. These brief remarks are offered as a step toward opening this line of inquiry.

Examining Alternatives

The trends in social development theory and policy noted here should encourage those who have been critical of technocratic solutions in international development policy. Stress on the transfer of technologies for basic needs and the role of social impact analysis in making technological innovations less destructive of human values can be effective in the struggle against the legacy of colonialism and imperialism.[3] Despite these strengths, the movement for appropriate technology transfer has some serious limitations. Lack of attention to questions of interest and control is curious because technologies are frequently proprietary entities. Certain industrial, medical, and military technologies can be transferred from one country to another yet ownership and rights associated with the technologies remain unchanged. This may or may not affect how and whether a transferred technology promotes social development. We can never know unless ownership is treated as a potentially important component of appropriateness. The problem has been ignored or suppressed both by the owners of proprietary modern technologies and by their critics in the appropriate technology and social impact movements. Owners might prefer to treat the question of ownership as moot or "academic" because it is, to their minds, a wholly settled issue (U.S. Department of State 1977; Mattill 1977). But it is not so easy to see why radical critics, even those in the Marxist tradition and contributors to the Lund Letter (1977-79), have not joined their Third World counterparts in calling into question the relationship between social development and ownership of the means of production.

Three Aproaches to Appropriate Technology Transfer

The various strategies proposed for appropriate technology transfer can be classed into three general categories: (1) the technological fix orientation; (2) the social responsibility movement; and (3) the social control movement (Lal 1977). Advocates of the technological fix stress that the negative impact of modern technology on social and physical environments can be neutralized either through the modification of currently available technologies or through new ones (Hetman 1977). The technological fix approach does not treat ownership as problematic, in part because its solutions to problems caused by the transfer of inappropriate technologies depend on increasingly profitable technologies. This account serves the interests of the owners of technologies and it is best to avoid examining the question of ownership from the point of view of these interests.

The social responsibility movement stresses the normative aspects of technology transfer (Goulet 1971; Illich 1973). From this perspective, the solution to problems of the unwelcome effects of technological innovation lies in educational programs that will permit Third World people to make wiser choices from a "shelf of world technologies." This position focuses on the importance of appeals to the conscience of scientists and technicians for a more humanistic type of inventiveness. The social responsibility approach does not address problems of ownership, in part because it tends to treat science and technology as free commodities. Proponents stress that it is people's choice mechanisms and the activities of innovators that require "education," presumably where there is none.

This presumption ignores the distinct possibility that people's choices are now being "educated" through advertising, propoganda, and other means to serve the interests of those who control the technology transfer process. The view that scientists can be expected to be more moral than, say capitalists, is often more a wish than a fact. Scientists, like everyone else, must face the powerful tendency to produce that which will bring wealth, prestige, and power. Whether this tendency is compatible with production of appropriate technologies is an open question. The answer depends in part on knowledge about the interests of those who control the national and international systems whereby social rewards are distributed, systems which, even for scientists, are increasingly influenced by the owners of technology.

Both proponents of the technological fix approach and supporters of the social responsibility movement see a need to design transferred technologies to suit local social and economic situations. They generally share the view that the negative effects of transferred technologies are caused by excessive scale and size. They recommend "descaling" of technology or steady-state, "no-growth" options as solutions. These and related views have been directly responsible for the coterie of value-laden terms like "alternate technology," "intermediate technology," and "low-level technology." Though the meaning of these terms is often context-specific, there is agreement on the basic premise that technologies have to be small enough and sufficiently labor-intensive to be used in the Third World with a minimum of disruption in present economic and social conditions.

The two-fold attempt to redesign the technologies that are "out of control" to a more human scale and to create new technologies limited in size or in terms of organizational and capital inputs they require constitutes the foundation of a worldwide program of reforms. The aim of this program is to replace or modify technologies that are not, in Ivan Illich's term, "convivial."

The developed countries' multinational corporations and state agencies are in search of indigenous technologies that can be perfected in their research and development divisions. Once they are made "convivial," they are better suited for diffusion back to the less-developed countries. Implicit in this program is one or both of two related assumptions: (1) that technologies are simply artifacts that increase efficiency and therefore are free from any social or political implications; and (2) that technologies are free goods, that is, they are not owned.

Lack of emphasis on control of technology in these two perspectives has promoted the rise of the third, social perspective. This movement takes a more radical approach in pointing out that technology has become the means for acquisition of power and privilege (Cooley 1977). Proponents of this position argue that fundamental changes are required in political and economic institutions to distribute more equally the benefits that technology has reaped for the few. Yet even this position does not confront the specific connection between ownership, interests, and control. While it stresses social (i.e., political) control of the innovation and diffusion processes, it does not acknowledge the de facto and de jure control vested in the property relationship itself, that is, the power to withhold, dismantle, and destroy. Such control cannot be altered without a fundamental change in this relationship. The emphasis of some authors on autonomous technology, and its antidote intellectual Luddism, has much appeal to radical critics of technology transfer. But, as noted in chapter 8, this perspective mystifies technology and deflects attention from the true "enemy," which is not technology but rather those who own and can thus choose to use it for promoting their own interests.

Fear of the consequences of such misplaced emphasis was well articulated by John Forge (1978:6) when he noted that he saw "the consequences of an alternate technology system as disguised sophisticated technology still in pregnancy. After birth, it will most likely develop into what we have already witnessed." While there may be several ways in which such deception could occur, unless we take into account the role of ownership, we may not be able to recognize such "disguised sophisticated technology" nor understand its conservative function in the transfer procedure.

Who Owns Technology?

International Patent System

Proprietary rights on technology are maintained by a well-organized system of awarding patents to technological innovations and discoveries. "Apart from the freely available and usually older technologies, one of the important determinants of the conditions governing access to technical knowledge is the nature, extent, and functioning of patent regulations at both national and international levels" (Patel 1974:3). An international body for the regulation of patents, the International Union for the Protection of Industrial Property, was instituted in 1883. Today, more than forty First, Second, and Third World countries are members. The organization supports principles such as equal treatment for foreigners and nationals of a member country and priority over all other applicants for the first applicant with a patentable invention. The assumption underlying patent law is that it is desirable to encourage invention for its own sake, and that granting certain rights over inventions to individual inventors is the best way of doing this. The original goals of the union were to reduce the social costs involved in granting patents to inventions developed abroad, prevent the exploitation of less-developed countries by developed countries in the technology market, and reduce the influence of the patents on the location of industrial activity. The strategies chosen to help achieve these goals include compulsory licensing and working of the patents (Penrose 1951).

Despite these international controls, almost all patents in less-developed countries are now owned by huge foreign companies. "In the case of the United States, Germany, Great Britian, Switzerland, and ten other countries, patents on local inventions exceed the number obtained in 'home nations' themselves."[4] Economic organs of the developed countries, such as MNCs, have secured patents in as many countries as possible and have emerged as monopolies in the world "innovation market." The situation has aided these corporations to minimize competition and disseminate certain kinds of technology, largely in keeping with their strategies of profit maximization and market expansion. Strategies adopted to achieve these ends include restrictive licensing and direct investment wherever possible (Vaiksos 1972; Lall 1976).

Patent holding by foreign governments and corporations has not led to the encouragement of domestic inventions. "The contradiction arises from the fact that, despite the theoretical significance of granting patent protection to private innovations, a very large proportion of patents actually taken out seems to be unimportant for innovative activity in the industry concerned" (Lall 1976:8). The assumption underlying patent laws, that patents will encourage innovation, is obsolete. The process of invention has transcended the individual level enterprise it was originally intended to serve and has become a programmed activity of multinational organizations.

Provisions in the patent laws — established by the Paris Convention of 1883 — for licensing patents have been used by developed countries and MNCs to their advantage, for profit maximization or political control. In the words of Win Straube (1974:478), an executive with Pegasus International, licensing does two things: "It makes somebody else in the international markets work with you instead of against you, and it may keep others from trying to copy or overcome your technological advantage — at least temporarily. . . and it gives you additional revenue for which you did not have to expend material or production labor." Licensing practices adopted by patent holders for strictly commercial reasons have not only supported the types of inappropriate technology transfer that have occurred, but they have also increased the social costs associated with the award of patents.

While licensing was originally conceived as a means to hasten the diffusion of technical know-how, a high concentration of patents in a few corporations and countries has resulted in discriminatory practices in the issuing of licenses. Where industrial development is very low, patent-holding organizations withhold licenses. When licenses are granted, a policy of cross-licensing is often adopted to avoid market segmentation. Cross-licensing is a method whereby patent holders pool patents and, through mutual agreement, monopoly privileges are integrated into a grand design to divide world markets. Under these market conditions, technologies (appropriate or otherwise) are transferred from the country of the patent holder to the Third World.

Licensing has now become a multibillion-dollar business: "In 1970, the U.S. received 2.4 billion dollars in royalties while paying about 230 million dollars for the acquistion of foreign technology" (Raffin 1974:472). Patenting can help realize a ten to one proportion for investments in technology transfer. This suggests that ownership can also mean a great deal in terms of the negative effects of technology on social develpment via foreign exchange deficits and other less subtle forms of expropriation. One of the key motives for securing patent rights is preservation of markets, especially those captured through exports and subsequently threatened by competition. The assertion of patent rights by corporations, during instances of threat against their monopoly privileges, is well documented. The case study of the pharmaceutical industry in Sri Lanka by Lall and Bibile (1977) is an especially good illustration.

Pharmaceutical Patents in Sri Lanka

The government of Sri Lanka formulated a scheme in 1972 to bring about a series of changes in the structure of production, distribution, and importation of pharmaceuticals. At the time the scheme was conceived, supply of imported drugs was directed by the Government Central Medical Store. This administration applied to public health care units and, through about eight hundred local and foreign firms, to private health care units. When a serious foreign exchange crisis arose, the government decided to pursue a program of social control. The aims of this program were: to reduce the number of drugs imported, channel all imports of processed pharmaceuticals through a state trading corporation, amend patent laws to obtain drugs from cheaper sources, replace brand names with generic names, and to subject local manufacturers to government decisions on rationalization and related industrial policies.

The local drug industry, which was dominated by foreign companies, reacted violently against the reforms. These firms vehemently refused the suggested rationalization policies and did not channel imports of intermediary chemicals through statutory trading corporations. At that time, the companies were securing intermediary chemicals from their parent firms at exorbitant rates, while the same chemicals were available at far lower prices on the world market. They held to their purchasing policies on the grounds that the quality of chemicals from other sources was doubtful. They also brought foreign governmental involvement and pressures to bear upon the local government. Under threat of immediate cessation of all forms of aid, the government of Sri Lanka relaxed the policies of rationalization in 1976.

The multinational pharmaceutical companies in Sri Lanka used their advantage as patent holders to continue to import intermediary chemicals on a selective basis and to overprice their products. In this instance and in general, the right to import intermediary products can give the patent holder leverage to produce products that are not themselves patented. Thus a company that has a patent for latex production may also produce rubber tires (which are nonpatented). "Patents do permit patentees and their licensees to charge higher prices in protected markets than those that other competitive suppliers might charge were they to enter the market" (Penrose 1973:777).

Technological Monopolies

The growth of these technological monopolies is further supported and accelerated by the "behavior of those groups in underdeveloped countries which, in search of individual advantage, obtain technology on less than optimum terms" (Turner 1977). Expansion of the modern-oriented middle class of consumers makes the situation even more profitable. Entrepreneurs in the Third World are very willing to invest in the manufacture of consumer items by obtaining licenses for them. Since the required technology in such instances necessitates heavy investments in research and development, entrepreneurs are forced to depend upon patented First and Second World technologies. These entrepreneurs, the new middle-class consumers, and patent holders constitute a formidable force in favor of foreign interests in local technological innovation policy.

Proprietary rights enjoyed by developed countries to technologies in the Third World have not only curbed the capacity of the Third World to innovate available technologies, but have also increased the social costs involved in their use. This is due to the unfair but widely practiced use of patent rights. Though such uses are well known, few Third World countries have taken any steps to challenge the patent system: "Many governments do not want to control them very strictly. They may tax them and bargain with them, but ultimately, the need for their technology, product, and capital is so pressing (and growing over time) that they will be willing to provide the necessary inducement for MNCs to operate them with relative freedom" (Lall 1976:5).

As foreign governments and corporations are able to evaluate the potential of domestic invention, they selectively offer rewards to local inventors and purchase their inventions. Once the proprietary rights are established, they "can utilize their research facilities to develop so-called intermediate technology; made to measure for the market size, climatic conditions, tastes, level of education, purchasing power, and locally available resources of the third world" (van Dam 1972:482). Such technologies, perfected and packaged to articulate with consumers' choices, may deal the final blow to the Third World in its effort to become technologically independent. It is doubtful whether technologies which can generate use values will ever become easily accessible and available under present conditions of proprietary rights.

In instances when the Third World has suggested changes in national patent laws, owners have reacted violently and proclaimed complete non-cooperation. On the question of appropriation of patents, a U.S. executive reacted thus:

> Another point of concern to us is the arbitrary stand taken by some countries on the free use of technology once it has transferred — even to the extent of selling it to someone else. Quite to the contrary, we view this as an intellectual property which we are merely disclosing for its licensing and use in each specific case. I submit that arbitrary laws of the buying country on royalty fees, length of time royalties can be paid, prices for know-how, etc., will stifle flow of technologies to these countries (Basehe and Duerr n.d.).

Such reactions are understandable in light of commerical benefits the developed countries are reaping from patents, though they are, ironically, quite contrary to the basic purpose of patent laws.

The Patent System Reconsidered

An important but latent countercurrent in the present situation is the consensus among almost all Third World countries that they are equally victimized by the current patent system. This has encouraged many to suggest partial to total rejection of patent laws as a strategy to break the monopolistic advantages that MNCs and foreign governments enjoy in their market. A general strategy seems to hinge on formulating plans of action consistent with the need that Third World countries have to maximize their profits and safeguard their interests in dealings among themselves and with developed countries. Though a great deal of technology transfer takes place among Third World countries and also among MNCs and developed countries, details of purchasing policies and corporate techniques are not well known. This may be one of the reasons why the bargaining capacity of Third World countries is very weak (Cooper 1972). Knowledge of strategies employed can substantially improve the capacity of Third World countries to negotiate license contracts.

Other reforms suggested to aid Third World countries in formulating new policies concerning proprietary technology include a new international convention on patents, technology-sharing arrangements, cooperative research and development programs, and the use of technological innovation purely for the purpose of training local engineers on-the-job and with no concern for immediate profitability. Although these are only a sample of suggestions for a more appropriate ownership system, they represent a sound and unexplored direction in appropriate technology transfer. The basis of this alternative requires considerable cooperation among Third World countries, skillful management practices, a reduction of the exploitative potential of technology, and a challenge to current proprietary rights on technology. There is reason to be skeptical about the potential for success of such alternatives. But insofar as they attack the problem at its roots by challenging ownership itself, this approach deserves more attention than it has received among social development theorists.

The question of ownership is an important avenue to explore in seeking to solve the problems that attend technology transfer. In the heading of this chapter, appropriateness and appropriation are presented as alternative strategies. But by the very nature of these issues, it should be clear that we do not really know whether terms of ownership need to be changed before technologies can be truly appropriate, that is, used to serve developmental ends. Perhaps it does not matter who owns technology; but perhaps it makes a big difference. We will not know this without an examination of the patent system as a key element in the complex relationship between technology transfer and social change in the Third World. Until this task is sufficiently under way, some burden of proof rests with the proponents of the technological fix, social responsibility, and social control perspectives to show that existing property relations do not stand as an essential factor mediating between technology transfer and lack of development.

Notes

1. A recent comparative social impact analysis that includes a variety of effects — positive and negative — is provided in Sofranko et al. (1977). For a current and fairly comprehensive bibliography of appropriate technology transfers, see VITA (1977).
2. The characterization of the question of ownership as "academic" can be attributed to proponents of quite different approaches. As a session chairman for the November 1976 meeting on UNCSTD at the State Department, Treasury Secretary G. William Miller (then chairman of the board of Textron Corporation) was asked what he thought of the discrepant claims made by Third World representatives, U.S. businessmen, and U.S. labor union executives about ownership of technology. His answer was that such claims were "academic." Former secretary of agriculture Orville Freeman concurred. See "Edited Transcript of November 17, 1976, Meeting in preparation for the 1979 Conference on Science and Technology for Development," U.S. Department of State, July 1977. Proponents of the autonomous technology thesis also stress that ownership and control are no longer serious considerations in understanding the social impact of technology.
3. For a review of this discussion see "Science Policy and Developing Countries" by Z. Sardar and D.G. Rosser-Owen, and "Science, Technology, and the International System" by Eugene B. Skolnikoff. Both are in Spiegel-Rösing and Price (1977). See also Goulet (1977:71).
4. Shipman (1967:59). Though outdated, this was a ground-breaking study in the politics of the international patent system.

14.

A Call for a Code of Ethics for the Transfer of Population Control Technology

During the past few years, a set of ethical issues associated with the transfer of population control aid from the United States to Third World countries has been identified by ethicists, social scientists, and policy specialists. This commentary has yet to be systematized, and, as policy is now formulated, it has no formal role in either the theory or pratice of population control technology transfer. In light of the recently demonstrated effectiveness of contraceptive chemicals, mechanical devices, and delivery systems in Third World field tests, this chapter presents a call to those involved in research, funding, policy administration, and other population control activities to consider ways for integrating knowledge of ethical impact with the technical aspects of their work. This project can be realized in the preparation of a code of ethics specifically attuned to the types of transfers favored in current aid programs. In pursuing the goal of formulating an ethical code, a forum can be established for the exchange of views among a diverse group of professionals. This exchange can be summarized in a systematic inventory and analysis of ethical issues associated with the transfer of population control technology, and the results communicated to a wide audience. Ultimately, these results should be used in formulating national and international policy on population control, research, technology transfer, and in stimulating scholarship in population ethics.

Current Demographic Issues

Since the early 1950s, well over one billion dollars have been spent by the U.S. government and by U.S.-based private research foundations in aiding Third World countries in the establishment and development of population control programs (Population Reference Bureau 1976; HEW 1978:6; Demerath 1976, ch. 1).[1] Historically, the countries of Central America, Thailand, Taiwan, Korea, and (until the early 1970s) India have been the main recipients of this highly advanced contraceptive know-how and equipment. Until recently, however, these programs have met with considerable skepticism from academic population specialists in the United States (Davis 1976; Matras 1977, ch. 13) and elsewhere (Mamdani 1972; USAID 1976). Much of this skepticism has been directed at the effectiveness of these programs in achieving their stated goals — to reduce birth rates to a degree sufficient to influence population growth in the face of declining mortality rates. Between 1950 and 1975, rates of natural increase in the Third World increased steadily, while crude birth rates showed no significant declines (Tapinos and Piotrow 1978; Weinstein 1976). Since about 1976, both USAID and the UN Office of Population have reported definite declines in birth and population growth rates for the Third World as a whole, and particularly for countries in Asia, Africa, and Central America in which U.S.-sponsored population control programs have been organized (UN Statistical Office 1978; Population Reference Bureau 1979; see also Population Index 1978).

In a recent series of USAID-sponsored studies, Teachman et al. (1978) concluded that a major share of these recently observed fertility declines is attributable to population control technology transferred from the United States. This conclusion is also supported by preliminary findings from the World Fertility Survey (Main 1978). While many questions about the effectiveness of current approaches to population control in the Third World remain unanswered, certain facts have muted the skepticism of a few years ago. First, the technical feasibility of cheap, safe, effective, and physically unobtrusive methods of fertility control has been demonstrated throughout the Third World. On-site trials with subdermal insertion of steroids have been especially successful (Ford Foundation Letter, 1 December 1978:8). Technology now exists to "vaccinate" women in rapidly growing populations of the Third World against conception in much the same way as vaccination against smallpox and other diseases took place during the past two decades. Second, the ability of family planning programs to contact and persuade, or otherwise induce, members of Third World populations to control their fertility has improved substantially. The work of Bogue and other social scientists, as well as reports from public and private population-development agencies, indicate that mass education through schools, the media, and household interviews, in conjunction with a variety of material and moral incentives, have greatly increased public awareness and acceptance of fertility control (National Academy of Sciences 1979).

One likely result of these successes is an increase in future levels of funding for public and private population transfer programs — subject to general economic constraints. This prognosis is supported by recent annual summaries by the Ford Foundation and the Federal Inventory of Population Research, both of which report record high expenditures for research and implementation of population control in Third World countries. If this outlook is correct, we are likely to see an increasing investment in types of transfers which have already proved effective (steroids, vasectomy, tubal ligation, and IUDs), and research and development along lines that will improve technical and attitudinal components of existing programs.

Because of the technical success of our contemporary population control aid programs abroad and because efforts are likely to be intensified in the near future, it is especially important that the ethical component of these programs be scrutinized. The state of population science and population control technology obviates the once-popular claim that family planning cannot work in the Third World.[2] Population policy administrators, researchers, and funding organizations have at their command a highly effective technology with potentially great economic and social impact. In light of this effectiveness, recognition of the need for appropriate technology transfers (U.S. House Committee on International Relations 1976) must be extended to population control technology. In these terms, ethical impact must now be viewed as a central aspect of the appropriateness of population control aid. What is required is the examination of the ethical component of contemporary population control technology transfer and application of these findings to the formulation of population control aid policies.

Current Ethical Issues

Stimulated by the same events and controversies which have concerned participants in population-development programs, the field of population ethics has, in the past few years, assumed a new shape and direction. Events associated with high rates of population growth in the Third World[3] and breakthroughs in U.S.-sponsored programs in curbing these high rates have given a renewed sense of urgency to these issues. While this urgency has usually been expressed, in the United States in any case, in terms of domestic population policy (Veatch 1977; *Report of the President's Commission on Population Growth and the American Future*), the attention of population ethicists has now turned to specific problems related to population control technology transfer.

The 1974 World Population Conference in Bucharest was among the most significant sources of these changes.[4] The *World Population Plan of Action,* adopted at Bucharest, is explicit in linking ethical with economic, medical, and demographic aspects of population control. The authors of the *Plan* stress that it "must be considered as an important component of the system of international strategies and as an instrument of the international community for the promotion of economic development, quality of life, human rights, and fundamental freedoms" (USAID 1976:4). In the presence of such sentiment, pointed questions have been raised in both popular and scholarly literature and in congressional hearings[5] concerning the transfer of population control technology from the United States to the Third World: Are U.S. population aid programs promoting human rights and fundamental freedoms (Bayles 1976)?

Despite this recent public and professional discussion, a systematic code of ethics for the international transfer of population control technology awaits formulation, though aspects of the problem have been addressed by philosophers, theologians, policy analysts, sociologists, and anthropologists (for an excellent summary see Reich 1979: 1215-1316). The result of this research is a modest but growing body of literature suggesting that U.S. population policy has avoided addressing questions of ethical impact in favor of other concerns viewed as more pressing: technical and economic feasibility, imminent Malthusian catastrophe, and at home, resistance from conservative anti-birth control interest groups. At present this commentary is too diffuse or fragmentary to be of much benefit to those who formulate policy or otherwise participate in U.S. population aid programs in Third World countries.

A code of ethics for population aid programs, overdue as it may seem, is not easy to devise. Several types of population control technologies, including IUDs, male contraceptives, subdermal "vaccinations," and incentives for vasectomy and tubal ligation have been transferred to scores of Third World countries. Each combination of a particular technology and culture has a unique ethical impact. Much of the information about population growth and control in Third World countries cannot be considered reliable. Since 1974, and perhaps as early as the days of Godwin and Malthus, the question of if and how the birth rates of the poor should be controlled by the wealthy has been imbedded in ideology, misunderstanding, and ignorance. Communication about actual demographic conditions and the impact of current U.S.-supported population control programs has been distorted by a range of often conflicting political interests.[6]

Another serious obstacle to the formulation of a systematic guide to the ethical impact of U.S.-transferred population control programs is the set of unresolved theoretical issues in social demography and related fields. It is not certain, for example, that high rates of population growth are inimical to social and economic development (Conroy 1977; cf. Coale 1978).[7] As has been well documented, Europe's development coincided with its second ("explosion") stage of demographic transition. One could argue with much evidence that development was a direct and indirect cause of increases in the use of fertility control in Europe.[8] Many Third World governments (including that of China) have at one time or another argued that a growing population ensures and reflects a growing economy, a view not uncommon in First and Second World countries until quite recently. In light of these claims and evidence, the "lifeboat ethic" propounded by Garrett Hardin (Hardin and Baden 1977; Hardin 1980), Paul Ehrlich, the authors of *The Limits to Growth,* and many others, as well as the more moderate versions of this view, appear to lack factual foundation. Because of such uncertainties, there is little if any consensus about what moral rules apply to population control. As Stanley Hauerwas (1977) has observed, "the problem with the population issue from an ethical perspective is that in spite of the huge literature and impassioned positions surrounding it, there is no immediate way of determining the kind of ethical problem that the issue should raise."[9]

Ethical Dimensions of Population Aid

As Hauerwas notes, recent work in the field of population ethics and related specialties has revealed a set of fundamental and unresolved value dilemmas in population aid activities. This work includes the Hastings Institute of Society, Ethics, and Life Science Program on Cultural Values and Population Policy,[10] recent research and theory in social demography, and related sociological and philosophical commentary.[11] Four dimensions of the ethical impact of population aid identified in this work are outlined here: (1) the role of cultural values; (2) the question of population crisis; (3) the ethical appropriateness of medical versus behavioral approaches to population control in Third World countries; and (4) the degree of participation of recipients of population aid in the planning, diffusion, and implementation of such programs.

Population Control Aid and Cultural Values

A central question raised in the Hastings Institute research relates to the possible discrepancy between the values that promote U.S.-supported fertility control efforts in Third World countries and the values held in those countries. Throughout the Third World, religious mores and social norms are strongly pronatalist. In many of these countries, public morality is opposed to the display, discussion, or even indirect reference to reproductive functions and organs. In attempting to foster widespread acceptance of modern fertility control techniques and attitudes, population aid programs can be morally offensive. For example the promotion of contraceptive use, especially in the Caribbean and South Asia, can easily be confused with the promotion of "illicit" sexual relations. From the point of view of those who accept local norms and values, population aid efforts can appear cynical, confusing, or even antagonistic.

Seen from another perspective, the sanctity of Third World values must be judged in relation to the consequences of the uncontrolled fertility they prescribe. In such a comparison, local culture may prove to be an inadequate basis for revising population aid efforts. As Warwick et al. (1977:18) argue:

> Cultural values are not written in stone, especially when adherence to them may be based not on rational choice but upon ignorance, fear, myth, superstition, and compulsion. . . . Values that are maintained in a society through ignorance, fear, uninformed choice and government fiat are not the best candidates for acceptance as moral values. . . in this case family planning information is clearly relevant and the persistence of existing cultural values is suspect.

To transcend the dilemma between donor and recipient values, an ethical code must take account of the multiplicity of value orientations entailed in population control technology transfer, while not necessarily subscribing to any of them. One key to this transcultural approach lies in the exploration of the other three underlying dimensions identified in previous research. This approach assumes that population control technology transferred from the United States is potentially effective. The most important ethical problems that arise in the transfer of population control know-how and equipment are related to social and demographic, as opposed to cultural, effects. The code of ethics must focus on the impact of effectively functioning population control programs, that is, the impact of rapid fertility decline.

If we assume that currently popular types of population control technology will continue to be developed and transferred to Third World countries, the examination of their social and ethical impact should proceed on the same basis as the study of the impact of any other well-functioning technology. We must ask about the effects on family, economy, and polity of an innovation that is engineered to change the person-environment relationship. In this instance, we must address the ethical problems that arise when families are constrained, encouraged, or forced to have fewer children than they ever thought desirable or possible.

The Question of Population Crisis

The work of population/development scholars and activists has often addressed the fundamental question of whether population growth rates observed in today's Third World imply imminent economic, environmental, or, in the classic Malthusian formulation, demographic crises.[12] Little consensus has been achieved as a result of this research. Gunnar Myrdal (1970a), Paul Ehrlich (1969), Kingsley Davis, Garrett Hardin (1968), and others have subscribed to the view that population growth constitutes a drain on scarce resources. Mahamood Mamdani (1972), Conroy et al. (1977), and the authors of the *World Population Plan of Action* represent those who hold a different view. From an economic and historico-demographic perspective, it is not certain that population growth is inimical to development. Under certain conditions, such as those that exist in the rural Third World, large families are viewed as a source of economic support in the event of crisis and on a day-to-day basis. In an attempt to assess both positions in relation to demographic transition theory, we have argued that the relationship between population growth and economic growth involves an inordinately large number of variable factors and an especially strong measure of intention, anticipation, and political interest (Weinstein 1976, 1978). It is likely that under these circumstances, the population/sustenance question cannot be answered with the usual range of scientific evidence.

If the uncertainties associated with population crisis are as serious as they appear to be, an ethical code to guide population control transfers must consider the risks and economic benefits associated with lower fertility. Such a code cannot, for example, be premised on the flat prognosis that "in order to preserve more precious freedoms, we must give up the freedom to breed — and that very soon" (Hardin 1968:1245). It is possible that high rates of population growth in a particular place and time represent a threat to economic well-being, liberty, or life. But no undisputed evidence exists to show that such a threat is a necessary or even likely result of the population/sustenance dynamic. To be ethically appropriate, a population control program that diminishes economic contributions ordinarily provided by children must be based on knowledge of economic losses that would be suffered if fertility were reduced. This is the case in every imposition of a specific population control technology in a given cultural environment. A code of ethics must address: (1) ways in which knowledge about specific economic and demographic factors can be incorporated into future population control technology transfers; (2) the crisis claim and its ethical implications, for developed as well as Third World countries; and (3) the ethical implications of procedures that estimate the cost of additional childbearing and fertility limitation. These are prerequisites to decision making affecting the intensity of population aid in specific societies.

Medical versus Behavioral Approaches

Programs that transfer population aid from the United States to Third World countries have been undertaken in an effort to control human reproduction through diffusion of modern mechanical and chemical contraceptives (Demerath 1976, refers to this as the "contraceptionist" strategy). The dominant approach guiding this work has been a medical, specifically epidemiological model of high fertility. A section of the pioneering Harvard School of Public Health Khanna Study plan of analysis quoted by Mamdani (1972:37) is explicit in labeling high fertility a disease that should be attacked from a public health perspective: "Overpopulation is a malady of society that produces wasted bodies, minds, and spirits just as surely as have other familiar scourges — leprosy, tuberculosis, cancer. . . . The problem in India is of epidemic proportions."

As a result of the domination of medical approaches, the ethical principles that have guided population technology transfers have been those incorporated into medical and legal codes and ethics governing experimentation with human subjects, codes that tend to focus on physical well-being. In this respect, population control work done in Third World countries appears to be medically-ethically sound: the health, nourishment, shelter, and sexual expression of recipients is not diminished — in many cases it is improved by the presence of medical personnel. Yet these ethical principles and the medical approach on which they are based do not adequately recognize that human reproduction is not merely a biological function.

While population aid programs prefer an epidemiological definition of high fertility, many demographic theorists take a behavioral approach (see Reich 1979:1225-41 for a review of this literature). From this perspective, fertility is viewed as a behavioral adaptation to social and economic contexts. Behaviorists point out that high fertility, far from being pathological, was the normal state for most of human history and is still normal in today's Third World. Human reproductive and social systems are viewed as parts of a complex whole. Without this recognition, no ethical issues beyond those that already relate to medical practice need be acknowledged. But as Ronald Freedman (1973:172) has observed, fertility reflects cultural needs:

Reproduction, whether at high or low levels, is so important to the family and society everywhere that its level is more or less controlled by cultural norms about family size and such related matters as marriage, timing of intercourse, and abortion. In each society, the cultural norms about these vital matters are consistent with social institutions in which they are deeply embedded. Changes in fertility are unlikely without prior or, at least, simultaneous changes in these institutions.

A similar observation was made by Roger Revelle (1976), according to whom people everywhere have two sets of needs which must be satisfied for the sake of survival. The first set, physiological needs such as food, shelter, and sexual expression, is characteristically of concern in current population/development programs. The second set of needs, specifically human, is frequently ignored. These needs arise from the evolution of a peculiarly human extended time sense. Lacking the sense of the "eternal now" characteristic of other animals, people require means whereby past, present, and future can be experienced as a whole. These means include the "intangibles" discussed in chapter 11 — the hope, control, security, and legacy which maintain continuity over a lifetime and between generations. For the vast majority of the human population now living in Third World countries and for all humanity for most of our history, the second set of needs has been satisfied not by development programs, with their stress on institutional sources of support, social services, etc., but by offspring. Until and unless things such as hope, control, security, and legacy can be provided by other means, Revelle concludes, the people of the Third World will continue to have large families.[13]

The ethical appropriateness of behavioral versus medical approaches must be explored in light of: (1) whether systematic differences exist between the ethical impact of contraceptive and sterilization programs on one hand, and economic and social change strategies on the other; and (2) the ways in which current types of population control technology transfers do or do not satisfy the kinds of cultural needs to which Freedman and Revelle refer, independent of the types of value issues discussed by the Hastings Institute. In light of the behavioral aspects of reproduction, it is clear that the ethical appropriateness of population control aid is only partly addressed by existing medical-legal codes. These codes do not ensure that specifically human needs are attended to, much less satisfied, by family planning clinics or other population control agencies in the Third World. In formulating a code of ethics, methods should be sought for accouting for the types of cultural provisions ordinarily expected from offspring. Guidelines need to be formulated for dispensing substitutes for the contributions of children at the same time that the know-how and equipment is dispensed to reduce the number of children. Existing ethical standards must be complemented with a set specific to the behavioral and medical components identified in research on population technology transfer.

Participation of Recipients

As is the case with other types of technology transfer, the transfer of population control technology is subject to criticism because it is so strongly controlled by the "supply side" (Hemmer and Ravenholt 1977). The extensive studies of the Ohio State University Center for the Study of Innovation Diffusion,[14] discussed in chapter 11, have demonstrated a consistent lack of attention to recipient input in population aid transfers. The sentiment expressed by the group of Third World representatives at the Bucharest World Population Conference indicates that U.S. population programs have been viewed as paternalistic and self-serving (Warwick 1974).[15]

Current procedures for including recipient participation in the planning, delivery, and evaluation of population control technology are deficient. There are no specific guidelines for consultation with those who are to be sterilized or given contraceptive devices before actual transfers occur. In light of the seriousness of the changes caused by lowered fertility, it is necessary to consider ways for broading the base of participation in decisions affecting the diffusion of population control technology. The social structure of most Third World societies is such that consultation between donors of population aid and local elites does not represent the values and interests of recipients. Local elites are unlikely to survey or even contact recipients as part of the transfer process — or in most circumstances.

To promote human rights in population aid programs, specific means for ensuring direct, preinnovation contact with members of target populations must be formulated. It is necessary to explore the ethical impact of various degrees of consultation with recipients of population aid: (1) the extent to which population/development programs formulated in the United States take account of local interests; and (2) specific procedures for incorporating recipient input. There are several possible ways in which broader participation can be incorporated in population aid planning, diffusion, and implementation; but those ways most supportive of the individual's right to know and to have a free choice remain to be identified.

Conclusions and Prospects

The four ethical dimensions of concern arise from the understanding that a successful program to limit fertility in the Third World will have far-reaching effects on family and society. Significant breakthroughs have been achieved in the development and diffusion of population control technologies in rural areas of Asia, Africa, and Latin America. Yet knowledge of the probable effects of lowered fertility on society and culture has neither been sought nor incorporated into current population aid programs. In addressing the question of population crisis, the medical and behavioral approaches to fertility control, the role of recipients in the transfer process, and related issues, such knowledge of social and ethical impact may be sought. By bringing these explorations together in a code of ethics and disseminating the code to parties of concern, an important step may be taken toward implementation of this knowledge in population/development research and policy.

These comments are meant to inspire a more systematic, policy-relevant approach than is currently available to the ethical problems that attend the international transfer of population control technology. The challenge today is to move beyond the investigation of cultural values, issues of cultural relativism, and assessment of religious views on contraception or other forms of fertility control as an academic exercise. Conditions now exist for the formulation of a code of ethics, complete with critical commentary by population ethicists, population policy administrators, and other key actors in population aid programs. Clearly such programs are of concern to the public and professionals in several fields. Because population aid is at once deeply personal and a public policy issue, it involves a substantial ethical component. And since Bucharest, at least, it is profoundly affected by political ideology. For these reasons, attention to formulating an applicable code of ethics can do much to increase understanding and communication with regard to this most serious and complex issue in the transfer of technology.

Notes

1. Warwick (1974) reports that in addition to direct transfers, more than one-half of the UN Population Program is funded by USAID, and that private individuals can obtain nonproject federal funds for fertility control work in the Third World. By far the largest single private supporter of population aid is the Ford Foundation. See also *Inventory of Private Agency Research*, prepared and distributed by the Interagency Committee on Population Research, Department of Health and Welfare (annual series), and World Health Organization, "Expanded Programme of Research, Development,

and Research Training in Human Reproduction" (annual series beginning in 1972).
2. Hauser (1967) argued that "it is not yet known whether a birth control communications program and a birth control clinic will bring about a more rapid decline in the birth rate than improved general education or new industry." Concepcion and Murphy (1967) propose "dispensing with the old evidence as unworkable or, at best, leading to unusable results."
3. These events reached a climax of sorts when, "in July, 1976, the Indian State of Maharashtra became the first political entity in the world to legislate population control by compulsory sterilization. Even without... final sanction... unofficial endorsement by Prime Minister Indira Gandhi, her son Sanjay, and other officials, has led to vigorous and sometimes fanatic efforts at mass sterilization" (Christiansen 1977). Christiansen suggests that ethical problems are presented in population controls of this sort "because they involve very basic deprivations in which ordinary people are subject to manipulation by elites. Compulsion is legitimized for purportedly moral ends, including survival." See also Donald Warwick (1975).
4. Ethical and political issues were of central concern at Bucharest. A panel on ethics was held at the World Population Tribune and a paper on "Ethics and the Implementation of Population Programs" was presented by Donald Warwick, organizer.
5. See "International Population Program Assistance Statement to the U.S. House of Representatives Appropriations Committee," delivered by R.T. Ravenholt, director, Office of Population, USAID, 24 June 1975 (Congressman Otto Passman, commitee chairman), Washington, D.C. The House Select Committee on Population (Congressman James Scheuer, committee chairman) has discussed aspects of international population and policies. Testimony has been presented by Kingsley Davis, George J. Stolnitz, and others. See *Science* 199 (24 February 1978:867).
6. For a review of this range of issues, See Warwick (1974) and Davis (1976).
7. See also Conroy (1974) and Conroy et al. (1977). Another recent study which contains a lucid reconsideration of the Malthusian question is "China's Population Problem: A May Fourth Period Debate," by Kathleen Lentz, M.A. Thesis, University of Iowa, July 1977.
8. Coale (1978), McNamara (1977), and others have argued that *development* may be too amorphous a term to be of much use in population control policy formulation. Nevertheless, when development also brings improved access to institutional social services, fertility does appear to respond (Weinstein 1978).
9. Hauerwas also considers some ethical dilemmas associated with population control in the United States. His investigation raises serious questions about the ethics of population aid.
10. A key work is Veatch (1977). See also *The Hastings Center Report*, especially vol. 7, no. 1 (February 1977). Also relevant is the set of internationally focused studies of the Hastings Institute's Project on Cultural Values and Population Policy (also see Sai 1977).
11. The excellent collection by Daniel Callahan (1971) is a landmark work. See also Ehrlich (1968); Martin Golding ("Obligations to Future Generations," *Monist* 56, January 1972:85-99); and Commoner (1971:133-39 and passism).

12. Two important reports issued a few years ago, ICP (1974) and World Bank (1974), give balanced reviews of the debate on the population/sustenance relationship. The Ford Foundation-sponsored study by Greep et al. is a central survey of recent population research which takes a strong position in viewing population growth as a definite drain on resources. In contrast, the following report appeared in the press during the Bucharest Conference: "An eight-country group led by Argentina kept up pressure on the United Nations Population Conference to endorse its view that popultion problems should be solved by economic development. . . . The new emphasis lauds population as the inexhaustible source of creativity and a determining factor of Progress" (Reuter News Service, 24 August 1974).
13. Revelle's argument, as presented here, is highly compressed. For a more complete statement see Weinstein (1978), where this relationship is expressed as a reciprocity (or tradeoff) between large families and institutional social services.
14. Most of the sixty papers in the Ohio State series are of interest. A comprehensive treatment of population aid diffusion is in L.A. Brown's (1981) recent theoretical summary.
15. The criticism has often been tied to a more fundamental anti-imperialism, as it was in the remarks of the Chinese delegation to Bucharest which argued that overpopulation is not the source of underdevelopment. "It is mainly due to aggression, plunder, and exploitation by the imperialists, particularly the superpowers" (quoted in Davis 1976:299).

PART IV

Sociology / Technology: Looking Ahead

Introduction

The social scientist in the next stage of social development will be actuated neither by bravado nor by idle ambition when he searches for methods which will do justice to the actual context in which the individual object exists.
                                            Karl Mannheim (1954:170)

In this part we return to our examination of the role of the postacademic social scientist. This discussion focuses on problems of political and philosophical commitment alluded to in earlier chapters. Thus far, we have reviewed intellectual and institutional trends that point to a renewed interest among professionals and the public in applied, interdisciplinary, and development-oriented approaches to technological innovation. It has been suggested that these postacademic approaches remain in a transitional state today: they are more than marginal aberrations from a strong academic status quo, but they are less than a fully established, clearly defined, and identifiable field unto itself. As we consider prospects for institutionalization of these postacademic approaches, key questions arise concerning the commitments of the practitioner which, depending on where they lay, could either improve or hurt these prospects.

If current trends toward integration of social-scientific knowledge, social policy, and technological innovation are to continue, further redistribution of material and symbolic resources will be necessary. Such a redistribution has never been known to have taken place, even in more prosperous times than ours, without concerted political effort. In this case, factors such as long-standing antagonisms fueled by mistrust and misunderstanding on both sides, and the fact that an effective social-scientific approach requires critical scrutiny and even reformulation of value-laden concepts such as "development" and "national interest," suggest that this political effort will be prolonged and difficult.

The effort to redistribute resources so that social science can make a positive contribution to the humanization of technology and policy will require cooperation and understanding of mutual interests and needs on the part of social scientists, engineers, planners, policymakers, and their publics. One can be hopeful yet skeptical about the prospects that technicians or the public are prepared or even able to contribute to the institutionalization of postacademic social science in its truest sense. Such caution is justified in light of the decades of misunderstanding and disdain which have characterized the relationship between the social sciences and technical fields. This disdain has served directly and indirectly to keep the technical fields and pure sociology subordinate to the interests of ruling economic and political elites.

Skepticism concerning this or any other type of progressive social change may be the appropriate end-product of academic sociological research. But it cannot be presented as a hard-and-fast conclusion and thus used as an excuse for inaction by those who share a postacademic orientation. Scholarly skepticism must be balanced by an activist commitment — even if it is only to learn, teach, and communicate about the reluctance of the technician to take sociology seriously.

Because of his activism and because he is aware that rationality is not merely a matter of technical adequacy, the postacademic social scientist requires a distinct political and philosophical point of view. In promoting the postacademic orientation in professional and public forums and in his daily work, he must have definite goals and priorities beyond mere technical efficiency; he must promote what Mannheim called "substantive rationality." As a sociologist or another type of disciplinary scholar, the postacademic social scientist is also heir to a tradition that speaks with many voices, especially concerning the "proper" political and philosophical principles upon which to base shorter- and longer-term goals and priorities. Within this tradition one can find proponents of virtually every conceivable ideology and theory of knowledge. While certain points of view have dominated in different eras, none has been established as definitive.

Postacademic social science thus faces a dilemma: while it needs substantive rationality, disciplinary social science tradition provides no clear positon on the matter. This dilemma is of direct concern in these concluding chapters: which, if any, of the various political and philosophical positions that have been put forward as appropriate by academic social scientists, and sociologists in particular, are the right ones for postacademic social science?

In pursuing this question, we must emphasize the heterogeneity characteristic of academic social sciences. Several aspects of these fields responsible for their diversity are shared with the non-social sciences, the humanities, and in some respects with all work in technological society. This includes increasing specialization and an abundance of "plausible" theories and methods with which to face an increasingly uncertain world. But by virtue of their subject matter, the social sciences have a unique set of dimensions along which disciplines like sociology can be internally differentiated (Ritzer 1975). One of these stems from the inherently ideological character of all social thought. The immense scope and complexity of social relations and the fact that the social investigator is himself a creature of social relations produce a situation in which the social research act, from conception to replication, is shaped by the investigator's personal interests, political commitments, and world view. Whether, for example, a sociologist decides to study class exploitation, social inequality, social stratification, or the status attainment process (four very different terms for the "same" phenomenon); how, once this decision is made, one defines key variables or chooses the appropriate theories, techniques, operationalizations, analytical tools, reporting styles, professional and public audiences — are all conditioned by such mundane factors as "Who pays the bills?" and/or such lofty considerations as "Whose side are we on?" This situation has produced a spectrum of social science styles parallel to the spectrum of Western political ideologies from ultrareactionary through ultraradical to the "end of ideology."

Another source of heterogeneity in sociology and related fields has roots in Vico's theories of historiography, in classic hermeneutics, and more familiarly, in Weber's debate with Willhelm Dilthey. Though given many names, internalist versus externalist, subjectivist versus objectivist, *Verstehen* versus positivist, these distinctions refer to the fact that social action as opposed to mere behavior entails actors who, accurately or not, impute meaning to what they do. The question then arises: Can we explain or understand social action without understanding the meanings, goals and motives of actors? And if not, are the observational and interpretive techniques of the non-social sciences appropriate for social science? How one answers these questions places one on the continuum that ranges from subjectivist (of which there are several types) to (one variant or another of) natural scientist.

While the bases upon which one type, school, or orthodoxy within sociology or related disciplines can be distinguished from another do not always correspond exactly with differences in ideological commitment or extent of subjectivism, distinctions between different schools can be traced to some combination of these two dimensions. This point will be further illustrated as we consider some perspectives identified as opposing ones in contemporary social science. Our major purpose is not merely to classify styles of academic sociology, but to examine the bearing of these styles on the prospects and conduct of postacademic research, teaching, and practice.

15.

Perspectives on Sociological Ideology

Let us begin with some direct questions. What are the politico-ideological commitments of postacademic social science? Does this approach have a basic world view? If so, is it conservative, Marxist, "value-free"? In seeking to identify a distinct postacademic orientation — based on critiques of academic sociology discussed earlier — this chapter outlines several responses to these questions. The major premise of these comments is that the effective pursuit of postacademic social science demands a review of traditional views and criticisms concerning sociology's proper political role. From this review, it is possible to discern a new — yet also old — set of commitments that accompany the kind of work in which the postacademic social scientist is likely to be involved (Bendix 1971; Merton 1971). This chapter begins with a brief statement on the "proper" ideology for social science represented in three of its most familiar variants: pluralist, conservative, and radical. In contrast to these, elements of a postacademic ideology are examined here and, in greater detail, in the next chapter.

The Pluralist Response

There is good reason to argue that diversity of political views, not a particular commitment, ought to characterize postacademic social science. To legislate that the social scientist must be liberal or radical, for example, is to break the long-standing and important precedent, at least in the United States and other First World countries, of keeping personal political views separate from evaluations of professional competence (not to say that this is always what happens in practice, but the ideal exists).

To politicize social science in this manner would be to invite tests of purity, including perhaps purges, punishment for deviation, and associated rituals. These could easily destroy social science by polarizing relations between the social scientist and the rest of society.

A distinct advantage of political diversity in social science is that it does not require any test or certification procedure beyond those that demonstrate professional competence. Whereas the commitment to maintain a specific political viewpoint, if meant seriously, would require that methods be implemented, at graduation and from time to time, to establish the extent to which the practitioner is adhering to "the line." By insisting that a specific political ideology be a prerequisite to professional practice, the field might be unnecessarily limiting its recruitment and training efforts to only the "true believers," thus letting the talents of skillful but "wrong-headed" people be absorbed into other fields.

If anything less than thoroughgoing political diversity were the rule in postacademic social science, agencies and procedures would have to be established to decide what orientation the field ought to have. Obviously such a decision cannot be made once and for all, since individuals and groups change, sometimes dramatically, in ways that affect their political orientation. Thus continual monitoring of ideological criteria would be required, which would subject the field to further polarization and ideological conflict. Who would make these decisions? Other sociologists? The public? Might it occur, with such criteria, that social scientists be elected on the basis of their ideological credentials by peers or public — or by all enfranchised citizens for that matter? None of these options seems very appealing or realistic. From a pluralist point of view, the very question of "correct" ideological commitment for social science is a contradiction in terms. Would not a better policy — if one must be formulated — be to establish ways for ensuring that the academic ideal of recruiting and promoting the best social scientists, regardless of their political views, be practiced? "Since no man knows all the answers, it is important to avoid the dictatorship of false answers" (Dahrendorf 1967:140).

The interaction between social and other factors in technological innovation is profoundly complex, extensive, often subtle, and always subject to uncertainty. The sociologist's ability to contribute meaningfully to innovation for development under any circumstances is strained by his capacity to absorb and relate relevant information. Limitations imposed by a political ideology to which he might be expected to adhere, and which must conform to that of all other practitioners, would exacerbate this situation beyond tolerable dimensions. The ideologue, even and especially the ideologue who is "ideology-free," is a fettered if not dangerous partner in the innovation process. Ideologies orient — one of their chief functions is to combat sources of disorientation, including the complex, extensive, subtle, and uncertain character of technology. But they orient by permitting us to see only part of the world before us and to follow only one or a few of the many paths that lie ahead. Such narrowing is what is most offensive, not to mention threatening to life and liberty, in the world view of the technician who has no use for social science.

In terms of this argument, political and ideological commitments of the postacademic social scientist should be pluralist. Only under such conditions can the scope of innovation be broadened. Lacking such a commitment, technology will continue to serve the interests of elites, and only incidentally will it contribute to development and the human interest. The pluralist's best antidote to this, according to one of its most articulate proponents, "is to see to it that at all times and everywhere more than one answer may be given. Conflict is liberty because by conflict alone the multitude and incompatibility of human interests and desires find adequate expression in the world of notorious uncertainty" (Dahrendorf 1967:40).

The Conservative Response

Can a postacademic social scientist be effective unless he is a conservative? After all, a key factor in the attempt to reconcile social scientific and technical approaches to innovation is the promotion of a willingness on the part of the technician and the public to take seriously social institutions, such as the state, economy, church, family, or other aspects of social and cultural traditions such as values and norms. One of the most important discoveries of the technical fields during the last few years, one that has provided an intellectual basis for the development of the postacademic approach, is the proliferation of unanticipated, unintended, and often ironic second- (and higher-) order consequences of technological innovation. In this discovery is the implicit, and increasingly explicit, call for the social scientist to bring to bear his understanding on the design, development, and diffusion of technology. A sociologist who lacks sufficient appreciation of the institutions, traditions, and dangers of tampering with parts of the complex social whole — one who is not sufficiently conservative — can hardly be expected to contribute productively to this task.

Perhaps this is a weak and harmless restriction. It does not entail very specific partisanship, for a "conservative" in this sense might be a supporter of any of several political parties or programs. Thus it might be granted that such conservatism is fundamental to the postacademic role, but as an adjunct rather than a substitute for the pluralist strategy just outlined. As long as the social scientist is sensitive to social forces, he would "qualify" politically. It may therefore be true that the postacademic social scientist must be conservative, but this seems to be something less than an unusual ideological commitment. Or is it?

Many conservative and radical commentators, including Daniel Boorstin (1961), Robert Lynd, and Christopher Lasch, have shown that academic social science is not conservative, even in this broad sense; that a characteristic "Whiggishness" typifies contemporary social thought and that unwarranted progressive evolutionary concepts pervade sociological analysis.[1] The effect is that the analyst's desire to equate improvement or at least substantial change with liberal institutional reforms distorts his perception of what is real, what persists, and how things may have degenerated. If a broad type of conservatism is inherent in any sociological perspective, as for instance Durkheim's arguments and his own work demonstrate (Durkheim 1964, pt.1), why is it not apparent in the views of those whom Boorstin, Lasch, and the others criticize?

If conservatism means an appreciation of all social experience and an awareness that the whole and its parts are intimately related, it probably does not include what is practiced today in the name of sociology. Abstracted empiricism, whose perfection is the hallmark of academic social science, suffers precisely because it violates the intimacy of the part/whole relationship in social life. It is decontextualized social thought par excellence, in which hypotheses drawn "from the literature" are "tested" in the timeless and spaceless world of model specification, operationalization of variables, regression analysis, and variance reduction ("while life goes on all around you").[2] If to be effective the postacademic social scientist requires a conservative sensitivity to the historical forces that produce social change and an openness to the gestalt of social relations, then a nonpluralist program of ideological reform may be necessary. For these are systematically excluded — as "exogenous" variables — in abstracted empiricist research.

The progressivist tendency to equate development with the accumulation of labor-saving, energy-using technologies that maximize corporate profits or the power of The Party is not peculiar to the technician. This and the consensus model of society that it promotes have become something of a general world view to which most academic social scientists adhere. This is one of the main conclusions that can be drawn from the critiques of academic social science of Veblen, the Frankfurt School, Mannheim, and Mills. This means that we cannot assume that because one is a sociologist, one will ipso facto be sympathetic to a more holistic view of development and a more critical appraisal of progress. If the successful pursuit of postacademic social science depends upon the nurturing of such sympathies, the approach must "purge" itself of this major ideological tendency of its predecessor.

The Radical Response

These observations suggest that some type of ideological discipline beyond pluralism may have its place in the postacademic era. But they do not necessarily support the case for conservatism. An equally strong argument can be marshalled in support of a radical perspective. Do not radicalism and the commitments of the postacademic social scientist converge at a very significant point — at which knowledge of social structure is sought so that meaningful change can occur? The very idea of reorienting technology so that its aims coincide, not with the interests of economic and political elites in present-day capitalist and command economies, but with the human interest, is thoroughly radical. The desire to replace technocratic (anti-) utopias with a humanized technological society is a characteristically radical program.

It is no coincidence that the most important critical commentary on technology produced during the academic era is that of Marx, Marxists such as members of the Frankfurt School, other radicals such as Veblen, or those with a radical mistrust of established state power such as Mannheim and Mills. The type of radicalism that encompasses this diverse list of intellectuals is partly defined by their approach to technology: the critical study of forces that have shaped the innovation process and social thought generated in technological society. The work of these and other radicals suggests that without mastery of the dialectical orientation and related components of Marxist, but also non-Marxist, critical social analysis, the postacademic social scientist cannot achieve his aims of understanding and improving the relationship between society and technology.

The state of technology today, the fact that technological innovation is a tool of ruling elites, the understanding that the technician can proceed without the help of the independent — nonpropogandist and nonmanipulative — input of social science, indicate that the only possible political role for the postacademic social scientist is a radical one. In redirecting the innovation process to serve the species, the social scientist must challenge the power of ruling elites. This means the power of corporations, ruling juntas, reigning religious "prophets," and party directorates. The relationship of such elites to technology today corresponds to the relationship of rulers in the days of old to their land, their gold, and their armies: that of owner, or at least controller of commodities that ensure hegemony. The aim of postacademic social science is to socialize the technology that has been privatized and nationalized during the course of the past two centuries. On these grounds, anything less than a radical commitment on the part of the sociologist would render his participation in the innovation process weak and superfluous.

A Postacademic Ideology

We have come to a paradox concerning the ideological commitment of postacademic social science. We are not suggesting that the arguments for pluralism, conservatism, and radicalism have been exhausted, or that effective arguments could not be framed in support of other ideological positions. While the pluralist position has considerable appeal, it cannot completely dispel the doubts that might be raised concerning the advantages of a more definite and less tolerant position. While the pluralist may argue against tests of ideological purity, these are administered anyway under labels such as mastery of "the" methods or "the" literature. One wonders how many decisions to grant Ph.D.s, promotions, tenure, accept research proposals and manuscripts, or to decide whether a social scientist is qualified to teach, are made without regard to one's political orientation.

What is distinctive about the current system of certification is not that it follows the pluralist model, but that ideological correctness is determined by such an indirect, hidden, and thus intrinsically undemocratic method of operation. In academic social science the "paradigm" of professional orthodoxy serves as a proxy for the ideological position it indirectly implies. One may never be rewarded or penalized for sharing or not the mildly technocratic managerial liberalism, doctrinaire academic Marxism, or the good-old-boy conservatism of one's peers. Yet these and far more subtle shades of ideological discrimination pervade the reward system in academic social science. They function, in part, because one can and should, in the view of most, be rewarded for being an effective abstracted empiricist, or in the margins of academe, for being an effective "Marxologist," celebrator of a mythical past, or other type of technical virtuoso.

A remedy for such duplicity is inherent in the conservative and radical responses outlined here. Ideological criteria should be made explicit, a certified social scientist, in addition to having other certifiable knowledge and skills, is someone known to have the appropriate political commitments. This suggestion does not answer all questions raised by the pluralist concerning what would constitute an "appropriate" commitment or who would decide this matter. Nor does it even necessarily promise to bring the desired results, such as an increasingly effective social-scientific contribution to technological innovation. Yet it is compatible with a frequent observation throughout the history of social science: that ideological commitments do matter in sociology and related fields and that some are better than others in inspiring an effective applied social science of technology. But are these commitments pluralist? Conservative? Radical? All these and more. There is in the perspective of the postacademic social scientist a set of ideological commitments which both encompass and transcend these specific political philosophies.

Postacademic social science is a shorthand way of referring to the interdisciplinary social-scientific study of the social antecedents and consequences of technology whose aim is to incorporate knowledge of social relations into the innovation process, the education of professionals, and other practical activities. In its preacademic origins and more current forms, this approach stresses the promotion of development, in the human interest, at every stage of technological innovation — from conception, to design, to diffusion, to evaluation, to reconception. In the social context of the postacademic era in the United States, the Third World, and elsewhere, such a study can neither be conceived nor pursued without a broad, even abstract, but distinct world view. In part, it is this world view that provides the "correct" ideological perspective for the postacademic social scientist.

If the relationship between technology and power were not what it is, if genocide and the destruction of the species were not now "mere" problems in the "engineering of death" (Horowitz 1980:6), if the craftsman had not been superseded by the routinized order-taker — postacademic social science might not have such an essential ideological dimension. But things are as they are, and social science therefore has an ideology that the pluralist, radical, and conservative must all address at the risk of irrelevance and ineffectiveness.

Relevance of Older Critiques

It should be evident from the commentary of Veblen and others highlighted in earlier chapters that there is nothing especially new about the sociological critique of sociology, in which distinctions are made between the way "they" do it and the "correct" way. Nor is the criticism raised or implied here novel. Why then should one want to repeat these points, or further suggest that they cohere in an "ideological orientation"? The answer is that though negative concepts such as pecuniary gain, abstracted empiricism, technical virtuosity, and scientism, and positive concepts, such as intellectual craftsmanship, the sociological imagination, reflexive theory, the new sociology, and the sociological way of looking at the world, have been employed before in critiques of academic social science, they take on a significant new meaning in the postacademic era.

When these observations were first formulated, their purpose was in part to instruct and by extension to contribute to reforms in the educational and professional development of the academic discipline. Critiques of social science provided by the line of sociologists of technology from Veblen to Mills and beyond are not isolated discourse. They are aspects of a more general criticism of the institutions and thought of technological society. When, for example in the cases of Mills and Mannheim, these critiques are also tied to classic perspectives in the sociology of knowledge, they go further to illustrate the nature of the relationship between social thought and institutions. These critiques were inspired by the promise of what sociology might be, and were guided by an optimism that made it seem worth the effort to attempt to reform the discipline, to close the gap between performance and promise.

Because such criticism of social science exists and is accessible to all, it has generated further debate, countercriticism, and elaboration in print and, perhaps more often, in the classroom. Without this criticism, the heritage of sociology would be much poorer than it is. Whatever its various motivations, it has kept alive the idea of an alternative to the narrow, presumptuous, and empty claims of academic disciplines. It has formed the basis of an idealism that has sustained many an academic sociologist who might otherwise have seen his professional options narrowed to a choice between leaving the field and giving in to pressures to conform to the canons of abstracted empiricism or another local orthodoxy. This criticism has kept the definition of legitimate sociology more encompassing and its practitioners more tolerant than would have been the case had it not become part of the professional culture. For the individual who, by temperament and training, cannot reconcile his understanding of sociology with the version(s) of the discipline that rule in university departments and other academic forums, it has provided comfort and inspiration.

Despite these significant benefits, this criticism has limits that stem from having been shaped within an academic frame of reference. It has been formulated by academic scholars, based on material drawn from academic literature and the empirical study of academic milieux, and addressed to academic audiences. One thing that makes the work of Veblen, the Frankfurt School, Mannheim, and Mills distinctive is its marginality in relation to mainstream academic styles and concerns (which might partly explain the relative popularity of these writers outside of academic circles). Yet it has been conditioned by the fact that no viable alternative to academic practice, no lively strain of nonacademic sociology, has been conceivable.[3] The most that the critics could have hoped for was a better type of academic sociology which, viewed realistically, would only be possible as a component of a "better" society.

Increasingly, professional sociologists and other social scientists are doing work, including the traditional scholarly work of research and teaching, outside of the standard academic setting, and they find that their needs differ in important ways from those of their academic counterparts. While mastery of theory, method, and current research trends are as vital as ever, for today's postacademic social scientists the purpose of such mastery is no longer to contribute to the "growth of the discipline" through the production of "new knowledge" certified as such by peers whose judgment is based on evidence of mastery and other less noble criteria. Their purposes now include making a contribution to the processes of technological design, innovation, diffusion, and evaluation in the role of sociologist at a nursing school, an anthropologist in government work, a political scientist in a private research organization, or as a historian at an institute of technology. For these purposes, the concerns and aspirations of the traditional academic discipline have proved inadequate.

The postacademic social scientist has been asked to "help move the earth" with the "lever" of his professional skills and knowledge, but he seeks in vain in the traditional disciplines for a place to stand. Such a point of intellectual anchorage has been developed in the critiques of academic sociologists of technology. In their diagnosis of the ills of the academic discipline, so profoundly influenced by pragmatist, radical, and conservative theories, they orient the postacademic toward the search for useful and substantial social knowledge. In their prescriptions, guarded as they must be, for curing these ills, they point the way to a new framework for organizing this type of social knowledge. And in their very existence, they permit the postacademic to benefit from the heritage of academic theory and method while remaining aware of assaults on one's moral and political sensibilities which so often accompany what academic sociologists and others are saying. These diagnoses and prescriptions are old; but for the first time since the fission, they may become part of the working knowledge of a group of practitioners in the full sense of the word.

Important as the critiques of Veblen and others have been in the academic era, their ideological value is only now being fully realized with the institutionalization of postacademic social science. They provide the postacademic social scientist with an independent basis for making the inevitable value judgments that attend his work. These critiques are true to the spirit, if not the letter, of the concerns of pluralism, radicalism, and conservatism. They were formulated by knowledgeable and concerned social scientists, with due awareness and sympathy for such ideological concerns, but also with attempted intellectual detachment from them.

The elements and themes that characterize the postacademic orientation, as originally expressed in the critical and programmatic views of the sociologists of technology and as they have been organized here into a single "doctrine," are not manifestly political or ideological. But in the present circumstances, their promotion and application are bound to encourage partisan responses such as opposition, indifference, and support. While not necessarily formulated or intended as a "position" in a struggle for power, the program of postacademic social science is, secondarily but no less authentically, just that. Among other things, this means that consideration must eventually be given to the nature of this political struggle and to the chances that the outcome will be satisfactory for the postacademic social scientist and society.

Notes

1. See Nisbet (1969), Lasch (1973), and our review in Weinstein (1974). For another recent critique of "Whiggishness," see the comments in Horowitz (1980, chs. 5-7 and passim).
2. An excellent example of the contrasts between abstracted empiricism and contextualized approaches in social science is the exchange between Featherman (1979) and Willhelm (1979). Not incidentally, this heated controversy centers on the study of social inequality. Willhelm points to the ideological and political implications of studying inequality in an ahistorical, aspatial, and asocial manner.
3. A distinction might be made between academic and transitional sociology of technology, since the degree to which a nonacademic alternative is viable is a relative matter. As the academic era drew to a close, it became easier for sociological critics of social science such as Mannheim and Mills to conceive of an institutional alternative to the academy.

16.

Ideological Principles of Sociology

While one cannot provide a detailed description of a professional role at such a very early stage in its evolution, it seems clear enough that the active and effective participation of the social scientist in technological innovation, professional education, and related activities carries with it a commitment to at least four broad working principles: a focus on technology, value explicitness, interdisciplinarity, and developmentalism.

The latter two are discussed at length in earlier chapters along with a brief indication of the manner in which they bear on postacademic social science's ideological orientation. This discussion continues here with a closer examination of the first two principles.

Focus on Technology

Interest in the technological dimension of human relations seems inherent to the general program of social science. It would be difficult enough for an anthropologist of tribal culture, for example, let alone a sociologist of Western society, to achieve anything but the most superficial insight into "his people" without attempting to understand how these people go about manipulating man and nature to make life longer, less burdensome, and more rewarding. The social and psychological processes that underlie invention, the criteria and procedures whereby people are certified as professional technicians, the character and history of what we call R&D, the uses to which technology is put, and the effects of its use and disuse on environment, population, culture, and social relations constitute, if not the essence of organized social life, a major component without which social life would be meaningless.

The vision of man without technology is terrifying. Even those who have been imprisoned and fettered, exiled, or otherwise deprived of access to tools and techniques for improving their lot, have kept their hopes alive by engineering imagined escapes in the dust of the prison yard and seeking cures for some illness of the body, mind, or social constitution. Technology makes freedom believable. If interest in technology is inherent in the idea of social science, the practice of academic social science and especially sociology has (with the important exceptions noted) barely spent its inheritance.

Technology As a Variable

While technology has been featured in sociological research and theory, explicitly by those like Veblen and Ogburn and implicitly by Mills, Durkheim, Weber, and others, it has not received the kind of attention devoted to other subjects such as urbanization, stratification, deviant behavior, and political relations.[1] For every collection that focuses on "the engineers and the social system" (Perucci and Gerstl 1969), one can find several that discuss religion and the social system, the family and the social system, and even more just The Social System. This is not to say that the study of technology and of "the engineers," on one hand, and urbanization, stratification, deviant behavior, political relations, religion, or the family, on the other, are disjointed. Technology is "implicated" in all these realms in a way paralleled by few other social phenomena. Yet the commitment to deal sociologically with these implications is in no way necessary to the academic discipline.

The reasons for this relative lacuna are many and varied, but the more obvious suggest why the sociological analysis of technology has definite ideological undertones. Academic sociology finds it difficult to deal with technology on the most elementary level: as a social variable. Instead, it is either excluded from explanatory schemes, treated as a Juggernaut from which our valued traditions cannot escape being destroyed, or included in a list of causes or (far more rarely) effects in place of an "etc." Thus technology is dealt with perfunctorily and as a "necessary factor" like gravity or the fact that man is a biped.

In a related way, academic sociology has chosen not to "tamper" with technology, but has instead tacitly agreed to accept its prevailing social definition in ways that it would never allow prevailing definitions of family, religion, or deviance to enter unscrutinized into its analysis. It has chosen to treat the existing technological innovation process as the noncontingent outcome of natural laws.[2] But technology is a variable. It is subject to conscious, semiconscious, and unconscious alteration by people and groups. Yet its variance is not felt to be the legitimate concern of the academic social scientist, either in the explanatory or "control" sense. In-depth knowledge of technology and technological change is treated as the special preserve of the engineer or other technician. And this is how the engineer, technician, and society-at-large traditionally prefer it. One result of such specialization is that a large share of the theory and practice of social change is left to those who, by training, temperament, and other forms of conditioning, are unconcerned with, or as Layton, Horowitz, and others have suggested, dead-set against such "side issues" as social conflict, social responsibility, let alone development or the human interest.[3]

Technology and Freedom

If man without technology is not free, the type of sociology which does not recognize technology as a crucial, variable, and eminently *social* force is a sociology of men who are not free. This may be the main flaw in the various styles of academic sociology: pluralist or radical, abstracted empiricist or grand-theoretical, new causal theory or ethnomethodological, functionalist or conflictualist. None of these styles has managed to be scientific about an animal that enjoys the mixed advantage of being able to transform himself and his environment through the application of technology. One reason why a science of free people is yet to be formulated is that the long-run course of technological innovation in societies where social scientists live and study has so often threatened and denied life and liberty. These effects of even this type of technological innovation are contingent upon our ability to see beyond particular outcomes, to appreciate the variable character of technology and, as social scientists, to communicate such observations.

Academic social science has been unable to develop a science of freedom because technology, an essential variable to such a science, has given every appearance that, far from being intimately associated with freedom, is grim necessity "depersonified." We often read about "technology versus man" or, less often, about "technology serving man," but rarely do we understand technology as an extension, a reflection, and in theory at least, a liberator of man, as in the theories of Harold Innes and Marshall McLuhann. The fact that technology's effects depend so much on what we believe them to be and that the social scientist is a professional "believer" in such things also support the conclusion that academic social science has been more often a part of the problem than a provider of the solution.

Questions concerning the relationship between technology and freedom have usually been excluded from consideration within academic social science or have been relegated to its margins in the work of a Veblen. Such exclusion and relegation have left the core of social science without a clue as to how to develop law-like generalizations about human behavior. Many social scientists and schools of thought have actively participated in the conscious depoliticization of their disciplines (for a discussion of the German case see Dahrendorf 1967, chs. 3,4) or, at other times, in their purification of "speculative" contaminants — as if atemporal, aspatial empiricism, grand theory, or price theory were not speculative. They had help from the inertia of academic institutions in which their science evolved, from their sources of economic and political support outside of academe, and from technicians who preferred, because it was in their interest, to retain for themselves specialized knowledge and control over technology's effects.

Postacademic social science seeks to rectify this imbalance, for it seeks to "specialize" in the applied study of technology in society and thus in the study of the problems of freedom. This search is not motivated by an interest in reforming academic disciplines, for this was the main intent of the critiques and programs of sociologists of technology. It is motivated by the need to contribute to the design and implementation of technology. If postacademic social scientists believe that the search for useful knowledge about the relationship between man, technology, and freedom does not have a definite ideological "loading" — critical yet neither pluralist, conservative, nor radical — they have not learned some important lessons from the history of their disciplines and society.

Values Explicitness and Neutrality

The prescription that the postacademic social scientist be critical and humanistic about his own values has serious ideological implications. These are closely related to a key distinction between neutrality and objectivity as they affect the performance of the social scientist. To suggest, as Mannheim, Mills, and others have done, that effective social analysis requires that one confront, criticize, and transcend interests of group, class, and nation to which one has been socialized in favor of commitments to life, liberty, and development in the human interest is to take issue with both sides of the neutralism debate in academic social science. Such a position challenges those who believe that the social scientist should avoid such self-confrontation and simply set aside secular interests. It also challenges partisans who, in a more honest but limited appraisal of the role of secular interests and values, hold that one may as well take "sides" and orient research and its reporting so that his side benefits. The neutralist and partisan positions exhaust academic responses to a root dilemma in social science. Thus to suggest a procedure that differs from both is to invite ideological conflict.

Whose Side Are We on?

Social phenomena are recalcitrant, they often defy analysis. Among the consequences of this is the special burden of the social scientist in coming to a judgment that enough information has been gathered to allow for proper conclusions (Rudner 1953). Because of the complexity of its subject matter and its theoretical lack, social science has special problems in achieving the degree of objectivity or intersubjective agreement that non-social sciences can attain. But it is a mistake for the postacademic to abandon the search for objectivity in favor of partisanship or disguise its elusiveness by keeping values implicit and feigning neutrality. Neutrality and partisanship must be superseded by an even more basic commitment to humanism.

In his address, "Whose Side Are We on?" Howard Becker (1967) stated the rudimentary case against neutral social science. According to Becker, because questions of traditional interest to social scientists inevitably involve conflict among group-related definitions of the situation — ruler/ruled, jailer/convict, manager/worker — the social scientist is constrained to take a side at the point of deciding or failing to decide regarding "enough" data. "There is no position from which sociological research can be done that is not biased in one way or another" (Becker 1967:116). Social science may shift its focus from traditionally interesting questions (this point is taken up in the following chapter, see also P. Park 1969), but within its established substantive concerns the only decision that must be made vis-à-vis neutrality is pointedly posed in the title of Becker's address.

In an extension of this argument, Becker (1971), his critics (Gouldner 1968; Riley 1971), and other commentators (Horowitz 1971) have stressed that one must not "regard partisanship as incompatible with objectivity" (Gouldner 1968:54; see also Becker's "Reply" and Riley's "Comments" in Riley 1974). The special problem of objectivity in social science arises in the relationship between its theoretical poverty, the complexity and dynamic character of its subject matter, and criteria employed in drawing (even provisional) conclusions. The partisan's critique of neutrality points to the relationship between the fact of social stratification and the act of making sufficiency judgments: that we will say "enough" when it serves either the interests of the ruler or those of the ruled, the poor or the rich, etc. But unless neutralists and partisans can agree about the character of stratification and its bearing on knowledge, the debate will never be resolved.[4]

By seeing the imperatives of social stratification as an impediment but not a substitute for objectivity, one can maintain commitments to other societal values over and above secular group interest, values toward which the postacademic social scientist cannot afford to be neutral. In a formal sense, the social scientist must take a side. But if this means that his conclusions invariably support a particular group, organization, party, or profession, he will be less committed to reaching scientifically based conclusions than one who interprets "side" to mean commitment to development and humanist values such as life, liberty, and the pursuit of happiness for all. Since one can never know when "enough" information exists in social science, the social scientist who takes a side by responding to group, class, or organizational loyalties is allowing that group to determine canons of sufficiency. For some, the decision may be conscious, akin to a "noble" lie or, in this case, the drawing of a "noble unwarranted conclusion."

The Noble Unwarranted Conclusion

It is difficult to come to conclusions on the nature of class, crime, inequality, exploitation, social order, or other aspects of human relations upon which social scientists have traditionally focused. Affiliation with a partisan group or similar attachment to canons of neutrality can be of service in lieu of theory and in the face of disorienting complexity. If a social scientist can assume that broadcasting a particular conclusion will have desirable or undesirable effects on his group, or that it will have no such effects on any group, he will at least be guided in his sufficiency judgments. For example a social scientist attempting to determine whether social inequality in a community has increased during a particular period may find that nothing he can say will be unqualified; that inequality and its dynamics are multidimensional phenomena and that he has asked a very difficult question. But if he believes that a conclusion to the effect that conditions have changed in a certain way — that inequality has increased — when made public or used in some other way will help his group, say in raising class consciousness, or that it will be "free" of any such bias, then he has some ground upon which to decide that his research has gone far enough.

To the extent that moral sentiments are attached to group interests, that some groups are good and others evil, the scientist even has an ethical basis for his decision. Even if his conclusions are unwarranted by scientific canons, he is right in using them to promote moral principles and utilitarian aims. Since Weber identified this problem of *Wertrational* without resolving it, sociologists have debated and, more often, quietly applied the principle that since one must conclude research at some logically arbitrary point, one may as well err: for some, on the side of "neutrality"; for others, on the side of their own favorite groups and the principles for which they stand. If the proper objects of social science research have already been more or less delimited in the traditional way mentioned, it appears equally important for those who favor the powerful, or the powerless, or neutrality to explicate and then distance themselves from their group-based biases when coming to social-scientific conclusions. Those who view affiliation with secular groups or with the ideal of neutrality as a substitute for value explicitness are disinclined to pursue objectivity; for them, wishes and fears replace understanding. In some cases, like that outlined by Becker, it is a question of intention; in others, like that of the conscious user of noble unwanted conclusions, it is a question of expediency. In any case, unscrutinized neutrality or partisanship is translated into a fatalistic course that can lead the social scientist further from the truth. Eventually this course can even do injury to the group with which the social scientist's interests lie.

The researcher who assumes that broadcasting a particular finding about social inequality, for example, will benefit his group might be mistaken. How can he know beforehand and avoid such unfortunate incidents, how can he determine the effects of broadcasting research findings on the welfare of a given group? How can he know, for instance, that the announcement of increased inequality will help the poor or the powerless? If social reality is too complex, in the absence of theory, to allow even approximate answers about changes in social equality in a community, it must also be too complex to allow definitive findings on the effects of broadcasting such conclusions.

As Peter Berger (1964) has noted, a spy must give accurate information to his side, regardless of what his side might like to hear. Even the most committed partisan must seek to know the world as it is, if he is to have a chance of influencing it rationally. If we cannot face and criticize our own secular commitments as social scientists, we are deluded into the belief that group interests and creeds are substitutes for scientific validation. In *Man's Fate* Andre Malraux stresses this distinction: One is a Marxist in order to conquer, not in order to be right. We are all Marxists, or conservatives, or pluralists to achieve political ends. The use of propaganda, advertising, persuasion, and other manipulative applications of knowledge have a place relative to these ends. But these are all different from social science, which is knowledge for transpolitical ends. We are social scientists to be right about improving man's lot.

Commitment and Self-Consciousness

"The physician, after all, is not necessarily less objective because he has made a partisan commitment with the patient and against the germ"

(Gouldner 1968:54). Similarly, it is not sensible to doubt findings about social relations simply because the researcher is committed to changing or maintaining these relations. But when the social scientist decides to say "enough" because he believes that at that point the conclusion, when broadcast, will serve the cause (including that of "neutrality") to which he is dedicated, he deceives himself. Certainty in this case is as misconstrued as in the case of a physician who believes that announcing a cure is tantamount to applying it. Knowledge about the effects on human behavior of the announcing of scientific findings is limited (Weinstein 1976, ch.1), since the production of such knowledge depends upon the state of social science, whose very weaknesses appear to justify the noble lie. What sort of guide to the "enough" judgment shall we employ in answering scientific questions about how, for example, the committed social scientist can raise class consciousness or achieve other aims? Here the "noble unwanted conclusion" entails an infinite regress.

The simple and abused position of self-conscious humanism is the most reliable guide to sufficiency judgments we have. We will never, as scientists, have that "conclusive" piece of evidence; and this is most emphatically so for social scientists. But the scientific method guided by humanist commitment and a critical awareness of how our biases distort our approach can serve as an appropriate guide to truth about the world.[5] The "unreflexive" social scientist, whether he is a partisan or neutralist, who cannot see the necessity of articulating his biases or pretends that they do not enter the social research act would do well to reexamine the traditional (and humanist) foundations of liberalism, conservatism, and radicalism.

Ideology and Postacademic Commitment

As an applied intellectual orientation, postacademic social science cannot afford to have no ideological definition. At the risk of becoming ineffective, some measure of intolerance — beyond that stemming from partisanship and personal interests — is necessary. However valuable in terms of its general cultural contributions, a social science that is neutral about humanist development, unprepared to explicate and criticize the practitioner's biases, unconcerned about problems of technology and freedom, and unable to relate to interdisciplinary activity is too costly for the postacademic to underwrite. In simple economic terms alone, resources invested in grand theory, Marxology, abstracted empiricism, or any other form of partisanship or technical virtuosity would always be better invested elsewhere — in light of the goal of orienting technology to human ends.

Interdisciplinarity is one such target for wiser investment. It is here that tolerance in its most vigorous application should reign. In the type of conflict that arises in the clash of interpretations generated in interdisciplinary research, teaching, and application, the canons of pluralism can be most faithfully and effectively applied. Another thing that should be encouraged is social knowledge for development. As an "impure" social science field, developmentalism is premised on the view that the pursuit of social knowledge with no immediate purpose other than accumulation per se or the display of technical virtuosity is epistemologically and morally questionable, and that some purposes such as development are better than others.

Having a purpose, even a developmentalist purpose, cannot stand alone as a guide to problem choice, selection of theories and techniques, and other critical decisions the social scientist must make. Purposefulness still requires a larger context for expression. But in the postacademic social scientist's activities, purposefulness and developmentalism are expressed through and, ideally, supported by other principles: a focus on technology — which by its very nature is relevant to development (Solomon 1977) — interdisciplinary pluralism, and value explicitness.

Each of the components of the postacademic orientation — developmentalism, interdisciplinarity, etc. — has an ideological "loading." Promotion of any is bound to meet resistance in traditional academic and technology-policy settings. Together, they constitute an integrated program that is subversive to the academic ideals of sociology and related disciplines. In the societal and professional contexts in which postacademic social science will be pursued and promoted, commitment to this program and its components is a political act. The political and ideological character of the program arises not because it promotes the interests of a particular class — it seeks to serve the human interest — nor because it is conservative, liberal, or radical. It arises because the object of study and application, technology, is so thoroughly politicized that to attempt to understand it social-scientifically and orient it to human ends is to become political "by conduction."

Since we shall return to this theme one more time, suffice here to refer to the question that opened the preceding chapter. Is a particular ideological position associated with postacademic social science? If this question is meant prescriptively, the answer is that it is inappropriate to establish that the field "should" be ideological in this or any other way. Such matters cannot be legislated. If the question is meant descriptively, there is no conclusive evidence that postacademic social scientists share a given ideology. But if the question is meant in a sense that is neither prescriptive nor descriptive, the answer is affirmative: the intellectual heritage and activities relating most closely to postacademic social science do orient it toward the applied study of technology, value explicitness, etc., which is intensely political and ideological.

Postacademic social science is a search for knowledge with a humane purpose, without regard to disciplinary or secular interests. It is not the only or necessarily the best way to conduct social inquiry, nor is adherence to the four components of its ideology sufficient to ensure that the practitioner abide by the other, and equally necessary, norms of social-scientific procedure. The pursuit of social knowledge without regard for these four components is neither less authentic nor necessarily immoral; nor is such a pursuit necessarily valueless to postacademic social science. It simply is no longer appropriate as the main preoccupation of social science. Even with this limited claim understood, postacademic social science is likely to be either offensive to those within and outside of academe whose interests are vested in academic sociology, or ineffective in relating to problems at hand: to understand and reorient technology as knowledge for development.

Notes

1. While a content analysis of appropriate samples of sociological literature has not been performed, it is safe to assume that such an analysis (especially if it excluded the exceptions noted) would reveal a relative neglect of technology. Rau's (1979) review of some of the work dealing with the subject (at least in name) shows, if not total neglect, a lack of willingness to confront technology with the degree of analytical seriousness exercised in other areas.
2. One important exception to the neglect of technology as a variable is in the concept of the "ecological complex" — population, organization, environment, and technology (Duncan 1959). Use of the concept has tended to be overly mechanistic (Willhelm 1964), and even then, the part played by technology in empirical research on the ecological complex has been minimal (Frisbie and Clarke 1979).
3. Innovation pursued purely in light of technical rationality is most susceptible to abuse by larger social control agencies, including the state. In failing to treat technology as a social variable, freedom denied on an intellectual plane can become freedom denied in political relations.
4. From its earliest days, academic sociology was fated to treat objectivity not, as in other scientific fields, as something taken for granted or at least as a desirable goal, but as something to be studied and debated — also from a sociological perspective. The tendency to focus on the most elemental bivariate level, which can serve the non-social scientist in the discovery of truth, may be quite independent of commitment to social justice, progress, or some other general ideal. But this same narrowing — and the accompany-

ing preoccupation with instruments that provide an increasingly precise focus — is so out of touch with the nature of social reality and the state of social theory that its support in social science is ethically partisan — it is a denial of moral purpose.

5. Discussion of the appropriate techniques for being self-conscious about one's values and incorporating the knowledge gained in such exercises into research activity can be found in the methodological literature on technology assessment cited, as well as in more traditional social science sources. Techniques of value explication have received the greatest attention in connection with policy-oriented work. Duncan MacRae's (1976) discussion of this and several related issues of interest is based upon his extensive experience in policy and applied technology. On value explicitness, he suggests three general guidelines:

- Each of the ethical systems at issue in a given evaluative discourse among social scientists shall be "specified in writing in advance" (MacRae 1976:92).

- Opponents will have equal opportunity to criticize one another's ethical system in terms of (a) the generality of hypotheses in the system; (b) internal consistency of the system; and (c) the consistency of hypotheses with "shared moral contributions."

- Each of his opponents shall have a right "to alter his ethical system" (MacRae 1976:93; Firey 1978:277).

17.

Subjectivism Versus Natural Science

The heterogeneity of disciplines such as sociology is evident not only in the variety of ideological positions from which it can be pursued, but also in terms of several other internal differences, some of which have been mentioned earlier. This chapter discusses a major one of these differences, focused on the subjectivism versus natural science debate. The thesis of this discussion is that while the debate has brought to light significant problems that attend the scientific study of social relations, it also illustrates the futility of most academic disputes. Though a sense of urgency has often been associated with both sides of the debate, its bearing on the aims and priorities of postacademic social science is minimal. This typifies the ways in which the concerns of academic disciplines are no longer as pressing or even as relevant as they once seemed.

The problem posed in the subjectivism/natural science debate is an ancient one, with roots in Greek philosophy and even more remote sources. Its first modern expression can be traced to the work of the Italian philosopher Giovanni Vico (1668–1744). Vico's argument is contained in his best known work, *The New Science* (1961), which, as much as any other book of its time, is a foundation document for modern social science. In it Vico draws a sharp distinction between the objects of the natural and of the historical (i.e., social) sciences. Because man does not create nature, he argues, the objects of natural science are comprehensible only in certain circumscribed, limited, and external ways. But the "world of nations," upon which the historical sciences focus, is built and maintained by men; thus in attempting to understand the motives and meanings of people as they create their realities, these sciences can hope to know what they study in a far more intimate and adequate way. The historical sciences have a different goal from the natural sciences, that must be pursued with a different method. The historian should, according to Vico, seek to understand men's actions not as mechanical responses to stimuli but in light of the subjective purposes these actions are meant to serve.

The Debate in Academic Social Science

This position entered academic social science largely through the writings of the German social philosopher Willhelm Dilthey (1833-1911) and commentary on Dilthey by Max Weber. Dilthey, a leading critic of Comte and other proponents of a positivist, natural science model for sociology, defined the social sciences as those studies whose "object becomes accessible to us through the attitude which is founded on the relation between life, expression, and understanding [*Verstehen*]" (1974:17). For him, this "relation" has no counterpart in the realm of physical and organic nature: only man is capable not merely of existing but also of incorporating meaning into his acts and products. To understand this meaningfulness in human society, the methods of natural science are of no benefit, for they cannot achieve, nor are they intended to achieve, what Dilthey calls a "reliving" of the events they explain. Natural science can produce only an external, and thus a deficient, understanding of man.

Based upon this distinction, Dilthey argues for a separate method in social science, one more like that of classic hermeneutics — the attempt to discern deeper levels of meaning in scriptural writings, as in the study of the *Zohar* and other cabalistic works — than of physics (Rickman 1961). The type of understanding of which Dilthey speaks requires an empathic, even transitive (in the psychoanalytic sense) relationship between the scientist and the object of his research: "It is through the process of understanding [*Verstehen*] that life in its depths is made clear to itself, and on the other hand, we understand ourselves and others only when we transfer our own lived experiences into every kind of expression of our own and other people's lives." With regard to studies that focus on man and his social relations, "all the leading concepts with which this group of studies operates are different from the corresponding ones in the field of natural science" (Dilthey 1974:77).

Since Dilthey's time, this argument for a separate method has found expression in several forums within academic social science: in Max Weber's discussion of methodology, in anthropological literature on ethnographic technique (Martin 1968), in the widely cited Abel/Wax debate of the late 1940s and early 1950s, in Peter Winch's *The Idea of a Social Science*, in the programmatic statements of phenomenological (Bien 1978) and ethnomethodological sociology, and most recently in sociology-as-poetics. While there is much that separates these various positions in the degree to which they subscribe to Dilthey's (and each other's) views, they all share an interest in exploring the extent to which positivist, natural science models of research procedure apply to the social sciences (Bauman 1980).

Much time could be spent in reviewing and critically assessing the various classic and recent positions in the subjectivism/natural science debate.[1] Our interest here is more limited; we wish to pose and attempt to answer the following questions: Is there a position that has emerged from this debate which is of particular relevance for postacademic social science? Is either the natural science or the subjectivist model of research method especially appropriate for the type of interdisciplinary research, teaching, and related activities characterized in the preceding chapters? That this distinction is important and even necessary in contemporary academic sociology (though perhaps less important today than in the early 1970s) is evident in the differences in substantive research approaches of pairs of leading sociologists such as Erving Goffman and Otis Dudley Duncan, or Aaron Cicourel and Hubert Blalock, or Peter Berger and Norval Glenn, in commentary on sociology's major factions in *The American Sociologist,* and in books on the sociology of sociology (especially Mullins 1973).

A review of this debate leaves room for serious doubt about its value for postacademic social science. In the preceding analysis of ideological perspectives it was concluded that while postacademic social science does have a political orientation, it is neither pluralist, radical, nor conservative in the usual sense of these terms. In a similar way, the choice between subjectivism and natural science is false and unnecessary outside of academic settings.

Subjectivism As a Critique of Positivism

The primacy of classic logical positivism, whose philosophical predecessor was Dilthey's original target, has been challenged on many fronts by philosophers, historians of science, and social theorists. A partial consensus exists in these fields and related studies such as linguistics that the positivist-operationalist models founded in the classic work of Russell and Whitehead (1925), Ludwig Wittgenstein's earlier writings (1923), those of George Bridgeman, George Hempel, and others, offer inadequate prescriptions and descriptions of science as process (the act of doing science as opposed to the body of knowledge accumulated in performing this act; see Rudner 1966, ch. 1). In terms of the sociology of knowledge, descriptions of positivist philosophies of science distort or are naive relative to the complex behavior and thought that constitute "doing science." The prescriptions of these philosophies instruct the scientist to behave in ways that, though perhaps necessary for certain extrascientific purposes, are inadequate for dealing with the choices that must be made in his professional role.

Some of these critiques apply to the descriptive and/or normative philosophy of any type of science, while others have special bearing on the philosophy of social science. The first category includes the later work of Wittgenstein (1953), whose argument against the "picture theory" of language inherent in positivism continues to have profound impact in academic philosophy and on Western culture. For him and for the ordinary language school that emerged around his ideas, the very noun-verb/object-event formulation on which classic positivism is based misconstrues the nature of the relationship between "descriptive" language and the reality to which it is applied. The second category, critiques referring to social science in particular, includes the works of Alfred Schutz, Claude Levi-Strauss — whose structuralist approach represents a lively alternative to positivism in European and, increasingly, in U.S. academic social science — Harold Garfinkel, and their students (see Coulter 1971 for a review of these positions). Proponents of these positions argue that the positivist stress on objectivity — intersubjectivity — and hypothesis testing is inappropriate and overly simplistic in light of the complexities of the social world and the complex ways in which knowledge about this world is attained.[2]

A set of works applying Wittgenstein's observations to social science problems combines characteristics of both critiques. Included among these is the argument by the contemporary English philosopher Peter Winch. Winch's *The Idea of a Social Science* (1958) contains arguments from Wittgenstein's *Philosophical Investigations* as well as Winch's interpretations of Weber's views on subjective explanation. He develops a model for social disciplines which, unlike positivist sociology, does not "decontextualize" social knowledge by attempting to express it in general theories and propositions. It treats social relations from within their historical and psychological contexts. Its purpose is not to describe and explain — in the usual scientific sense — the "game" of social life, but rather to grasp and convey the meaning and experience of what it is like "to play" this game. Like Dilthey, Winch (1958:135) holds that a separate method is required for social science: "Whereas in natural science it is your theoretical knowledge which enables you to explain occurrences you have not previously met, a knowledge of logical theory, on the other hand, will not enable you to understand a piece of reasoning in an unkown language; you will have to learn that language, and that in itself *may* suffice to enable you to grasp the connections between the various parts of arguments in that language."

From this perspective, the formal, decontextualized types of language appropriate to the natural sciences may be adequate for understanding their objects, since these are not language users themselves. For the social sciences, such languages are inadequate in comparison to those spoken and understood by their objects, human beings. Natural or ordinary languages, their rules, and their other verbal and nonverbal aspects provide social actors with effective means to organize their worlds. They permit users to perceive and act, to know their environments and other people well enough to survive and even prosper. But these rules and vocabularies of ordinary language are remote from rules (laws, generalizations, propositions) used in empirical science. The meaning of words in ordinary language is much more complex than in formal languages (in Wittgenstein's terms, the former have a "family" relationship as opposed to a one-to-one correspondence between symbol and object). Thus in order to discover social knowledge, the social scientist must be prepared to abandon the externalist perspective and the decontextualized language of the natural sciences and enter society and learn the rules that actors address (and create) as they behave in orderly ways.

A Strange Idea of a Social Science

While the point of these remarks is not to criticize but to evaluate the usefulness of arguments such as Winch's for postacademic social science, one observation will be made before continuing this review. It relates to Winch's interpretation of Weber. Winch (like many others) believes that one source of the separatist position in sociology is Max Weber's commentary on meaningful action. As he points out, Weber was insistent that human *action* derives its meaning from "within" and that it cannot be adequately understood as entirely (externally) caused *behavior*. Winch invokes Weber on this issue to dispute the views of the positivist philosophers Karl Popper and Morris Ginsberg. For Winch (1958:45) the categories of social sciences are those "of which we can sensibly say that they have a *meaning,* a *symbolic* character. In the words of Max Weber, we are concerned with human behavior if and insofar as the agent or agents associate a subjective sense *(sinn)* with it."

The argument has been made (by Popper and Ginsberg, for example) that despite this stress, Weber was not a proponent of a separate method. According to this interpretation, *Verstehen* was meant by Weber to be used as a supplement to the scientific method, not a substitute for it. In particular, an orientation to the symbolic and meaningful character of social life is intended to aid in the formulation of hypotheses, not in their testing where the method of natural science remains applicable even in the study of social relations. Subjective understanding applies to the context of discovery where hypotheses get posed, as opposed to, as Winch argues, the context of validation where hypotheses are tested (Rudner 1966, ch. 2). Winch (1958:112) is aware of this interpretation, but believes that because "Weber never gives a clear account of the logical character of the subjective approach" or *Verstehen,* many writers have been led to "allege that Weber confuses what is simply a technique for framing hypotheses with the logical character of the evidence for such hypotheses." Winch (1958:113) wants "to question Weber's implied suggestion that *Verstehen* is something which is logically incomplete and needs supplementing by a different method altogether, namely, the collection of statistics."[3]

Winch is right in observing that Weber was neither as clear nor as consistent as he might have been in discussions of *Verstehen.* Weber often used *Verstehen* as a vehicle to smuggle his strong ideological views into arguments intended to be purely scientific. One well-known example is his demonstration of the "elective affinity" between the Protestant ethic — as Weber understood it — and the spirit of capitalism — as Weber understood it. But Weber was clear enough about the role he prescribed for *Verstehen.* His commentary in the collection entitled *The Methodology of the Social Sciences* (Weber 1949) and elsewhere is explicitly mounted against Dilthey and Rickert's separatism (Truzzi 1974:18-19). In his programmatic views as a whole, Weber was committed to natural science models, at least in the context of validation. On the specific issue of the role of subjective understanding, Weber's statement leaves one wondering what Winch had in mind when he mentioned Weber's "confusion" between hypothesis formulation and testing, or that Weber gave only an "implied suggestion" that *Verstehen* must be supplemented by statistics. Weber (1974:26-27) held that

> every interpretation strives to achieve utmost verifiability. But even the most verifiable interpretation cannot claim the character of being casually valid. It will remain only a particularly plausible hypothesis. Thus, what appears [with our subjective understanding] to be a conscious motivation to the individual involved may only serve to hide the deeper lying motives. . . . For example, the proposition known as Gresham's Law is a rational interpretation of human conduct within a given context and on the basis of an ideal hypothesis of a purely rational course. To what extent such behavior really follows Gresham's Law can only be ascertained on the basis of statistical information.

An Aesthetic Response to Natural Science

Many critiques of positivism put forth by Winch, Wittgenstein, Lévi-Strauss, and Garfinkel have recently been drawn together in a new movement for a separate method which views social-scientific knowledge as poetry (Brown 1977). This sociology-as-poetics approach consists of observations on the great use made of irony, metaphor, point of view, and other literary devices in social theory. Proponents of this approach criticize, both on stylistic grounds and for more substantive ideological reasons, overly prosaic and methodologically transfixed types of social thought. How many people, they argue, including academic sociologists, would choose the collected works of Herbert M. Blalock over the collected works of Shakespeare if allowed to select only one book for companionship on the proverbial desert isle? In this approach, a sociological tradition is identified in which literary, dramatic, and aesthetic insight are given due respect. But like others who argue the subjectivist position — from Dilthey to Winch, proponents also seriously err in attributing excesses like scientism, obscurantism, and outright opportunism (to which academic social science is clearly prone) to a fault in the scientific method itself.

At many points in the poetics approach, observations on social theory are sound, but it also overestimates the extent to which a call for a poetic for sociology damages the natural-science position. The approach proposes to outline an "aesthetic" as a form of subjective understanding to take the place of the theory of knowledge upon which positivist science is based. What is suggested is not a substitute for natural science; the approach offers correctives and guidelines for formulating hypotheses. In the absence of a well-established and widely agreed upon body of theory, such correctives and guidelines may be most valuable. But even their assiduous application hardly constitutes an alternative to the (prosaic) process of comparing hypothetical expectations — even if formulated with due aesthetic sensitivity — to experience and combining well-warranted hypotheses to form theories.

Very little attention is paid, in either this recent variant of the subjectivist position or in more classic versions of it, to science as a social activity. Like the philosophies it seeks to invalidate, the poetic approach seriously underestimates how much is involved in science-as-process even beyond poetics, the scientific method, or other criteria for the generation of knowledge. In this approach, the focus is exclusively on the end-product of scientific activities and on the general ways in which social science as product (the ideas produced by social scientists and their representations in speech and writing) can be distorted. As a result this new approach falls prey to an excessive and obsolete type of idealism.

Idealism and Separatism

While for Dilthey, Immanuel Kant "is a pathetic rationalist" (Dahrendorf 1967:148), the philosophical basis for social science of which the poetics approach speaks, draws heavily on Kant's analysis of aesthetic judgment. According to this analysis, in order to be comprehensible, observations about social relations must not only be true but also clear, economical, and in accord with other principles of aesthetic judgment. In this way, it is argued, aesthetics guides scientific understanding. The social scientist's observations must eventually be communicated or at least formulated. To this end ideas must be expressed in words or other symbolic systems. But the principles that guide verbalization, communication, and the use of symbols generally are those of literature, drama, music, and other arts rather than those of science and logic.

Even if we grant a master role to aesthetics, the implication drawn by the poetics approach that art and science are one is unwarranted. It may be true that a certain type of creative intellect is required of all (good) scientists, as well as moralists and artists; but it does not necessarily follow that science, moralizing, and art are not still distinct creative intellectual activities. This is an indiscriminate blurring of distinctions between types of activity whose purposes differ significantly. It is little more than an overreaction to the equally unreflective argument of positivist philosophers of social science that one must keep morals, science, and aesthetics from scientific research. Such overreaction and such heavy reliance on Kantian aesthetics forces the new separatists to defend an idealist philosophy that is as stilted as the classic positivism it attacks.

Kant believed that concepts like "fitness to perception" and "autotelic cogency" — characteristics of objects which "allow them to be understood" — were more than mere metaphors. He developed an ontology in which scientific understanding was so powerful that it corresponded (a priori) to an ordered but unknowable noumenal world. Kantian aesthetics apply only in this kind of dual reality: an underlying but incomprehensible substratum and the world of phenomena man knows. More recent philosophers, especially pragmatists on whose work, since the days of the Chicago School, contemporary social science is based, do not agree with Kant. Most contemporary social scientists do not accept the noumenal level as part of their considerations since by definition it can never be observed or studied.

Like Dilthey and others who have seen a close kinship between history and hermeneutics, the new separatists support an elaborate set of criteria for "adequate" social theory. In distinction to the positivist position, these criteria are aesthetic as well as semantic and syntactic. Consequently it can easily be shown that few if any "adequate" theories have ever been produced. This type of adequacy can be predicated to theories only in a Kantian world wherein things like "fitness to perception" are at stake or a Diltheanian world in which "reliving" is a possibility. Since Comte's law of three stages, theories have been treated simply as linguistic entities: sentences believed to be true that have a logical relationship to one another. But separatists both old and new seem compelled to require extralinguistic standards by which the adequacy of theories is to be judged — that they be beautiful as well as true. The argument that an aesthetic philosophy of social science is required rests on a view of the relationship between scientific language and reality that results in a proliferation of muddles in the name of integration between social science and poetry.

Unmediated Effects and Thinkerless Thoughts

The poetics approach suggests that ideas and theories can have unmediated effects on social life (Brown 1977:229ff). Ideas and other symbolic constructs can be very powerful tools; they act as factors in shaping social relations. But neither they nor any other tools are powerful enough to use by themselves. In such instances and in other contexts, the new separatists appear genuinely confused (rather than merely reflective) about the relationship between people's actions and their ideas before, during, and after they act. We find in the work of these new separatists an almost mystical quality: "Unreflective thought imagines itself to be subjectivity at work on its object. Yet, by raising thought to a higher power through an aesthetic of self-consciousness, this initial subject-object dualism may be dissolved" (Brown 1977:23).

One century after Dilthey, thinkerless thoughts, subjectless verbs, and actorless social science continue to pervade the subjectivist program. In the absence of a sociological analysis of the norms, roles, behaviors, cognitive states, political factors, and economic constraints that inform social science, the approach glosses the complexity of scientific activity. It proposes to achieve a goal it cannot achieve: to make sociology and related disciplines nonscientific fields. Sociologists could and should be metaphoric, ironic, reflective, and more poetic about the people and behavior they seek to understand. Sociology suffers when styles like abstracted empiricism can dominate the field without their practitioners having to take account of such intensely human activities as intent, anticipation, and the creation of their own realities. But it remains to be demonstrated that all of this, and the many other things that social scientists can do to be more effective, cannot be done with due respect for scientific method.

The poetics approach appears to be mystified by its own depiction of models, theories, sentences, and even science itself as metaphors. Proponents admit that even highly "elaborated" metaphors can be shown to be false, but they fail to reflect on their own root metaphor, "poetics." It is one thing to be sensitive to poetics when formulating hypotheses and quite another to substitute aesthetics for logic when seeking to establish whether hypotheses are true. When the separatists suggest the second option, they overextend their metaphor. As a collection of useful and often insightful observations on social thought, the poetics approach contains much truth and beauty; but as a program for transforming social science, it fails from the outset. Proponents should be credited for their insistence that social scientists must have a sense of the poetic and that they pay due attention to the meaningfulness of human action. But the position suffers seriously from a lack of receptivity to a sociology of science and a lack of reflectiveness about the magnitude of the task of overthrowing scientific sociology.

Hermeneuticism As Technical Virtuosity

From the time of Dilthey to our own, a period that spans the era of academic social science, critics of positivism have sought to replace the scientific method with hermeneutics, *Verstehen,* poetics, inner awareness, or some other method felt to be more faithful to the nature of man and more likely to produce a "deeper" level of understanding of human action. Within sociology itself, this program has been developed self-consciously as a counterforce to the prevailing abstracted empiricist school (Mullins 1973; Coulter 1971). In its phenomenological and ethnomethodological variants (Cicourel 1964) and as a "poetic for sociology" of the new separatists, it has posed an important challenge to sociological positivism — to pursue a homocentric social science. Whatever its weakness, methodological separatism (in sociology at least) is based on correct assessment of the field: the ahistorical, atemporal, methodological world of long-linear modeling, causal path analysis, etc., is wanting in several crucial ways in terms of the world of human relations to which such models are applied.

From the perspective of postacademic social science, subjectivist programs overreact to what are not faults in the scientific method but rather exaggerations on the part of positivists about the relationship between scientific method and social knowledge. Subjectivism is as excessive in its claims as positivism. Because it fails to grasp the nub of the issue, it is equally prone to the chief foible of academic sociology — technical virtuosity. Once the battle for legitimacy for academic sociology is resolved, scientific method will be of use in answering only a very small portion of the questions a social scientist must answer. Critical decisions related to problem choice, the "enough" judgment, and other issues both in the context of discovery and validation are more difficult than those faced by scientists whose subject matter is less complex and whose theoretical repertoire is richer. The social scientist must be a scientist, moralist, and aesthetician all at once. The scientific method cannot help us render better moral or aesthetic judgments, nor is it meant to perform these roles.

Those who have used the arguments of classic positivism and the methods of abstracted empiricism in battles for personal or disciplinary legitimacy have left the strong impression that since the purposes of scientific method are limited, social science must shape its purposes accordingly; it must exclude as unscientific a concern with morality and aesthetics. Some of these people have used their power and influence over the means of certification, reward, and legitimation to exclude any but the narrowest sociological charicature of "real" scientific research and writing. Because of their influence, displays of certain logical and quantitative techniques have become the dominant criterion for judgment, and with the capture of the discipline by the profession, the dominant type of sociology practiced in academe.

At the same time, promotion of strict positivist models has given impetus to its long-time foe, subjectivism, in new and more elaborate forms. In the (usually implicit) claim that morality and aesthetics have no place in social science, positivists attempt to legislate away something that is inherent in the enterprise. They thereby offend those who believe otherwise and leave themselves open to the charge of hypocrisy (since positivist social scientists do make moral an aesthetic judgments anyway). In response, subjectivists — tolerated in liberal professional circles — attempt to legislate away the scientific method and with it the scientific part of the sociological enterprise. This leaves the subjectivists also open to charges of hypocrisy (since they use the scientific method anyway), and leaves them vulnerable to the critique of positivists, rationalists, and pluralists.

Positivism presumes that the moral and aesthetic needs of social science can be "satisfied" — ignored — by attending only to descriptive and explanatory issues. This is met by the complementary claim that descriptive and explanatory needs can only be "satisfied" — again, ignored — by subjective understanding, by a "poetic." "The 'Sciences' that emerge from [the latter] methodological preferences may be poetic and pleasing, profound and thought-provoking, but they are lacking in definition and reality, that is, they are lacking in necessity" (Dahrendorf 1967:148). The result in this case, as with abstracted empiricism (and for ironically similar reasons), is that the criteria used for certification, reward, and legitimation center not on how much truth one reveals but on technical virtuosity. Since scientific canons such as the ability to reproduce findings and the formulation of hypotheses that can be proved false are irrelevant to subjectivists, specialized expertise and, at the extreme, masterful triviality are of special importance. Subjectivism "allows the individual scholar to fence himself in and thus further the subdivision of the world into innumberable little front gardens, one of which is allotted to each scholar with his Ph.D. subject, to put up his garden dwarfs in, after he has made sure in the proper registry of the history of science that nobody else could possibly advance a claim for his particular plot" (Dahrendorf 1967:149).

As Dahrendorf goes on to note, "it is no accident that this appearance reminds one of ideologies of political conflict. The two notions of [positivist and subjectivist] science. . .have political consequences as well as political implications" (150). To be effective, social science must take account of the subjectivity of actors, be alert to the moral aspects of social action, be sensitive to poetics, and be capable of applying the scientific method where appropriate. But in light of the exclusivism, the categorical character, and the extravagant claims of both positivism and subjectivism, they are polemics. The two positions need and feed off each other as do any pair of ideologies employed in the struggle for power and legitimacy. At stake is the power to determine what shall and shall not count as "authentic" sociology. Lost in this exchange is any sense of purpose for social science other than academic: perpetuation of knowledge for its own sake, personal aggrandizement, development of expertise, and outright vanity. The debate remains — appropriately — "academic."

Method in Postacademic Social Science

The method of natural science has an important role to play in postacademic social science; so too do the critiques of subjectivists. But neither can be treated as ends in themselves. By virtue of its commitment to interdisciplinarity, development, value explicitness, and to an applied focus on technology, postacademic social science has the means to correct the narrowness and lack of homocentricity of abstracted empiricism and to counter other exaggerations of the scope of natural science. Yet it need not dispense with scientific procedure, but can remain faithful to the rational, public, and pluralist character of natural science. Positivism is mute and separatism is shrill and excessive about the special problems (and opportunities) of a science about people conducted by people. The postacademic orientation seeks to humanize technology and its impact and thereby put to the test any approach or understanding derived in social science.

The final judgment concerning whether the scientific method or some other is the best for sociology need no longer be sought in academic forums, where it can never be resolved intellectually but only politically. No test exists, beyond debate, concerning the exercise of either type of virtuosity. In postacademic activities, the scientific method, poetics, *Verstehen,* and anything else that works can be used and evaluated in proper perspective. These are not ends in themselves nor means to equally elusive ends such as "improving the knowledge base"; they are to be judged insofar as they contribute to the task at hand: solving problems in the design, innovation, and diffusion of technologies promoting development. From this perspective, it is unlikely that this final judgment, if ever rendered, will favor either strict positivism or strict subjectivism. Social life is too complex to be apprehended, in a manner relevant to the pursuit of postacademic social science, by any single road to knowledge.

Notes

1. Dilthey's views have been laid to rest, only to be revived many times. See Holborn (1950), Masur (1952), and Rickman (1960); also cited in Truzzi (1974:9). English translation of Dilthey's writings can be found in Rickman (1961) and Hodges (1944, 1952).
2. Here one can place the Abel/Wax debate in context. Wax argued that *Verstehen* is a separate and necessary operation in social science for ontological and epistemological reasons. Abel held that there is nothing in the "operation called *Verstehen*" not encompassed by the positivist model. See Abel (1948, 1967), Munch (1957), and Wax (1967). The resolution of this and related arguments on this issue depends on how rigorously one wishes to separate the context of discovery (hypothesis formulation) from that of validation (hypothesis testing). This is discussed further in the text.
3. Something of the character of Winch's (1958:118) ideological program can be discerned in the invidious distinction he makes between Weber's "generally correct" approach and those portions of Weber's work where he too (implicitly) affects distance from and "imposes" categories on social action. Here Weber is accused — along with several others — of playing God: "What is dangerous is that the user of these [distancing] devices should come to think of his way of looking at things as somehow more real than the usual way. One suspects that Brecht may sometimes have adopted this God-like attitude (as would be consistent with his Marxism); it is certainly involved in Pareto's treatment of "residues"; and although it is an attitude which is, on the whole, very uncharacteristic of Weber, it nevertheless follows very naturally from his methodological account."

18.

Technology, Society, and Social Science

The very idea of a non-repressive civilization. . .appears frivolous. Even if one admits this possibility on theoretical grounds, as an extension of the achievements of science and technology, one must be aware of the fact that these same achievements are being used to the contrary, namely to serve the interests of continued domination. The modes of domination have changed: they have become increasingly technological.

Herbert Marcuse (1955:vii)

Postacademic Social Science and Varieties of Academic Sociology

In the preceding discussion of the ideological, subjectivist, natural science, and other dimensions of sociology, the many styles and varieties that characterize the academic discipline were classified and analyzed. These have not been explored in sufficient depth or with the detail of citation that they perhaps deserve. Nor have we considered some dimensions that might be even more revealing of sociology's heterogeneity. Other distinctions, discussed by Comte, Veblen, Mills, and the Frankfurt School, are also worth investigating — including some very useful ones such as craft versus profession, Sociology versus sociology, and *sociologie* versus social physics. From the little that has been said, it may be possible to understand the bearing of sociology's heterogeneity on postacademic social science (Reynolds and Reynolds 1970).

The distinctions that underlie the division of sociology into schools, camps, and styles, arose and were nurtured in response to specific social conditions within and outside of academe. As they have performed polemical functions, these styles have also served to motivate much interesting and important debate. A large portion of sociology produced in the name of some ideological or methodological orthodoxy is of value for the postacademic social scientist. These approaches and their products deserve careful and critical study. Knowledge of the main positions and counterpositions can only help sharpen one's own critical awareness. These distinctions and the choices between schools and styles which they reveal were not meant to apply to the type of activity in which postacademic social scientists are likely to be involved. For such purposes, these distinctions and associated choices can be taken too seriously, so that one becomes caught between the horns of dilemmas that should be kept at a distance.

The various styles and schools have persisted in part because, in the face of inordinate complexity of subject matter and in the absense of widely agreed-upon bodies of theory and method, they serve as criteria for judgment in academic certification, reward, and legitimation rituals. They have helped nuture the peculiar stress on a combination of internal-political loyalty and virtuoso performance around which the profession is organized. They are creatures of academic settings and they function best in such settings. Postacademic social science has an additional public, different peers, and different purposes from those of academic sociology. Additional criteria apply to the evaluation of its work. The postacademic social scientist cannot be effective simply by exercising virtuosity with positivist or separatist methodologies, by arguing a consistent ideological line, or solely by grasping the poetic and "deeper" meaning of social relations. When all these performances have been perfected, the question still remains: What use is sociology in humanizing technology? The answer points beyond all distinctions, schools, styles, and orthodoxies yet to evolve.

Like academic fields, postacademic social science remains subject to disturbing influences from organizations that support it and other sources, and it is not immune to all the "fads and foibles" that P.A. Sorokin (1956) once catalogued. The practical orientation of postacademic social research, teaching, and related activities, and the nature of the organizations likely to support such activities also expose the practitioner to a whole new range of vulnerabilities and to a set of relatively unfamiliar ethical choices (Wax and Cassell 1979). This demands considerable caution and moral fortitude. The ideal of an independent professional social science community, free of pressure to conform to secular interests, is as valid in the postacademic era as it ever was. In the absence of institutional protection afforded by universities, with their built-in (though archaic and elitist) moral codes, the postacademic social scientist must be especially self-conscious and protective of his specific rights in relation to those of his employers.

He should not equate the achievement of such ideals as independence with what is or is imagined to be academic reality. The freedom enjoyed by academic social science was purchased at a very high price. Within the (not always subtle) political and economic constraints imposed by professionalization and the drive for legitimacy, the academic social scientist is free to be a "Yogi or a Commissar" (Koestler 1967), conservative, pluralist, or radical, a positivist or a subjectivist. But he is not free, as a social scientist, to ignore disciplinary boundaries and propriety, nor to respond to pressing moral and technological problems of our times because they are pressing, regardless of how the profession chooses to relate to them. The academic social scientist is not as free as he should be to use his craft to help in the reconstruction of the social world, a world in which nuclear destruction and genocide get treated as technical inputs in equations for reckoning the costs and benefits of development or, more nakedly, for the perpetuation of state power.

Technology and Totalitarianism

Today

This "technological fact" makes contemporary styles of political oppression all the more heinous and frightening, and the inability of even radical academic social science to face this fact all the more irresponsible. The Frankfurt School insisted that the empirical side of social life should be studied with constant reference to the possibility of, on the hopeful side, the "other" society, and on the desperate side, Nazi Germany. But for all the critical acumen that members of the school applied to the study of the relationship between society and technology in Germany and elsewhere, their own ideological commitments blinded them to a crucial "possibility" so grimly realized in our time: That the state in antifascist countries, particularly those organized around Soviet-style communism, could equal fascism in its calculated elimination of dissidence; not for ideological reasons, but for technological ones.

The technology of the Holocaust was crude, and toward the end of World War II, R&D even in the technologies of death could not be afforded. As time went by, the slaughter of Jews and others whose culture and social life did not fit the image of the Reich grew to wholesale, unselective, and wasteful (by any standards) proportions. With the defeat of fascism and in the wake of the postwar boom, states whose existence depended upon shaping the national culture, society, and population to their own narrow ends, which had taken a firm and cynically hypocritical stand in favor of an "ideal" over human life and liberty, would perfect the technologies of oppression to a finely tuned, microcomputerized, solid-state, modular, integrated, fail-safe system.

As much as man without technology is not free, man in a society in which high technology is the exclusive possession of the state and is combined with ideological zealotry (even to the point at which ideology itself can be replaced by fear and the pursuit of narrow self-interest) is an endangered species. In the contemporary Soviet version of technological society, those deemed unfit can eliminated with precision — even without being killed through "nonexistent" status — more cheaply, efficiently, and discretely than ever before. An R&D program of proportions unthinkable in the last days of the Third Reich has been carried out long after the collapse of Nazism. The problem is not that Soviet communism is any better or worse on moral or ideological grounds than German Nazism, but that such distinctions are irrelevant today compared to the simple difference between their respective levels of technological sophistication. We have traveled a long way since Saint-Simon and even Marx dreamed of the day when technology would at last allow people to live full, free, and happy lives.

Mastery of the craft of academic social science will always deserve the respect and support of society, but today it must assume a role of servant to practical concerns. Mastery can no longer be an end in itself, for the stakes are far too high for such indulgence. Virtuoso performances with the manipulation of social theories and methods — be they bourgeois, Marxist, positivist, aestheticist — ultimately unconcerned with the social antecedents and impact of technology, oblivious to the ideological uses of development, and cast within specifically sociological, economic, historical, and other disciplinary frameworks, may have a value in themselves. But such isolated performances were rendered meaningless in the death camps of the Third Reich, in the "technological advances" incorporated into the Gulag Archipelago, and in the "engineering" of the killing fields strategy in Kampuchea. Looking ahead from the darkness of these events, and

> after a long search for abstract principles the time must come when more people will realize, even if they have to learn it from their own distress, that the reconstruction of society is a matter of life and death for every citizen, and that most of our calamities can only be removed once we have understood that politics form a set of problems which can never be solved by prejudice, but only by a gradual and conscientious study of society.
> (Mannheim 1954:366)

Tomorrow

The incredible growth of technology and the expansion of the innovation process today have obliterated differences between fascist, nonfascist, and antifascist totalitarianisms. Even in formally democratic societies, the battle between public interest and the will of those "who nurse the big machine" — in the face of nuclear weaponry and the available stock of environment-destroying, privacy-invading, control-extending technologies — is all but lost. In these circumstances, politics has been narrowed to a choice between two paths.

On one hand, we can continue as we have since the mid-nineteenth century: defining development in ways that ensure elite rule, perfecting the types of technologies that promote such development, and maintaining R&D with these technologies as a prerogative of engineering, non-social science, corporate executives, and political elites. With this course we can also expect our understanding of technology to remain at the current polemical level: with one side convinced that technology must ultimately make everybody happy, the other side equally convinced that technology has become a Frankenstein's monster bent on destroying its inventor. At best, the outcome will be a haphazard type of social progress for some people at the expense of others and at costs that could easily be avoided with the proper uses of technology. At worst, the outcome is the unthinkable.

On the other hand, we can decide to pay the price of a fundamental reorientation of the innovation process: defining development in ways that serve the human interest; designing and perfecting technologies for peace, long life, and freedom, regardless of their profitability to economic or political elites. With such a commitment, it would be possible to forge an understanding of technology as a tool for freedom and to live with it as the powerful servant that the amateur social scientists once dreamed it to be. To think of it not as the stolen treasure of Prometheus for which man has to pay forever, but as the "forbidden fruit," knowledge to do great good or great evil, for whose theft the guilty parties have paid their debt to society in full long ago.

The path that we do follow will be determined not by technology, but by people, as they formulate official policies and as they make their day-to-day decisions that affect how, how much, and to whom resources get distributed: money, manpower, and political power. If such decisions continue to be made in the same way as in the past century, the likely future path is the one of more antihuman priorities for innovation and increasing mystification of technology. But things have been changing lately: in all three worlds of development people are reexamining their concepts of technology and progress, a general public awareness of technology's many side effects and unintended consequences is growing rapidly, and with these changes there has arisen a new postacademic commitment in sociology and other social science disciplines. These trends indicate that some type of reorientation of the innovation process is occurring, that resources are being redistributed (to a small extent) so that the social dimensions of technology are being attended to. If these trends are to continue and the future path to the "other" society of technology-for-freedom is to be followed, it is going to take money, manpower, political power, and a struggle on the part of all interested parties. Social scientists, engineers, and everyone whose life is affected by technology should be prepared to work hard and long for a humanized technology and against the momentum in the other direction that technological society has been gathering for over a century.

It seems likely that these trends toward the humanization of technology will continue. We have argued and tried to illustrate that there is good reason to promote the continuation of these trends. The additional effort it will take to get social science and technology back together is well worth it; as Mannheim noted, it may be a matter of life and death. Technology has narrowed the distance that separates people from both totalitarianism and the "other" society. How we decide to treat this new and sobering fact and to proceed from here will make a very big difference in the quality, length, and meaning of life of this and all future generations.

References

AARON, DANIEL
    1951   "Thorstein Veblen: Moralist and Rhetorician." In *Men of Good Hope*. New York: Oxford University Press.

ABEL, THEODORE
    1929   *Systematic Sociology in Germany*. New York: Columbia University Press.
    1948   "The Operation Called *Verstehen*." *American Journal of Sociology* 34:211-18.
    1967   "A Reply to Professor Wax." *Sociology and Social Research* 51:334-36.

ADDAMS, JANE
    1965   *The Social Thought of Jane Addams*. Ed. Christopher Lasch. New York: Bobbs-Merrill.

ADORNO, THEODOR W.
    1941   "On Popular Music." With George Simpson. *Studies in Philosophy and Social Sciences* 9(1):17-48.
    1945   "A Social Critique of Radio Music." *Kenyon Review* 7(2):208-17.
    1967   *Prisms*. Trans. Samuel Weber and Sherry Weber. London: Spearman.

ADORNO, THEODOR W., ELSE FRENKEL-BRUNSWICK, DANIEL J. LEVINSON, and R. NEVITT SANFORD
    1950   *The Authoritarian Personality*. New York: Basic Books.

ANDERSON, MARY B., and PETER BUCK
    1980   "Scientific Development: The Development of Science, Science and Development, and the Science of Development." *Social Studies of Science* 10:215-30.

ANDREWS, FRANK M., and S.B. WITHEY
    1975   *Development and Measurement of Social Indicators*. Ann Arbor: University of Michigan, Social Science Archives.

ARNSTEIN, SHERRY, and ALEXANDER CHRISTAKIS (eds.)
    1975   *Perspectives on Technology Assessment*. Jerusalem: Science and Technology.

BARAN, PAUL A.
1957   *The Political Economy of Growth.* New York: Monthly Review.

BARAN, PAUL A., and PAUL M. SWEEZY
1966   *Monopoly Capital.* New York: Monthly Review.

BARNET, RICHARD J., and RONALD E. MULLER
1974   *Global Reach: The Power of Multinational Corporations.* New York: Simon & Schuster.

BARTH, R.T., and R. STECK (eds.)
1979   *Interdisciplinary Research Groups: Their Management and Organization.* Proceedings of the First International Congress of Interdisciplinary Research Groups. Schloss, Reisenberg (Germany), April 22-28.

BASEHE, JAMES R., and MICHAEL G. DUERR
(n.d.)   "International Transfer of Technology: A Worldwide Survey of Chief Executives." Report of the Conference Board (no date, no place).

BAUER, RAYMOND A.
1966   *Social Indicators.* Cambridge, Mass.: MIT Press.

BAUER, RAYMOND A., RICHARD S. ROSENBLOOM, and LAURE SHARP
1969   *Second-Order Consequences: A Methodological Essay on the Impact of Technology.* Cambridge, Mass.: MIT Press.

BAUMAN, ZYGMUNT
1980   *Hermeneutics and Social Science.* New York: Columbia University Press.

BAYLES, MICHAEL
1976   *Ethics and Population.* New Brunswick, N.J.: Transaction.

BECKER, HOWARD
1967   "Whose Side Are We on?" *Social Problems* 14 (Winter):239-47.
1971   "Reply to Riley's 'Partisanship and Objectivity.'" *American Sociologist* 6 (February):13.

BEN-DAVID, JOSEPH
1971   *The Scientist's Role in Society: A Comparative Study.* Englewood Cliffs, N.J.: Prentice-Hall.

BENDIX, REINHARD
1960   *Max Weber: An Intellectual Portrait.* New York: Doubleday.
1971   "Sociology and Ideology." In E.A. Tiryakian, *The Phenomenon of Sociology.* New York: Appleton-Century-Crofts.

BENJAMIN, WALTER
    1968   *Illumination: Essays and Reflections.* Ed. Hannah Arendt, New York: Viking.

BERGER, PETER
    1964   *Invitation to Sociology.* New York: Doubleday.

BERGER, PETER, BRIGITTE BERGER, and HANSFRIED KELLNER
    1973   *The Homeless Mind: Modernization and Consciousness.* New York: Random House.

BERNARD, L.L., and JESSIE BERNARD
    1965   *Origins of American Sociology.* New York: Russell and Russell.

BERRY, BRIAN J.L.
    1975   *The Human Consequences of Urbanization.* New York: St. Martin's.

BERRY, BRIAN J.L., and JOHN D. KASARDA
    1976   *Contemporary Urban Ecology.* New York: MacMillan.

BERNSTEIN, BASIL
    1972   *Class, Code, and Control,* vol. 1. *Theoretical Studies Towards a Sociology of Language.* London: Routledge & Kegan Paul.

BIEN, JOSEPH (ed.)
    1978   *Phenomenology and the Social Sciences: A Dialogue.* The Hague: Nijhoff.

BLACK, MAX (ed.)
    1961   *The Social Theories of Talcott Parsons: A Critical Examination.* Englewood Cliffs, N.J.: Prentice-Hall.

BOGUSLAW, ROBERT
    1965   *The Utopians: A Study of System Design and Social Structure.* Englewood Cliffs, N.J.: Prentice-Hall.

BOORSTIN, DANIEL J.
    1961   *The Image: A Guide to Pseudo-Events in America.* New York: Atheneum.

BOTA, KOFI, JAY WEINSTEIN, and J.D. WALTON (eds.)
    1979   *Proceedings of the African Solar Energy Workshop,* vol. 1, *Summary Report.* Atlanta: Georgia Institute of Technology.

BOTTOMORE, THOMAS B., and MAXIMILLAN RUBEL (eds.)
    1961   *Karl Marx: Selected Writings in Sociology and Social Philosophy.* London: Watts.

BRAMSON, LEON
- 1961 *The Political Context of Sociology.* Princeton, N.J.: Princeton University Press.
- 1971 "The Rise of American Sociology." In E. Tiryakian (ed.), *The Phenomenon of Sociology.* New York: Appleton-Century-Crofts.

BRAUDEL, F.
- 1966 "European Expansion and Capitalism: 1450-1650." In Columbia University (ed.), *Chapters in Western Civilization.* New York: Columbia University Press.
- 1973 *The Mediterranean.* New York: Harper & Row.

BRITTAIN, JAMES, and ROBERT McMATH
- 1976 "Engineers and the New South Creed: The Formation and Early Development of Georgia Tech." *Technology and Culture* 18(2):175-201.

BROWN, LAWRENCE A.
- 1981 *Innovation Diffusion: A New Perspective.* New York: Methuen.

BROWN, NORMAN
- 1959 *Life against Death.* New York: Vintage.

BROWN, RICHARD H.
- 1973 "L'Ironie dans la théorie sociologique." *Epistémologie Sociologique* 15-16:63-96.
- 1977 *A Poetic for Sociology.* New York: Cambridge University Press.

BUNGE, MARIO
- 1966 "Technology as Applied Science." *Technology and Culture* 7(3):329-47.

BURKE, JOHN G. (ed.)
- 1972 *The New Technology and Human Values.* Belmont, Calif.: Wadsworth.

BURKE, JOHN G., and MARSHALL C. EAKIN (eds.)
- 1979 *Technology and Change.* San Francisco: Boyd & Fraser.

BURNHAM, JAMES
- 1941 *The Managerial Revolution.* New York: John Day.

CALDER, NIGEL
- 1969 *Technopolis: Social Control of the Uses of Science.* New York: Simon & Schuster.

CALLAHAN, DANIEL (ed.)
- 1971 *The American Population Debate.* New York: Anchor.

CAMPBELL, ANGUS, PHILIP CONVERSE, and WILLARD RODGERS
    1976   *The Quality of American Life: Perceptions, Evaluations, and Satisfactions.* New York: Russell Sage.

CARPENTER, STANLEY
    1978   "Autonomous Technology by Langdon Winner." *Technology and Culture* 19(1): 142-44.

CHAMBLISS, WILLIAM, and ROBERT B. SEIDMAN
    1971   *Law, Order, and Power.* Reading, Mass.: Addison-Wesley.

CHASE-DUNN, CHRISTOPHER
    1975   "The Effects of International Economic Dependence on Development Inequality: A Cross-National Study." *American Sociological Review* 40(6):720-38.

CHODAK, SZYMON
    1973   *Societal Development.* New York: Oxford University Press.

CHRISTIANSEN, D.
    1977   "Ethics in Compulsory Population Control." *Hastings Center Report* 5(February):30-33.

CHUBIN, DARYL E.
    1976   "The Conceptualization of Scientific Specialties." *Sociological Quarterly* 17:448-76.

CHUBIN, DARYL E., F.A. ROSSINI, A.L. PORTER, and IAN I. MITROFF
    1979   "Experimental Technology Assessment: Explorations in Processes of Interdisciplinary Research." In *Technological Forecasting and Social Change* 15:87-94.

CICOUREL, AARON
    1964   *Method and Measurement in Sociology.* New York: Free Press.

CIPOLLA, CARLO
    1965   *Guns and Sails in the Early Phase of European Expansion, 1400-1700.* Cleveland, Ohio: Collins.

COALE, A.J.
    1978   "Population Growth and Economic Development: The Case of Mexico." *Foreign Affairs* 56(2):415-29

COHEN, BENJAMIN J.
    1973   *The Question of Imperialism: The Political Economy of Dominance and Dependence.* New York: Basic Books.

COLEMAN, JAMES
    1972   *Policy Research in the Social Sciences.* Morristown, N.J.: General Learning.

COMMONER, BARRY
    1971   *The Closing Circle: Nature, Man, and Technology.* New York: Knopf.

COMTE, AUGUSTE
    1896   *The Positive Philosophy of Auguste Comte,* 3 vols. Trans. H. Martineau. London: George Bell.
    1912   *Système de politique positive,* 4 vols. Paris: Crès.

CONCEPCION, M., and G.M. MURPHY
    1967   "Wanted: A Theory of the Demographic Transition." International Union for the Scientific Study of Population Conference, Sydney (Australia).

CONROY, MICHAEL
    1974   "Recent Research in Economic Demography Related to Latin America." *Latin American Research Review* 9(2):3-27.
    1977   "Population Growth as a Deterrent to Economic Growth: A Reappraisal of the Evidence." Hastings-on-Hudson, N.Y.: Institute of Society, Ethics, and the Life Sciences.

CONROY, MICHAEL, K. KELLEHER, and R. VILLAMIZAR
    1977   "The Role of Population Growth in Third World Theories of Underdevelopment." Hastings-on-Hudson, N.Y.: Institute of Society, Ethics, and the Life Sciences.

COOLEY, MIKE
    1977   "Contradictions of Science and Technology." In G. Boyle et al. (eds.), *Politics of Technology.* New York: Longmans.

COOPER, CHARLES
    1972   "Science, Technology, and Production in the Underdeveloped Countries: An Introduction." *Journal of Development Studies* 9(1):1-18.

COSER, LEWIS A.
    1977   *Masters of Sociological Thought,* 2nd ed. New York: Harcourt-Brace-Jovanovich.

COTTRELL, WILLIAM F.
    1972   *Technology, Man, and Progress.* Columbus, Ohio: Merrill.

CROSLAND, MANUEL
    1967   *The Society of Arcueil.* New York: Heinemann.

COULTER, JEFF
    1971   "Decontextualized Meanings: Current Approaches to *Verstehende* Investigations." *Sociological Review* 19(3):301-23.

Current Biography
    1955   "William Fielding Ogburn." Ed. Marjorie Dent Candee. New York: H.W. Wilson.

DAHRENDORF, RALF
    1967   *Society and Democracy in Germany.* New York: W.W. Norton.

DAVIS, KINGSLEY
    1976   "The World's Population Crisis." In R.K. Merton and R. Nisbet (eds.), *Contemporary Social Problems*, 4th ed. New York: Harcourt, Brace, Jovanovich.

DELACROIX, JACQUES, and CHARLES RAGIN
    1978   "Modernizing Institutions, Mobilization, and Third World Development," *American Journal of Sociology* 84 (July): 123-50.

DEMERATH, NICHOLAS J.
    1976   *Birth Control and Foreign Policy.* New York: Harper & Row.

DE NEVERS, NOEL
    1972   *Technology and Society.* Reading, Mass.: Addison-Wesley.

DIGGONS, JOHN P.
    1968   *The Bard of Savagery: Thorstein Veblen and Modern Social Theory.* New York: Seabury.

DILTHEY, WILLHELM
    1974   "On the Special Character of the Human Sciences." In M. Truzzi (ed.), *Verstehen.* Reading, Mass.: Addison-Wesley.

DOBEL, J. PATRICK
    1978   "The Corruption of a State." *American Political Science Review* 72(3):958-73.

DOBIANSKY, L.E.
    1957   *Veblenism: A New Critique.* Washington: Public Affairs.

DORFMAN, JOSEPH
    1940   *Thorstein Veblen and His America.* New York: Viking.
    1973   "New Light on Veblen." In Joseph Dorfman (ed.), *Thorstein Veblen: Essays, Reviews, and Reports.* New York: Kelly.

DOWD, DOUGLAS F. (ed.)
    1958   *Thorstein Veblen: A Critical Reappraisal.* Ithaca, N.Y.: Cornell University Press.

DUMONT, LOUIS
    1977   *From Mandeville to Marx.* Chicago: University of Chicago Press.

DUNCAN, OTIS D.
    1959   "Human Ecology and Population Studies." In P.M. Hauser and O.D. Duncan (eds.), *The Study of Population.* Chicago: University of Chicago Press.

DURKHEIM, EMILE
    (1890)   "The Principles of 1786 and Sociology." Reprinted in E.A.
    1971   Tiryakian (ed.), *The Phenomenon of Sociology.* New York: Appleton-Century-Crofts.
    1958   *Socialism.* Ed. Alvin Gouldner. Yellow Springs, Ohio: Antioch.
    1962   *The Division of Labor in Society.* Glencoe, Ill. Free Press.
    1964   *The Rules of Sociological Method.* Glencoe, Ill. Free Press.

EHRLICH, PAUL R.
    1968   *The Population Bomb.* New York: Ballantine.

ELDRIDGE, J.E.T. (ed.)
    1980   *Max Weber: The Interpretation of Social Reality.* New York: Schocken.

ELLUL, JACQUES
    1964   *The Technological Society.* New York: Knopf.

ENGELS, FRIEDRICH
    1892   *Socialism, Utopian and Scientific.* Trans. E. Aveling. London: Sonnenschein.

EVANS, DONALD D., and LAURI NOGG ADLER
    1979   *Appropriate Technology for Development: A Discussion and Case Histories.* Boulder, Colo.: Westview.

FANON, FRANZ
    1963   *The Wretched of the Earth.* New York: Grove.

FARIS, ROBERT E.L.
    1945   "American Sociology." In G. Gurvitch and W. Moore (eds.), *Twentieth Century Sociology.* New York: Philosophical Library.

FAUNCE, WILLIAM A.
    1968   *Problems of an Industrial Society.* New York: McGraw-Hill.

FEATHERMAN, DAVID
    1979   "Opportunities are Expanding . . . ." *Society* 16(3):4, 6-11.

FERGUSON, ADAM
    1825   *The History of the Progress and the Termination of the Roman Republic.* London: Jones.
    1979   *An Essay on the History of Civil Society.* New Brunswick, N.J.: Transaction.

FIREY, WALTER
   1978   "Theory and Valuation in the Social Sciences." *Contemporary Sociology* 7(3):275-79.

FORGE, JOHN
   1978   "Commentary." *Lund Letter on Science, Technology, and Basic Human Needs* 5(May):6.

FRANK, ANDRE G.
   1967   *Capitalism or Underdevelopment in Latin America.* New York: Monthly Review.
   1969   "The Development of Underdevelopment." In A. Frank (ed.), *Latin America: Underdevelopment or Revolution.* New York: Monthly Review.

Frankfurt Institute of Social Research, The
   1972   *Aspects of Sociology.* Ed. M. Horkheimer and T.W. Adorno. Boston: Beacon.

FREEMAN, DAVID
   1974   *Technology and Society: Issues in Assessment, Conflict, and Choice.* Chicago: Markham.

FREEDMAN, RONALD
   1973   "Norms for Family Size in Underdeveloped Areas." In M. Micklin (ed.), *Population, Environment, and Social Organization.* Hinsdale, Ill.: Dryden.

FRISBIE, W. PARKER, and CLIFFORD J. CLARKE
   1979   "Technology and Ecological Perspective." Austin: University of Texas, Department of Sociology (ms.).

FROMM, ERICH
   1927   "Der Shabat," *Image* 13:2-4.
   1947   *Man for Himself.* New York: Harper & Row.
   1968   *The Revolution of Hope: Toward a Humanized Technology.* New York: Harper & Row.

GALBRAITH, JOHN K.
   1968   *The New Industrial Age.* New York: New American Library.

GALTUNG, JOHANN
   1971   "A Structural Theory of Imperialism." *Journal of Peace Research* 2:81-117.

GARFINKEL, HAROLD
   1967   *Studies in Ethnomethodology.* Englewood Cliffs, N.J.: Prentice-Hall.

GELLAR, SHELDON
  1979   "Administration of Development: Part I." *Social Development Issues* 3(1):45-60.

GENDRON, BERNARD
  1977   *Technology and the Human Condition.* New York: St. Martin's Press.

GERTH, HANS, and C. WRIGHT MILLS (eds.)
  1946   *From Max Weber.* New York: Oxford University Press.

GILLESPIE, D.F., and P.H. BIRNBAUM
  1977   "Status Concordance, Coordination, and Success in Interdisciplinary Research Teams, Project Report 10." Seattle, Wash.: University of Washington, Research Management Improvement Project.

GOLDING, MARTIN
  1972   "Obligations to Future Generations." *Monist* 56:85-99.

GOLDSCHMIDT, ARTHUR
  1963   "The Development of the U.S. South." In Scientific American (ed.), *Technology and Economic Development.* New York: Knopf.

GOULDNER, ALVIN W.
  1954   *Patterns of Industrial Bureaucracy.* New York: Free Press.
  1962   "Anti-Minotaur: The Myth of Value-Free Sociology." *Social Problems* 9 (Winter):199-213.
  1968   "The Sociologist as Partisan: Sociology and the Welfare State." *American Sociologist* 3 (May):103-16.
  1975-76   "Prologue to a Theory of Revolutionary Intellectuals." *Telos* 26 (Winter):3-36.
  1976   *The Dialectic of Ideology and Technology.* New York: Seabury.
  1980   *The Two Marxisms: Contradictions and Anomalies in the Development of Theory.* New York: Seabury.

GOULET, DENNIS
  1977   *The Cruel Choice.* New York: Atheneum.

GREEP, R., M. KOBLINSKY, and F. JAFFE
  1976   *Reproduction and Human Welfare: A Challenge to Research.* Cambridge, Mass. MIT Press.

GRIMSHAW, ALAN
  1979   "Editor's Page." *American Sociologist* 14(November).

GURR, TED R.
  1970   *Why Men Rebel.* Princeton, N.J.: Princeton University Press.

GURVITCH, GEORGE
    1971   "The Social Settings of Sociological Knowledge." In E.A. Tiryakian (ed.), *The Phenomenon of Sociology.* New York: Appleton-Century-Crofts.

HABERMAS, JURGEN
    1968   *Antworten auf Herbert Marcuse.* Ed. J. Habermas. Frankfurt: Suhrkamp.
    1970   *Toward a Rational Society.* Boston: Beacon.
    1971   *Knowledge and Human Interests.* Boston: Beacon.

HAGEN, EVERETT E.
    1962   *On the Theory of Social Change.* Homewood, Ill.: Dorsey.

HALTY-CARRERE, MAXIMO
    1980   *Technological Development Strategies for Developing Countries.* Montreal: Institute for Research in Public Policy.

HARDIN, GARRETT
    1968   "The Tragedy of the Commons." *Science* 162:1243-48.
    1980   "The Life Boat Ethic." In Melvin Kranzberg, T. Hall, and J. Scheiber (eds.), *Energy and the Way We Live.* San Francisco: Boyd & Fraser.

HARDIN, GARRETT, and J. BADEN (eds.)
    1977   *Managing the Commons.* San Francisco: W.H. Freeman.

HAUERWAS, S.
    1977   "The Moral Limits of Population Control." In S. Hauerwas and Richard Bondi, *Truthfulness and Tragedy: Further Investiations in Christian Ethics.* Notre Dame, Ind.: University of Notre Dame Press.

HAUSER, PHILLIP
    1967   "Family Planning and Population Programs: A Book Review Article." *Demography* 4(1):397.

HAWLEY, AMOS
    1968   "Human Ecology." In D. Sills (ed.), *International Encyclopedia of Social Sciences* 4:328-37.

HAYEK, F. A.
    1952   *The Counter-Revolution of Science.* New York: Free Press.

HEILBRONER, ROBERT
    1967   "Do Machines Make History?" *Technology and Culture* 8(3):335-45.

HEITOWIT, EZRA D.
    1977   *Science, Technology, and Society; A Survey Analysis of Academic Activities in the U.S.* Ithaca, N.Y.: Cornell University Program on Science, Technology, and Society.

HEMMER, C., and R.T. RAVENHOLT
    1977   "Curbing the Population Explosion" (review). In *War on Poverty.* Washington, D.C.: USAID.

HETMAN, FRANÇOIS
    1973   *Society and the Assessment of Technology.* Paris: Organization for Economic Cooperation and Development.
    1977   "Technology on Trial." In G. Boyle et al. (eds.) *Politics of Technology.* New York: Longmans.

HIGGOTT, RICHARD
    1980   "From Modernization Theory to Public Policy: Continuity and Change in the Political Science of Political Development." *Studies in Comparative International Development* 15 (Winter).

HINKLE, ROSCOE C., and GISELA HINKLE
    1954   *The Development of Modern Society.* Garden City, N.Y.: Doubleday.

HIRSCHMAN, ALBERT O.
    1958   *The Strategy of Economic Development.* New Haven: Yale University Press.

HIRSCHMAN, ALBERT O., and C. LINDBLOOM
    1962   "Economic Development, Research, and Development Policy Making: Some Converging Views." *Behavioral Science* 7(2):211-22.

HODGES, H.A.
    1944   *Willhelm Dilthey: An Introduction.* New York: Oxford University Press.
    1952   *The Philosophy of Willhelm Dilthey.* New York: Humanities.

HOFFSTADER, RICHARD
    1955   *Social Darwinism in American Thought.* Boston: Beacon.

HOLBORN, HAJO
    1950   "Willhelm Dilthey and the Critique of Historical Reason." *Journal of the History of Ideas* 11:93-118.

HORKHEIMER, MAX
    1974   *Eclipse of Reason.* New York: Seabury.

HORKHEIMER, MAX, and T.W. ADORNO
    1973   *Dialectic of Enlightenment.* London: Allen Lane.

HORNE, THOMAS A.
1978 *The Social Thought of Bernard Mandeville: Virtue and Commerce in Early Eighteenth-Century England.* New York: Columbia University Press.

HOROWITZ, IRVING LOUIS
1963 "An Introduction to C. Wright Mills." In C. Wright Mills, *Power, Politics, and People.* New York: Oxford University Press.
1966 *Three Worlds of Development,* 1st ed. New York: Oxford University Press.
1969 "Engineering and Sociological Perspectives on Development: Interdisciplinary Constraints in Social Forecasting." *International Social Science Journal* 21(4):545-56.
1971 (ed.) *The Use and Abuse of Social Science.* New Brunswick, N.J.: Transaction.
1972 *Three Worlds of Development,* 2nd ed. New York: Oxford University Press.
1974 "Capitalism, Communism, and Multinationalism." *Society* 12(40):32-45.
1978 "The Sociology of Development and the Ideology of Sociology." Paper Presented at the 73rd annual meeting of The American Sociological Association, San Francisco (September).
1980 *Taking Lives: Genocide and State Power.* New Brunswick, N.J.: Transaction.

HOUGH, GRANVILLE W.
1979 *Technology Diffusion: Federal Programs and Procedures.* Mt. Airy, Md.: Lomond.

HUGHES, H. STUART
1958 *Consciousness and Society.* New York: Knopf.

HYMER, STEPHEN H.
1970 "The Efficiency (Contradictions) of Multinational Corporations." *American Economic Review* 60(May):441-48.

ICP
1974 *The Policy Relevance of Recent Social Research on Fertility.* Washington, D.C.: Interdisciplinary Communications Program Staff Report. Smithsonian Institution.

ILLICH, IVAN
1973 *Tools for Conviviality.* New York: Harper & Row.

INNES, HAROLD
1951 *The Bias of Communication.* Toronto: University of Toronto Press.

JAY, MARTIN
    1973   *The Dialectical Imagination.* Boston: Little, Brown.

JANIS, IRVING L.
    1972   *Victims of Group Think.* Boston: Houghton-Mifflin.

JOHNSTON, JON, D.E. CHUBIN, R.C. McMATH, D. RAY AND F.A. ROSSINI
    forthcoming   *The Role of the Humanities and the Social Sciences in Undergraduate Engineering Curricula.* Atlanta: Report to the Exxon Educational Foundation. Georgia Institute of Technology.

KASH, D.E.
    1977   "Observations on Interdisciplinary Studies and Government Roles." In R. Sribner and R. Chalk (eds.), *Adapting Science to Social Needs.* Washington: American Association for the Advancment of Science.

KAUTSKY, JOHN H.
    1969   "Revolutionary and Managerial Elites in Modernizing Regimes." *Comparative Politics* 1(July):441-67.

KELLMAN, M., and D. LANDAU
    1978   "Economic Development and Population: Further Empirical Tests of The Malthusian Hypothesis." *Studies in Comparative International Development* 13(Fall):3-27.

KIDD, CHARLES V.
    1980   *Manpower Policies for the Use of Science and Technology for Development.* New York: Pergamon.

KLASS, MORTON
    1978   *From Field to Factory.* New York: ISHI.

KOESTLER, ARTHUR
    1967   *The Yogi and the Commissar.* New York: MacMillan.

KORNHAUSER, WILLIAM
    1959   *The Politics of Mass Society.* Glencoe, Ill.: Free Press.

KRANZBERG, MELVIN
    1967   "The Unity of Science-Technology."*American Scientist* 55:48-66.
    1968   "The Disunity of Science-Technology." *American Scientist* 56:21-34.
    1979   "The Science-Technology Complex." *Society* 15 (Jan.-Feb.):54-55.

KRANZBERG, MELVIN, and JOSEPH GIES
    1975   *By the Sweat of Thy Brow: Work in the Western World.* New York: G.P. Putnam's Sons.

KRANZBERG, MELVIN, and T. HALL (eds.)
    1980   *Energy and the Way We Live.* With Jane Scheiber. San Francisco: Boyd and Fraser.

KUHN, THOMAS S.
    1970   *The Structure of Scientific Revolution,* 2nd ed. Chicago, Ill.: University of Chicago Press.

LAKATOS, IMRE, and A.L. MUSGRAVE (eds.)
    1970   *Criticism and the Growth of Knowledge.* Cambridge (Eng.): Cambridge University Press.

LAKE, ANTHONY
    1976   *The Tar Baby Option: American Policy toward Southern Rhodesia.* New York: Columbia University Press.

LAL, SHIVWAJI
    1977   "Social Control and the Social Responsibility in Science and Technology." In G. Boyle et al. (eds.), *Politics of Technology.* New York: Longmans.

LALL, SANJAYYA
    1976   "The Patent System and the Transfer of Technology to Less-Developed Countries." *Journal of World Trade Law* 10, 6:8.

LALL, SANJAYYA, and SENAKA BIBILE
    1977   "The Political Economy of Controlling Transnationals: The Pharmaceutical Industry of Sri Lanka (1972-76)." *World Development* 5 (8): 677-97.

LANDES, DAVID S.
    1969   *The Unbound Prometheus: Technological Change and Industrial Development in Europe from 1750 to the Present.* New York: Cambridge University Press.

LANDMAN, DAVID
    1980   "University of Illinois Research Led to Oil-Eating 'Bugs': Geneticist." *Alumni News* (University of Illinois): 59 (July): 1, 3.

LANGE, OSKAR
    1965   *Wholes and Parts.* New York: Pergamon.

LASCH, CHRISTOPHER
    1965   *The Social Thought of Jane Addams.* New York: Bobbs-Merrill.
    1973   *The World of Nations.* New York: Knopf.

LAWLESS, EDWARD
    1977   *Technology and Social Shock.* New Brunswick, N.J.: Rutgers University Press.

LAYTON, EDWIN JR.
    1971   *The Revolt of the Engineers.* Cleveland, Ohio: Case Western Reserve Press.

LEHMANN, W.C.
    1930   *Adam Ferguson and the Beginnings of Modern Sociology.* New York: Columbia University Press.

LENGERMANN, PATRICIA MADOO
    1979   "The Founding of *The American Sociological Review:* The Anatomy of a Rebellion." *American Sociological Review* 44(April): 185-98.

LERNER, DANIEL (ed.)
    1963   *Parts and Wholes.* Glencoe, Ill. Free Press.

LERNER, MAX (ed.)
    1948   *The Portable Veblen.* New York: Viking.

LENTZ, KATHLEEN
    1977   "China's Population Problem: A May 4th Period Debate." M.A. thesis, University of Iowa.

LEWIS, W. ARTHUR
    1968   "Development Planning," In D. Sills (ed.), *International Encyclopedia of Social Sciences.* 12:118-25.

LILLEY, SAMUEL
    1966   *Men, Machines, and History.* London: Cobbett.

LINDBLOM, CHARLES
    1959   "The Science of Muddling Through." *Public Administration Review* 19:79-88.

LLOYD, IRIS
    1978   "Don't Define the Problem." *Public Administration Review* 38(3):283-86.

LRIS
    1977-79   *Science and Technology for Development: International Conflict and Cooperation.* A Bibliography of Studies and Documents related to UNCSTED. Consortium for International Studies-Education. New York.

LUCE, PHILIP
    1972   *The New Left Today: America's Trojan Horse.* Washington: Capitol Hill Press.

Lund Conference
    1977   *Science, Technology, and Basic Human Needs.* Conference on International Conflict and Cooperation, Lund University, May 10-11.

LUNDBERG, GEORGE, C.C. SCHRAG, O.N. LARSEN (eds.)
    1968   *Sociology.* New York: Harper & Row.

LYMAN, STANFORD M., and MARVIN B. SCOTT
    1970   *A Sociology of the Absurd.* New York: Appleton-Century-Crofts.

LYND, ROBERT S.
    1939   *Knowledge for What?* Princeton, N.J.: Princeton University Press.

MacRAE, DUNCAN Jr.
    1976   *The Social Function of Social Science.* New Haven: Conn. Yale University Press.

MAINE, D.
    1978   "IPPF Survey: World's Contraceptors Increased by 35 Million: 5 Million More in FP Programs." *International Family Planning Perspectives and Digest* 4(Spring):21.

MAMDANI, MAHAMOOD
    1972   *The Myth of Population Control.* New York: Monthly Review.

MANDEVILLE, BERNARD
    1934   *The Fable of the Bees: Or Private Vices, Public Benefits.* London: Wishart.

MANN, THOMAS
    1960   "The Art of the Novel." In H. Brock and H. Salinger (eds.), *The Creative Vision.* New York: Grove.

MANNHEIM, KARL
    1944   *Diagnosis of Our Time.* New York: Oxford University Press.
    1950   *Freedom, Power, and Democratic Planning.* New York: Oxford University Press.
    1952   *Essays on the Sociology of Knowledge.* London: Routledge & Kegan Paul.
    1953   "German Sociology." In *Essays on Sociology and Social Psychology.* London: Oxford University Press.
    1954   *Man and Society in an Age of Reconstruction.* New York: Harcourt, Brace, Jovanovich.
    1956   *Essays on the Sociology of Culture.* London: Routledge & Kegan Paul.
    1968   *Ideology and Utopia.* New York: Harcourt, Brace, & World.

MANUEL, FRANK
    1956   *The New World of Henri St.-Simon.* Cambridge, Mass. Harvard University Press.

MARCUSE, HERBERT
1941 "Social Science Implications of Modern Technology." *Studies in Philosophy and Social Sciences.* 9(3): 414-39.
1955 *Eros and Civilization.* Boston: Beacon.
1964 *One Dimensional Man: Studies in the Ideology of Advanced Industrial Society.* Boston: Beacon.
1968 *Negotiations: Essays in Critical Theory.* Boston: Beacon.
1970 *Five Lectures.* Trans. Jeremy J. Shapiro and Sherry M. Weber. Boston: Beacon.
1972 *Studies in Critical Philosophy.* Trans. Joris de Bres. London: NLB.

MARINI, MARGARET M.
1978 "The Transition to Adulthood: Sex Differences in Educational Attainment and Age at Marriage." *American Sociological Review* 43(August): 487-507.

MARTIN, MICHAEL
1968 "Understanding and Participant Observation in Cultural and Social Anthropology." In Robert S. Cohen and Max Wartofsky (eds.), *Boston Studies in the Philosophy of Science,* vol.4. Dordrecht (Holland): D. Reidel.

MARX, KARL
1904 *Contribution to the Critique of the Political Economy.* Chicago, Ill.: Kerr.
1964 *Capital.* Moscow: Foreign Languages.
1973 *Grundisse.* New York: Vintage.

MARX, KARL, and FRIEDRICH ENGELS
1960 *The German Ideology.* New York: International Publishers.

MARX, LEO
1964 *The Machine in the Garden: Technology and the Pastoral Ideal in America.* New York: Oxford University Press.

MASUR, GERHARD
1952 "Willhelm Dilthey and the History of Ideas." *Journal of the History of Ideas* 13:94-107.

MATRAS, JUDAH
1977 *Introduction to Population.* Englewood Cliffs, N.J.: Prentice-Hall.

MATTHEWS, FRED H.
1977 *Quest for an American Sociology: Robert E. Park and the Chicago School.* Toronto: McGill-Queen's University Press.

MATTILL, JOHN I.
1977 "Technology as an Instrument of Power." *Technology Review* 79 (March-April):78-80.

McLELLAN, D. DAVID (ed.)
1973 *Marx's* Grundisse. New York: Harper & Row.

McLUHAN, MARSHALL
1964 *Understanding Media: Extensions of Man.* New York: McGraw-Hill.

McLUHAN, MARSHALL, and QUENTIN FIORE
1968 *War and Peace in the Global Village.* New York: Betram.

McNAMARA, ROBERT S.
1977 "An Address on the Population Problem." MIT Program on World Change and World Security (April 28).

MEADOWS, D.H., D.L. MEADOWS, J. RANDERS, and W. BEHRENS
1972 *The Limits to Growth.* New York: New American Library.

MERTON, ROBERT K.
1949 *Social Theory and Social Structure.* Glencoe, Ill. Free Press.
1970 *Science, Technology, and Society in Seventeenth-Century England.* New York: Harper & Row.
1971 "The Precarious Foundations of Detachment in Sociology: Observations on Bendix's Sociology and Ideology." In E.A. Tiryakian (ed.), *The Phenomenon of Sociology.* New York: Appleton-Century-Crofts.

MESTHENE, EMMANUEL G.
1968 "How Technology Will Shape the Future." In W.R. Ewald (ed.), *Environment and Change: The Next Fifty Years.* Bloomington: Indiana University Press.
1970 *Technological Change: Its Impact on Man and Society.* Cambridge, Mass.: Harvard University Press.

MILLAR, JOHN
1930 *A Historical View of the English Government,* 4 vols.
(1803) London: Mawman.

MILLS, C. WRIGHT
1940 "Situated Actions and Vocabularlies of Motive." *American Sociological Review* 5(6):904-13.
1943 "The Professional Ideology of Social Pathologists." *American Journal of Sociology* 49 (September):65-180.
1951 *White Collar.* London: Oxford University Press.
1956 *The Power Elite.* London: Oxford Universtiy Press.
1958 *The Causes of World War Three.* New York: Simon & Schuster.

1959 *The Sociological Imagination.* London: Oxford University Press.
1960 *Listen Yankee.* New York: McGraw-Hill.
1962 *The Marxists.* New York: Dell.
1963 *Power, Politics, and People.* Ed. Irving L. Horowitz. London: Oxford University Press.
1969 *Sociology and Pragmatism: The Higher Learning in America.* Ed. Irving L. Horowitz. London: Oxford University Press.

MITROFF, I.I.
1974 *The Subjective Side of Science: A Philosophical Enquiry into the Psychology of the Apollo Moon Scientists.* New York: Elsevier.

MITROFF, I.I., and RALPH H. KILMANN
1978 *Methodological Approaches in the Social Sciences.* San Francisco: Jossey-Bass.

MOORE, WILBERT E.
1969 *The Sociology of Ideology.* Englewood Cliffs, N.J.: Prentice-Hall

MORAVCSIK, MICHAEL J.
1975 *Science Development.* Bloomington, Ind.: Pasitam.

MUECKE, D.C.
1969 *The Compass of Irony.* London: Methuen.
1970 *Irony.* London: Methuen.

MULLINS, NICHOLAS J.
1973 *Theory and Theory Groups in Contemporary American Sociology.* New York: Harper & Row.

MUMFORD, LEWIS
1963 *Technics and Civilization.* New York: Harcourt, Brace, & World.
1966 "Technics and the Nature of Man." *Technology and Culture* 7(3):303-17.
1967 *Technics and Human Development: The Myth of the Machine.* New York: Harcourt, Brace, Jovanovich.

MUNCH, PETER A.
1957 "Empirical Science and Max Weber's *Verstehende Soziologie.*" *American Sociological Review* 22:26-32.

MYRDAL, GUNNAR
1968 *Asian Drama: An Inquiry into the Poverty of Nations.* Harmondsworth (Eng.): Penguin.

1970a   *The Challenge of World Poverty: A World Anti-Poverty Program in Outline.* Harmondsworth (Eng.): Penguin.

1970b   "The 'Soft State' in Underdeveloped Countries." In Paul Streeten (ed.), *Unfashionable Economics: Essays in Honour of Lord Balogh.* London: Weindenfield & Nicholson.

National Academy of Sciences
- 1979   *Contraception: Science, Technology, and Application.* Proceedings of a Symposium. Washington, D.C.

NELKIN, DOROTHY
- 1977   "Technology, and Public Policy." In Ina Spiegel-Rösing and Derek S. Price (eds.), *Science, Technology, and Society.* Beverly Hills, Calif.: Sage.
- 1979   (ed.) *Controversy: Politics and Technical Decisions.* Beverly Hills, Calif.: Sage.

NEUMANN, FRANZ
- 1944   *Behemoth: The Structure and Practice of National Socialism, 1933-1944.* New York: Viking.
- 1957   *The Democratic and the Authoritarian State: Essays in Political and Legal Theory.* New York: Viking.

NEWELL, W.T., B.O. SAXBERG. and P.H. BIRNBAUM
- 1975   *Management of Interdisciplinary Research in Universities: An Overview.* Project Report no. 1, Research Management Improvement Project. Seattle: University of Washington.

NISBET, ROBERT
- 1969   *Social Change and History.* New York: Oxford University Press.
- 1971 (1943)   "The French Revolution and the Rise of Sociology in France." In E.A. Tiryakian (ed.), *The Phenomenon of Sociology.* New York: Appleton-Century-Crofts.

NOBLE, DAVID F.
- 1977   *American by Design.* New York: Knopf.

OGBURN, WILLIAM F.
- 1922   *Social Change.* New York: B.W. Heubsch.
- 1934a   *Recent Social Trends.* With S.C. Gilfillan. New York: McGraw-Hill.
- 1934b   *You and Machines.* Chicago: University of Chicago Press.
- 1946   *The Social Effects of Aviation.* With Jean L. Adams and S.C. Gilfillan. New York: Houghton-Mifflin.
- 1957   "The Meaning of Technology." In F. Allen et al. (eds.), *Technology and Social Change.* New York: Appleton-Century-Crofts.

OGBURN, WILLIAM F., and M.F. NIMKOFF
    1940   *Sociology.* New York: Houghton-Mifflin.

O'CONNOR, JAMES G.
    1976   *The Fiscal Crisis of the State.* New York: St. Martin's Press.

Organization for Economic Cooperation and Development (OECD)
    1975   *Methodological Guidelines for Social Assessment of Technology.* Paris.

ORWELL, GEORGE
    1946   "James Burnham and the Industrial Revolution." London: Privately published.
    1968a   *In Front of Your Nose.* Ed. Sonia Orwell and Ian Angus. New York: Harcourt, Brace, & World.
    1968b   "Not Counting Niggers." In Sonia Orwell and Ian Angus (eds.), *An Age Like This.* New York: Harcourt, Brace, & World.
    1978   *As I Please.* Ed. Sonia Orwell and Ian Angus. Harmondsworth (Eng.): Penguin.

PALMER, R.R.
    1962   *The Age of Democratic Revolution.* vol. 1. Princeton, N.J. Princeton University Press.

PARK, PETER
    1969   *Sociology Tomorrow.* New York: Pegasus.

PARK, ROBERT E.
    1969   "The City: Suggestions for the Investigation of Human Behavior in the Urban Environment (1915)." In Richard M. Sennett (ed.), *Classic Essays on the Culture of Cities.* Reading, Mass.: Addison-Wesley.

PARSONS, TALCOTT
    1951   *The Social System.* Glencoe, Ill.: Free Press.
    1970   "The Impact of Technology on Culture and Emerging New Modes of Behavior." *International Science Journal* 12:607-27.

PATEL, SURENDRA
    1974   "The Patent System and the Third World." *World Development* 2(9):3-12.

PELTO, PERTTI J.
    1973   *The Snowmobile Revolution: Technology and Social Change in the Arctic.* Menlo Park, Calif.: Cummings.

PENROSE, EDITH TILTON
    1951   *The Economics of the International Patent System.* Baltimore: Johns Hopkins University Press.

1973 "International Patenting and the Less-Developed Countries." *Economic Journal* (September):768-86.

PERRUCCI, ROBERT, and JOEL GERSTL (eds.)
1969 *The Engineers and the Social System.* New York: Wiley.

PETERS, J., and S. WELSH
1978 "Political Corruption in America: A Search for Definitions and a Theory." *American Political Science Review* 72(3):974-84.

PETTIT, JOSEPH M., and JAMES M. GERE
1963 "Evolution of Graduate Education in Engineering." *Journal of Engineering Education* 54(October): 57-62.

PODHORETZ, NORMAN
1971 "Doomsday Fears and Modern Life." *Commentary* 52 (October): 4-6.

POLLACK, FRIEDRICH
1957 *The Economic and Social Consequences of Automation.* London: Oxford University Press.

Population Index
1978 44(October):376-84.

Population Reference Bureau
1976 *World Population Growth and Response.* Washington, D.C.
1979 *World Population Data Sheet.* Washington, D.C.

PORTER, ALAN, F.A. ROSSINI, S.R. CARPENTER, and A.T. ROPER
1980 *A Guidebook for Technology Assessment and Impact Analysis.* New York: North-Holland.

PRICE, DEREK S.
1963 *Little Science, Big Science.* New York: Columbia University Press.

PRICE, DON K.
1965 *The Scientific Estate.* Cambridge: Harvard University Press.

RAFFIN, JACQUES
1974 "Transfer of Technology through International Licensing." In Harold F. Davidson et al. (eds.), *Technology Transfer.* Leiden (Holland): Noordhoff International.

RAU, WILLIAM
1979 "Contratechnology: Or the Misplaced Metaphors of Macro-Sociology." Albany: SUNY, Department of Sociology (Ms.).

RAUSHENBUSH, WINIFRED
   1979   *Robert E. Park: Biography of a Sociologist.* Durham: University of North Carolina Press.

RAVETZ, JEROME R.
   1978   "Technology as Master." *Science* 200 (May 12):642-43.

REICH, CHARLES
   1970   *The Greening of America.* New York: Random House.

REICH, WARREN T. (ed.)
   1979   *Enclopedia of Bioethics,* vol. 3. New York: Free Press.

REVELLE, ROGER
   1977   Presentation at November 17 meeting in Preparation for 1979 UN Conference on Science and Technology for Development. Edited transcript, U.S. Department of State.

REYNOLDS, LARRY T., and JANICE M. REYNOLDS
   1970   *The Sociology of Sociology.* New York: Mckay.

RHOADES, LAWRENCE
   1980a   "Federal Funding Level for Social/Behavioral Sciences Indicates Low National Priority." *ASA Footnotes* (Washington, D.C.) 8(3):4-5.
   1980b   "The History of the American Sociological Association." Series in *ASA Footnotes* (Washington, D.C.).

RICHARDS, TUDOR
   1974   *Problem-Solving through Creative Analysis.* New York: Wiley.

RICHARDSON, JACQUES
   1979   *Integrated Technology Transfer.* Mt. Airy, Md.: Lomond.

RICKMAN, H.P.
   1960   "The Reaction against Positivism and Dilthey's Concept of Understanding." *British Journal of Sociology* 11:307-18.
   1961   *Meaning in History: W. Dilthey's Thoughts on History.* London: Allen & Unwin.

RIESMAN, DAVID, RUEL DENNY, and NATHAN GLAZER
   1950   *The Lonely Crowd.* New Haven: Yale University Press.

RIGGS, FRED W.
   1964   *Administration in Developing Countries: The Theory of Prismatic Society.* Boston: Houghton-Mifflin.

RILEY, GRESHAM
   1971   "Partisanship and Objectivity in the Social Sciences." *American Sociologist* 6(February):6-12.

1974 *Objectivity and Values in the Social Sciences.* Ed. G. Riley. Reading, Mass.: Addison-Wesley.

RITZER, GEORGE
1975 "Sociology: A Multiple Paradigm Science." *American Sociologist* 10(August):156-67.

ROBINSON, AUSTIN
1979 *Appropriate Technology for Third World Development.* New York: St. Martin's Press.

RODNEY, WALTER
1972 *How Europe Underdeveloped Africa.* London: Howard University Press.

ROSE, GILLIAN
1980 *The Melancholy Science: An Introduction to the Thought of Theodor W. Adorno.* New York: Columbia University Press.

ROSE, HILARY, and STEPHEN ROSE
1969 *Science and Society.* Harmondsworth (Eng.): Penguin.

ROSENBERG, BERNARD
1956 *The Values of Veblen: A Critical Reppraisal.* Washington, D.C.: Public Affairs.

ROSSINI, F.A., and A.L. PORTER
1978 "Frameworks for Intergrating Interdisciplinary Research." *Research Policy* 8:70-9.

ROSSINI, F.A., A.L. PORTER, and E. ZUCKER
1977 "Multiple Technology Assessments." *Journal of the International Society for Technology Assessment* 2:21-8.

ROSSINI, F.A., and BARRY BOZEMAN
1977 "National Strategies for Technological Innovations." *Administration and Society* 9(May):81-110.

ROSSINI, F.A., A.L. PORTER, P. KELLY, and D.E.CHUBIN
1978 "Frameworks and Factors Affecting Integration within Technology Assessments." Atlanta: Georgia Institute of Technology, Report to the National Science Foundation. Grant ERS 76-04474.

ROSSINI, F.A., J.E. BRITTAIN, S.R. CARPENTER, D.E. CHUBIN, T.A. HALL, D. RAY, and J. WEINSTEIN
1980 "Technology and Science Policy: A Proposal for a Graduate Program." Atlanta: Georgia Institute of Technology, Department of Social Sciences.

ROSTOW, W.W.
    1960   *The Stages of Economic Growth.* London: Cambridge University Press.

ROSZAK, THEODORE
    1969   *The Making of a Counterculture.* New York: Doubleday.

ROTHMAN, HARRY
    1978   "Technology Assessment and the Unanticipated Consequences of Technology." In K.D. Sharma and M.A. Quereshi (eds.), *Science, Technology, and Development.* New Delhi: Sterling.

RUDNER, RICHARD
    1953   "The Scientist Qua Scientist Makes Value Judgments." *Philosophy of Science* 20(January):1-6.
    1966   *The Philosophy of Social Science.* Englewood Cliffs, N.J.: Prentice-Hall.

RUSSELL, BERTRAND, and ALFRED NORTH WHITEHEAD
    1925   *Principia Mathematica.* Cambridge, (Eng.): Cambridge University Press.

SAI, FRED T.
    1977   "Some Ethical Issues in Family Planning." Occasional Essay 1. London: International Planned Parenthood Federation.

SAINT-SIMON, CLAUDE HENRI
    1952   *Selected Writings.* Trans. F. Markham. Oxford: Blackwell.

SALAMON, ALBERT
    1960   "The Messianic Bohemians." In G.B. de Huszar (ed.), *The Intellectuals: A Controversial Portrait.* New York: Free Press.

SAMPSON, ANTHONY
    1974   *The Sovereign State of ITT.* Greenwich, Conn.: Fawcett-Crest.

SARDAR, Z., and D.G. ROSSER-OWEN
    1977   "Science, Policy, and Developed Countries." In Ina Spiegel-Rösing and Derek S. Price (eds.), *Science, Technology, and Society.* Beverly Hills, Calif.: Sage.

SCHELER, MAX
    1960   *Die Wissenformen und die Gessellschaft.* Bern: Francke.

SCHNEIDER, LOUIS
    1964   "Toward Assessment of Sorokin's View of Change." In G. Zollschan and W. Hirsch (eds.), *Explorations in Social Change.* Boston: Houghton-Mifflin.
    1967   *The Scottish Moralists.* Chicago: University of Chicago Press.

1971	"Dialectic in Sociology." *American Sociological Review* 35 (August): 667-77.
1975	*The Sociological Way of Looking at the World.* New York: McGraw-Hill.
1976	*Classical Theories of Social Change.* Morristown, N.J. General Learning.
1979	"Introduction" to *An Essay on the History of Civil Society, by Adam Ferguson.* New Brunswick, N.J.: Transaction.

SCHUMPETER, JOSEPH
    1950   *Capitalism, Socialism, and Democracy.* New York: Harper & Row.

SCHUTZ, ALFRED
    1970   *On Phenomenology and Social Relations.* Ed. Helmut R. Wagner. Chicago: University of Chicago Press.
    1971   *Collected Papers.* Ed. Maurice Natansen. The Hague: Nijhoff.

Scientific American (ed.)
    1963   *Technology and Economic Development.* New York: Knopf.

SCOTT, JAMES
    1972   *Comparative Political Corruption.* Englewood Cliffs, N.J. Prentice-Hall

SCOTT, ROBERT A., and ARNOLD R. SHORE
    1979   *Why Sociology Does Not Apply: A Study of the Use of Sociology in Public Policy.* New York: Elsevier-North Holland.

SECKLER, DAVID
    1975   *Thorstein Veblen and the Institutionalists.* Denver, Colo. Associated University Press.

SEIDMAN, ROBERT B.
    1979   "Development Planning and the Legal Order in Black Anglophonic Africa." *Studies in Comparative International Development* 14 (Summer):3-27.

SELZNICK, PHILLIP
    1949   *TVA and the Grass Roots.* Berkeley: University of California Press.

SENNETT, RICHARD M. (ed.)
    1969   *Classic Essays on the Culture of Cities.* Reading, Mass.:Addison-Wesley.

SHARMA, K.D., and M.A. QUERESHI (eds.)
    1978   *Science, Technology, and Development.* New Delhi: Sterling.

SHELDON, ELEANOR B., and WILBERT E. MOORE
- 1968 *Indicators of Social Change.* New York: Russell Sage.

SHIPMAN, JOHN R.
- 1967 "International Patent Planning." *Harvard Business Review* 24 (March-April):56-72.

SIMONDS, A.P.
- 1978 *Karl Mannheim's Sociology of Knowledge.* New York: Oxford University Press.

SINGER, HANS
- 1976 *Technologies for Basic Needs.* Geneva: International Labour Office.

SKOLIMOWSKY, HENRYK
- 1966 "The Structure of Thinking in Technology." *Technology and Culture* 7:371-83.

SLATER, PHIL
- 1977 *Origin and Significance of the Frankfurt School: A Marxist Perspective.* London: Routledge & Kegan Paul.

SMALL, ALBION
- 1903 "What is a Sociologist?" In *American Journal of Sociology* 8 (January):468-77. Reprinted in E. Tiryakian (ed.), *The*
- 1971 *Phenomenon of Sociology.* New York: Appleton-Century-Crofts.

SMALL, H.C.
- 1977 "A Co-Citation Model of a Scientific Specialty: A Longitudinal Study of Collagen Research." *Social Studies of Science* 7 (May): 139-66.
- 1978 "Cited Documents as Concept Symbols." *Social Studies of Science* 8(August):327-40.

SMITH, ADAM
- 1759 *Theory of Moral Sentiments.* London: Millar.
- 1776 *The Wealth of Nations,* 1937 ed. New York: Random House.

SMITH, DUSKY LEE
- 1970 "Sociology and the Rise of Corporate Capitalism." In J.L. Reynolds (ed.), *The Sociology of Sociology.* New York: McKay.

SNOW, C.P.
- 1964 *The Two Cultures: And a Second Look.* New York: New American Library.

Social Development Issues
- 1978 *Quality of Life* 2(Fall).

SOFRANKO, ANDREW J., F. C. FLIEGEL, and N.C. SHARMA
    1977   "A Comparative Analysis of the Social Impacts of the Technological Delivery System." *Human Organizations* 36(Summer):193-97.

SOLOMON, JEAN-JACQUES
    1977   "Science Policy Studies and the Development of Science Policy." In Ina Spiegel-Rösing and Derek S. Price (eds.), *Science, Technology, and Society*. Beverly Hills, Calif.: Sage.
    1978   "Science Policy and Social Objectives." In K.D. Sharma and M.A. Quereshi (eds.) *Science, Technology, and Development*. New Delhi: Sterling.

SOROKIN, PITIRIM A.
    1929   "Some Contrasts of Contemporary European and American Sociology." *Social Forces* 8:57-62.
    1956   *Fads and Foibles in Modern Sociology and Related Sciences*. Chicago: Regney.
    1964   "Comments on Schneider's Observation and Criticisms," In G. Zollschan and W. Hirsch (eds.), *Explorations in Social Change*. Boston: Houghton-Mifflin.

SPENCER, HERBERT
    1896   *The Principles of Sociology*. New York: Appleton-Century-Crofts.

SPIEGEL-RÖSING, INA, and D.S. PRICE
    1977   *Science, Technology, and Society*. Beverly Hills, Calif.: Sage.

STANLEY, MANFRED
    1972   "Technicism, Liberalism, and Development." In *Social Development*. New York: Basic Books.

STEINER, GEORGE
    1970   *Language and Silence*. New York: Atheneum.

STRAUBE, WIN
    1974   "How to Obtain Higher Financial Rewards from International Technology Transfer." In Harold F. Davidson et al. (eds), *Technology Transfer*. Leiden (Holland): Noordhoff.

SUPPE, FREDERICK (ed.)
    1974   *The Structure of Scientific Theories*. Urbana: University of Illinois Press.

SYPHER, WYLIE
    1971   *Literature and Technology: The Alien Vision*. New York: Random House.

TAPINOS, G., and P.T. PIOTROW
 1979 *Six Billion People.* New York: McGraw-Hill.

TARR, JOEL A.
 1977 *Retrospective Technology Assessment 1976.* San Francisco: San Francisco Press.

TAYLOR, J.B.
 1975 "Building an Interdisciplinary Team." In S.R. Arnstein and A.N. Christakis (eds.), *Perspectives on Technology Assessment.* Jerusalem: Science and Technology.

TEACHMAN, J., D.P. HOGAN, and D.J. BOGUE
 1978 "A Components Method for Measuring the Impact of a Family Planning Program on Birth Rates." *Demography* 15(1):1-13.

THOMPSON, A.R.
 1948 *The Dry Mock: A Study of Irony in Drama.* Berkeley: University of California Press.

THOMPSON, VICTOR
 1971 *Decision Theory: Pure and Applied.* Morristown, N.J.: General Learning.
 1974 *Without Compassion or Enthusiasm.* Tuscaloosa: University of Alabama Press.

TIMASHEFF, NICHOLAS
 1967 *Sociological Theory: Its Nature and Growth.* New York: Random House.

TIRYAKIAN, EDWARD
 1971 *The Phenomenon of Sociology.* New York: Appleton-Century-Crofts.

TODD, RALPH A.
 1977 "A City Index." *Review of Applied Urban Research* (special ed.) 15(7):1-20.
 1978 "A City Index: Measurement of a City's Attractiveness." *Social Development Issues* 2(2):8-22.

TOFFLER, ALVIN
 1970 *Future Shock.* New York: Random House.

The Trend
 1974 26(2). Seattle: University of Washington, College of Engineering

TRUZZI, MARCELLO (ed.)
 1974 Verstehen: *Subjective Understanding in the Social Sciences.* Reading, Mass.: Addison-Wesley.

TURNER, TERESA
    1977   "Two Refineries: A Comparative Study of Technology Transfer to the Nigerian Refining Industry." *World Development* 5(3):236.

UNESCO
    Impact of Science on Society. Paris.

UN Statistical Office
    1978   *Statistical Papers,* Series A. New York.

U.S. Agency for International Development (USAID)
    1976   *World Population Plan of Action.* Washington, D.C. (August).

U.S. Department of Health, Education, and Welfare
    1978   *Inventory and Analysis of Federally-Funded Population Research.* Washington, D.C.: HEW.

U.S. Department of State
    1977   Edited Transcript of 17 November 1976 Meeting in Preparation for the 1979 Conference on Science and Technology for Development. Washington, D.C. (July).

U.S. House of Representatives Committee on International Relations
    1976   "Proposal for an Appropriate Technology." Transmitted by USAID (July 27).

VAIKSOS, CONSTANTINE
    1972   "Patents Revisited: Their Function in Developing Countries." *Journal of Development Studies* 9(October):71-98.

van DAM, ANDRE
    1973   "The Multinational Corporation vis-à-vis Societies in Transformation: The Case for Intermediate Technology in the Developing Countries." *Technology Forecasting and Social Change* 5:236-82.

VEATCH, ROBERT
    1977   *Population Policy and Ethics: The American Experience.* New York: Irvington.

VEBLEN, THORSTEIN
    1919   *The Place of Science in Modern Civilization and Other Essays.* New York: Viking.
    1934   *The Theory of the Leisure Class.* New York: Modern Library.

1939   *Imperial Germany.* New York: Viking.
1953   *The Theory of the Leisure Class.* Introduction by C. Wright Mills. New York: New American Library.
1957   *The Higher Learning in America.* New York: Sagamore.

VICO, GIOVANNI BATTISTA
(1744)   *The New Science.* Trans. T.G. Bergin and M.H. Fisch.
1961   Ithaca, N.Y.: Cornell University Press.

VIDICH, ARTHUR J., and JOSEPH BENSMAN
1958   *Small Town in Mass Society.* Princeton, N.J.: Princeton University Press.

VITA
1977   *Newsletter on Appropriate Technology.* Rangley, Me.: VITA.

WALLERSTEIN, IMMANUEL
1961   *Africa: The Politics of Independence.* New York: Vintage.
1966   (ed.) *Social Change: The Colonial Situation.* New York: Wiley.
1974   *The Modern World System.* New York: Academic Press.
1979   *The Capitalist World-Economy.* Cambridge, Eng.: Cambridge University Press.

WALSH, W.B., G.L. SMITH, and M. LANDON
1975   "Developing an Interface between Engineering and the Social Sciences: An Interdisciplinary Team Approach to Solving Societal Problems." *American Psychologist* 30(November):1067-171.

WARWICK, D.
1974   "Ethics and Population Control: The Case of the Developing Countries." Hastings-on-Hudson, N.Y.: Institute of Society, Ethics, and the Life Sciences.
1975   "Contraception in the Third World." *Hastings Center Report* 5(August):9-12.

WARWICK, D., T.W. MERRICK, and A. CAPLAN
1977   "Population Programs: Should They Change Local Values?" *Hastings Center Report* (October):17-18.

WAX, MURRAY
1965   "The Tree of Social Knowledge." *Psychiatry* 28(May):99-106.
1967   "On Misunderstanding *Verstehen:* A Reply to Abel." *Sociology and Social Research* 51:323-33.

WAX, MURRAY, and JOAN CASSELL (eds.)
1979   *Federal Regulations: Ethical Issues and Social Research.* Boulder, Colo.: Westview.

WEBER, MAX
- 1923 *General Economic History.* Trans. Franklin Knight. London: Allen & Unwin.
- 1946a "Politics As a Vocation." In H. Gerth and C. Wright Mills, (eds.), *From Max Weber.* New York: Oxford University Press.
- 1946b "Science As a Vocation." In H. Gerth and C. Wright Mills, (eds.), *From Max Weber.* New York: Oxford University Press.
- 1947 *The Theory of Social and Economic Organization.* Trans. A.M. Henderson and Talcott Parsons. New York: Free Press.
- 1940 *The Methodology of the Social Sciences.* New York: Free Press.
- 1958a *The Protestant Ethic and the Spirit of Capitalism.* Trans. Talcott Parsons. New York: Scribner.
- 1958b *The City.* New York: Free Press.
- 1958c *The Religions of India.* New York: Free Press.
- 1974 "On Subjective Interpretation in the Social Sciences." In M. Truzzi (ed.), Verstehen: *Subjective Understanding in the Social Sciences.* Reading, Mass.: Addison-Wesley.

WEINSTEIN, JAY
- 1974 "Contributions to the Theory and Measurement of the Soft State in Comparative Political Sociology." *Third World Review* 1(Fall):39-51.
- 1975 "Rethinking Modernization" (review). *Rural Sociology* 40:337-40.
- 1978 "Fertility Decline and Social Service Access: Reconciling Behavioral and Medical Models." *Studies in Comparative International Development* 13 (Spring):71-99.

WENK, EDWARD, Jr.
- 1979 *Margins for Survival.* New York: Pergamon.

WHITE, LYNN, Jr.
- 1962 *Medieval Technology and Social Change.* London: Oxford University Press.
- 1968 *Dynamo and Virgin Reconsidered: Essays in the Dynamism of Western Culture.* Cambridge, Mass. MIT Press.

WHITE, M.K.
- 1973 "Bureaucracy and Modernization in China: The Maoist Critique." *American Sociological Review* 38:149-63.

WHITEHEAD, ALFRED NORTH
- 1967 *Science and the Modern World.* New York: Free Press.

WILLHELM, SIDNEY
- 1964 "The Concept of the Ecological Complex: A Critique." *American Journal of Economics and Sociology* 23(3):241-48.
- 1979 "Opportunities are Diminishing. . ." *Society* 16(March-April): 7, 12-17.

WINCH, PETER
- 1958 *The Idea of a Social Science and Its Relation to Philosophy.* Boston: Routledge & Kegan Paul.

WINNER, LANGDON
- 1972 "On Criticizing Technology." *Public Policy* 20:35-59.
- 1977 *Autonomous Technology: Technics-Out-of-Control As a Theme in Political Thought.* Cambridge, Mass.: MIT Press.

WIONCZEK, MIGUEL S.
- 1977 "Science and Technology for LDCs." *Science* 196 (May 20):837.

WIRTH, LOUIS
- 1948 "American Sociology, 1915-1947." *American Journal of Sociology Index* to vol. 1-52.
- 1969 "Urbanism as a Way of Life." Reprinted in Richard M. Sennett (ed.), *Classic Essays on the Culture of Cities.* Reading, Mass.: Addison-Wesley.

WITTFOGEL, KARL A.
- 1957 *Oriental Despotism: A Comparative Study of Total Power.* New Haven: Yale University Press.

WITTGENSTEIN, LUDWIG
- 1923 *Tractatus Logico-philosophicus.* London: Routledge & Kegan Paul.
- 1953 *Philosophical Investigations.* London: Blackwell.

WOLF, C.P., and JOHN H. PETERSON, Jr.
- 1977 "Social Impact Assessment in a Cross-Cultural Perspective." In Peter Suedfeld et al. (eds.), *The Behavioral Basis of Design.* Stoudsberg, Pa.: Dowden, Hutchinson, & Ross.

WOLFF, KURT (ed.)
- 1971 *From Karl Mannheim.* Oxford: Oxford University Press.

WOLFF, ROBERT, and BARRINGTON W. MOORE (eds.)
- 1967 *A Critique of Pure Tolerance.* Boston: Beacon.

World Bank
- 1974 *Population Policies and Economic Development.* World Bank Staff Report. Baltimore, Md.: John Hopkins University Press.

World Health Organization
        "Expanded Programme of Research, Development, and Research Training in Human Reproduction." Annual Series.

ZEITLIN, IRVING M.
   1968   *Ideology and the Development of Sociological Theory.* Reading, Mass.: Prentice-Hall.

ZELDITCH, MORRIS, JR.
   1979   "Why Was the ASR So Atheoretical?" *Contemporary Sociology* 8(6):808-13.

INDEX

Aaron, Daniel, 65n
Abel, Theodore, 43, 282, 293n
Abel/Wax debate, 282, 293n
Abstracted empiricism (*see also* empiricism), 62, 76, 95-97, 263, 265, 266, 267, 269n, 272, 277, 289, 290, 291, 292; methodological creed, 175; and pure sociology, 173-77
Academic milieux, 179
Academic science (*see also* science): commitment to development, 29; era of, 3, 24, 26, 28-29, 38, 42, 43, 59, 126, 128, 170, 188, 195, 197, 206; first systematic sociology in era of, 35; Frankfurt School's role in, 101; and purity, 171; transition from, 124
Academic social science (*see also* social science): approaches to technology, 4, 271; contrasts with postacademic, 178, 258; critiques of, 266; and development, 198; domination by the pure physics model, 109; Frankfurt School's effects on, 112; future role, 126, 297; and "methodologyism," 143; Mills' comments on, 93, 263; as not conservative, 263; programs for technology, 110; pure/applied, 170, 172-79; radical, 296; reward system, 265; and a science of freedom, 272; and secular interests, 136-37; stress on criticism and application, 100; and the subjectivism/natural science debate, 282-83; U.S., 284; Veblen's critique of, 49-51, 263; Veblen's proposed reforms of, 48; Weber as founder, 39n
Academic sociology (*see also* sociology), 31, 33, 35, 38, 172, 177, 267; authentic, 176; circulation of elites in, 174; as dispensable, 180; effects of Americanization on, 44; European, 44, 65n; focus on technology, 271; and freedom, 295-96; growth in the U.S., 51; ideological dimensions, 294; as ideology, 61; as irresponsible, 108; limitations, 109; and Mannheim's later work, 68; Mills' observations on, 82, 91-96; moral weaknesses, 93; and objectivity, 137, 279n; and Ogburn, 57; in a period of decline, 64; and postacademic social science, 123, 258, 279, 295; as a pseudoscience, 94; and purity, 177; quest for legitimacy, 59, 173, 291, 296; research in, 258; styles of, 271; support for, 126; and technical virtuosity, 291; in the U.S., 41, 64n, 81n, 173
Academic system: and departments, 29; in Europe, 24; German, 29, 30, 38n; Veblen's views on, 48-50
Action versus behavior, 285
Addams, Jane, 64n
Adler, Laurie Nogg, 218
Administrators and the soft state, 213
Adorno, Theodore (*see also* Frankfurt School), 100, 103, 107, 116n, 151n; *The Authoritarian Personality*, 113n; quoted on poetry, 114, 115n
Aesthetics and science (*see also* separate method debate), 287
Afghanistan, 223
Africa: development planning 216; political independence movements, 195, 219, 222; population control, 243, 252
Age of democratic revolution, 18
Agricultural economics, 133
Allende, Salvador, 232
Alternation, 185
Altruism, Mandeville's argument

335

against, 10
Amateur science, 36, 39n
Amateur social science (*see also* preacademic social science): approaches to technology, 4; concern with morality, 31; explicit applied orientation, 172; revolutionary programs, 38; separated from development, 26; and technology for freedom, 296
American Economics Association, 64n
American Journal of Sociology, 63; first U.S. journal, 44; Ogburn as supporter, 55; replaced by *American Sociological Review*, 66n; Small as founder, 64n
American Sociological Association (successor to American Sociological Society), 97, 190n
American Sociological Review, 66n, 174-75
American Sociological Society: its first presidents, 45; first U.S. association, 44; founded in 1895, 64n; Ogburn as president, 55; second-generation presidents, 51; in support of the discipline, 63
American Sociologist, The, 139n, 283
Anderson, Mary, 173, 190n
Andrews, Frank, 167n
Année Sociologique, 39n
Anomie, 135
Anthropological component in SIA (*see also* SIA *and* social impact analysis), 133
Anthropology (ist), 29, 48, 108, 268; Chicago School, 63n; nonacademic activity for, 169; and population ethics, 243
Anti–birth control, 245
Antifascist (*see also* fascism): countries, 296; scholars and the Frankfurt School, 100; totalitarianisms, 297
Antipositivists (*see also* positivism *and* positivist), 163-66
Antitechnocrats (*see also* technocracy *and* technocrat), 163-66
Appropriate technology, 36, 131, 133, 226, 231-41; experimental centers, 231; movement, 210; planning, 228; production of, 234; transfer, 240, 241n, 244
Arbeit Macht Frei (work makes [one] free), 106
Architects, 126
Architecture, 21
Archiv für Sozialwissenschaft und Sozialpolitik, 39n
Argentina, 254n
Argonne National Laboratory, 190n
Aristotle, 81n
Arnstein, Sherry, 136, 140n, 166
Art and science (*see also* separate method debate), 166
Asia, political independence movements, 195, 219, 222; population control, 243, 252; South, 247
Atlanta, Georgia, 154, 161
Atomic bomb, 171
Auschwitz, 114
Automation, 102
Automobile: as exemplary technology, xiv-xvi; Ogburn's study of, 56, 135
Autonomous technology (*see also* technics-out-of control), 104, 120, 141-52, 202; and the City Index Approach, 153; and the Frankenstein's Monster theme, 145-46; history of the idea, 141-44; as ideology, 146, 148, 150-51; and political action, 149-151
Authoritarian (ism), 102; anti-, 124; in a democratic system, 163; elites, 103; personality, 101
Authoritarian Personality, The (Adorno et al.), 102

Bacon, (Sir) Francis, 16, 127
Baden, John, 164, 246
Baran, Paul A., 38n, 199, 222
Barnet, Richard, 222
Barth, R.T., 139n
Basehe, James, 239
Bauer, Raymond, 134, 153
Bauman, Zygmunt, 283
Baxter, Richard (Minister), 8
Bayles, Michael, 245
Becker, Howard, 274, 275

Bell, Daniel, 146
Bellamy, Edward, 50
Ben-David, Joseph, 3, 9, 13, 38n, 67, 126, 138, 139n, 170, 171, 173
Bendix Reinhard, 35, 260
Benin (Africa), 190n
Benjamin, Walter (*see also*, Frankfurt School), 105-6, 108, 115n
Bensman, Joseph, 70
Berger, Peter, 141, 151, 185, 276, 283
Bernard, Jessie, 45
Bernard, L.L., 45
Bernstein, Basil, 78
Berry, Brian J.L., 52, 65n, 217
Bhagavad Gita, 81n
Bibile, Senaka, 237
Bien, Joseph, 282
Big science, 129, 132
Biochemistry, 128, 190n
Bioethics, 128
Birnbaum, P.H., 186
Black, Max, 95
Blalock, Hubert M., 283, 287
Boas, Franz, 50, 65n
Bogue, Donald, 243
Boguslaw, Robert, 128
Bohemian: Messianic, 16; versus Philistine, 147
Bombay, India, 162
Boorstin, Daniel, 263
Bota, Kofi B., 36, 190n
Bottomore, Thomas B., 23n
Bourgeois: decadence, 111; development models, 203, 205; king, 18; mentality, 175; proletarian polarization, 84; social theories and methods, 297
Bourgeoisie, 147
Bossuet, Jacques, 16
Bozeman, Barry, 133
Brahmanism, 171
Bramson, Leon, 45, 46, 51, 52, 53, 64n
Braudel, F., 209n
Brecht, Bertolt, 293n
Bribery, 212, 220, 222, 226
Bridgeman, George, 283
Brittain, James, 29
Brooks, Harvey, 132
Brown, L.A., 210, 254n
Brown, Norman, 107, 115n; quoted on Marx and Freud, 116n
Brown, Richard H., 78, 81n, 164, 287, 289
Buck, Peter, 173, 190n
Bunge, Mario, 140n
Bureaucracy: and class as variables, 84; and the command economy, 33; and compassionate behavior, 220; development, 216; government, 130; and the "knowledge business," 92; Maoist critique of, 228; "mock" (*see* mock bureaucracy), monopoly of resources, 220-21; and prosperity, 39n; public, in the Third World, 211-12; as a sociological concept, 31; and the soft state, 211-14, 217; state, in the Second World, 222; and technology, 36
Bureaucratic: accountability, 221; conduct and the soft state, 224; control and the soft state, 220; officials, 79; organizations, 79; reorganization of research, 126; requirements and intellectuals, 77; research ethics, 174; system, 83
Bureaucratic society, 31, 32-35, 88, 98; role of sociology in, 98; Weber's theory of, 88
Bureaucratization, 25, 32, 34; of intellectual functions, 92; in Mills' and Weber's work, 84; of the sociologist, 94; and technology in Veblen's work, 47
Bureaucrats (*see also* technocrats), 79-149
Burgess, E.A. (*see also* Chicago School of Sociology), 53
Burke, John G., 140n
Burnham, James, 23n

Calder, Nigel, 141, 148
California, University of, 146
Callahan, Daniel, 253n
Cal Tech (California Institute of Technology), 140n
Campbell, Angus, 153, 157
Capital (Marx), 18, 23n
Capital accumulation, 33; and development (*see* development, as

338 Sociology/Technology

capital accumulation); and machine accumulation, 53; model, 44; and the state, 102
Capitalist, class, 163; countries, 84, 102, 130; economics, 264; post-, 208; society(ies), 83, 196, 202, 208
Capitalists: versus technocrats, 79; versus working class, 83
Captains of industry, 47; and development, 199; in a power struggle with technicians, 48
Caribbean, 247
Carlebach, (Rabbi) Shlomo, 115n
Carpenter, Stanley, 151n
Cassell, Joan, 130, 295
Causes of World War Three, The (Mills), 89
Central America, 224, 243
Chakrabarty, Amanda (*see also* "oil eating" bacteria), 190n
Chambliss, William, 212
Chase-Dunn, Christopher, 222
Chicago, Illinois, 52
Chicago School of Anthropology (*see also* anthropology *and* Boas, Franz), 65n
Chicago School of Sociology (*see also* Burgess, E.A.; Ogburn, William F.; Park, Robert E.; sociology, U.S.; Thomas, W.I.; *and* Wirth, Louis), 38n, 39n, 42-44, 51-53, 79, 288; challenge from the "Eastern School," 62; challenge from the positivists, 59-64; influence on Mills, 82, 85; Ogburn's promotion of, 59; rise to dominance, 51, 60; post- and anti-, 83, 96; social problems focus, 52, 173-74; views on autonomy, 86, 115n; waning, 62, 174
Chicago, University of, 44; Jane Addams' influence, 64n; "fringe" character, 46; Ogburn's career, 55; social gospel orientation, 51; Veblen's career, 50
Chile, 232
China: antisoftness policy, 228-29, 230n; communism, 228, 230n; development program, 228-29; population policies, 246, 254n; sociology declared illegal in, 177; in Weber's work, 36-37
"China's Population Problem: A May Fourth Period Debate" (Lentz), 153n
Chodak, Szymon, 199
Christakis, Alexander, 136, 140n, 166
Christiansen, O., 253n
Chubin, Daryl, 29, 139n, 186
Cicourel, Aaron, 283, 290
Cipolla, Carlo, 8
Citation analysis, 140n; and cocitation analysis, 130; meaning of, 140n
Cities (U.S.), 153-54, 155 (Table 1), 157ff, 160
City Index, 153-68; group, 154, 158, 161, 162, 164-66; literature, 160, 168n
Civil disobedience, 219
Civilization and Its Discontents (Freud), 135
Clarke, Clifford, 279n
Classicist versus romanticist, 147
Class struggle, 19
Closure, 123
Coale, A.J., 253n
Cohen, Benjamin, 222
Cold-war mentality, 89
Coleman, James, 130, 132, 138
Colonial era, 217
Colonialism (*see also* imperialism *and* neocolonialism): legacy, 232; and the soft state, 219-21, 223; Western, 89
Columbia University, Ogburn's career, 55; Mills' career, 97; temporary home of the Frankfurt School, 100-01, 112
Command economy, 33, 39n, 209n, 264
Command state model of development, 200
Commoner, Barry, 253n
Communication(s): in Chicago School's work, 51, 79; and the city, 51; control-extending innovations, 86, 102; in Frankfurt School's work, 79; industry, 73; infrastructure, 224; mass, 78; in Ogburn's work, 56; and public participation, 69; its study,

29, 38n; technologies, 204
Communication gaps, 120, 130
Communism(ist), 18-19; Chinese, 228, 230n; critics of Mills, 86; and Mannheim, 68; Marxist ideal, 105; movements, 24; Soviet style, 296-97; and technology, 72; utopian views, 70
Communist Manifesto, The (Marx and Engels), 18
"Compassionate behavior" of governments, 220
Comte, Auguste, 14, 15, 17, 22, 23n, 31, 64n, 94, 107, 131n, 172, 208, 294; Dilthey as a critic of, 282; as an ideologue, 28; his law of three stages, 15, 289; quoted on liberty of thought, 15
Concepcion, M., 253n
Condorcet, Jean Antoine de, 8, 14, 16
Conflict resolution in interdisciplinary research, 187
"Connections" (Public Broadcasting System Series), 125
Conroy, Michael, 248, 253n
Consensus model, 139, 263
Conservative(ism): and Grand Theory, 95; and humanism, 277; ideology, 79; interests, 74, 137; Ogburn's critique of, 56; and political ends, 276; strategies, 146; world view in social science, 260, 262-64, 265, 266, 268, 269, 273, 278, 283, 296
Considerant, Victor, 16
Context of discovery, 286, 291, 293n
Context of validation, 286, 291, 293n
Contribution to the Critique of the Political Economy (Marx), 23n
Contraceptive(s), 242, 247, 249, 250, 251
Control-extending: aims of the state, 102; technology, 86, 298
Converse, Philip, 153, 157
Cooley, Charles H., 51
Cooley, Mike, 235
Cooper, Charles, 240
Cooptation, 228
Corruption, 211-14, 220, 224, 225, 226-27
Coser, Lewis A., 16, 22n, 39n, 45, 50, 57, 65n, 68, 81n
Cottrell, William F., 128
Counterculture, 146, 147-48, 150; -al intellectual, 151
Coulter, Jeffrey, 284, 290
Craftsmanship, 266; intellectual, 83, 99n, 267; Mills on, 87-88; and the scientists' rights and power, 140n; and social science, 98
Crosland, Manuel, 13
Cross-disciplinary (*see also* interdisciplinary *and* multidisciplinary), 124, 128, 136
Critical theory, 108, 110; in the Frankfurt School's work, 104-5; lack of major role for the humanities in, 113; versus positivism, 108-10; versus the pure science orientation, 106
"Cultural apparatus," 79, 91
Cultural relativism, 113, 252
Culture and personality, 103, 131
Culture lag, xvi, 90; as an evolutionary concept, 65n
Cumulative causation, 213

Daddario, Emilo Q. (U.S. congressman), 133
Dahrendorf, Ralf, 112, 188, 273, 288; quoted on pluralism, 261, 262; quoted on subjectivist science, 291-92
Davis, Kingsley, 243, 248, 253n, 254n
de Bonald, Louis, 16, 22n
de Condillac, Etienne, 16
Dehumanization, 106; of technology, 103 (*see also* technology, dehumanized)
Delacroix, Jacques, 216
de Maistre, Joseph, 16, 22n
Demerath, Nicholas J., 211, 216, 243, 249
Democracy, 79; loss of, 88; in Mills' work, 83, 88, 89, 91-92; social scientist's commitment to, 188
Democratic: access to technology, 149; institutions, atrophy, 87; participation, 90; procedure, 102; processes in interdisciplinary research, 186-87, 188; system, 102

Democratic society, 69, 80; and elite planning, 226; and Mannheim's educational reforms, 73; Marcuse's study of, 103; in Mills' work, 84; and public interest, 297; and theory of the power elite, 86
Demographic transition, 246
Demography, 29-30, 131; social, 245
de Nevers, Noel, 128
Denny, Ruel, 70
Dependency theory, 89, 209n
Descartes, René, 16
Des Moines, Iowa, 162
"Despecialize," 130
Detachment, 80n; as a goal in social science, 73; Grimshaw quoted on, 121; its lack and ideology, 74; in Mannheim's work, 73-75; Mannheim quoted on, 74, 120
Developing countries, science and technology in, 129
Development: administration, 214-5; as accumulation of capital, 26, 28, 29, 32, 36, 45, 47, 53, 60, 96, 138-39, 263; asymmetric, 34; based on technological innovation, 12, 36-37, 196, 261; bourgeois, Marxist, and idealist models, 203-5; and bureaucratic efficiency, 221; Chicago School views on, 53; commitment, 274, 292; as a concern of social science, 13, 15, 16, 30; and corruption, 213; costs and benefits, 296; current societal definitions, 104; domestication, 28, 32, 85-86, 138, 199, 202, 204, 205, 207; elite, 86, 298; First and Second World concepts, 199-204; Frankfurt School's research on, 110; as historically operationalized, 200-201; in the human interest, 115, 266, 272, 273, 298; humanist, 277; ideals and realities, 199-207; impact on work and life, 9; institutionalization of, 25; and irony, 197; its lack and technology transfer, 240; machine- and capital-oriented, 61; in Marx's work, 19; Mills' views on, 99n; mixed strategies of, 200; "mock," 218; models in the Third World, 90; as a modern cultural goal, 7, 22; morality, 205; nondomesticated definitions of, 37; partial and uneven, 202; "per se," 199; planning, 211, 214-19, 225, 226, 229; planning organizations, 221; plans, future of, 215; policy, 224-29; policymakers, 123; as policy-saturated social change, 225; and population control, 245, 253n; and population growth, 246, 248; postacademic alternative to, 206; and postacademic social science, 198-209; programs and traditional authority, 225; promoted by applied social science, 70, 71; and pure science, 178; R&D approach to, 196; radical edge, 99n; and the rise of the Third World, 195-97, 202-3; and secularization, 10; social, 114, 136, 231, 233, 240; socialist, 26; sociology as knowledge for, 92, 184, 278; and the soft state, 211-30; as a state prerogative, 202; technological, 129; and technological and social change, 35; technology for, 90, 195-97; and technology, mutual influence, 202-3; theorists, 231; and the three births, 21; trivialized, 87-88, 98, 99n; and universalism, 205-7; and U.S. sociologists, 46; as a value-laden concept, 257; violation of its core ideals, 205
Developmentalism, 180, 184, 195, 199, 270, 278
Developmental orientation: classic, in sociology, 64; and irony, 72; neoclassical, 62; in U.S. sociology, 45; in Veblen's work, 49
"Development game," 195, 205
Deviant behavior, 174, 271
Dewey, John, 39n, 52; his concern with democracy, 91; his influence on Mills, 82, 84-85, 91
Diagnosis of Our Time (Mannheim), 68
Diagnostic dimension (*see also* therapeutic): of ideology, 76-77, 146, 147, 150, 173, 204; in Mills'

theory of craftsmanship, 88
Dial (magazine), 50
Dialectic: and autonomous technology, 145, 151; of ideology and technology, 75, 77, 136, 146; as a technique for discovery, 140n
Dialectical, 163; approaches, 163; orientation, 264; relationship between technology and society, 146
Dialectical materialism, 19
Diffusion of technology (*see also* technology, diffusion; *and* technological innovation): and licensing, 237; one-way and two-way models, 201-11; and population control, 246; and science, 231; and social planning, 218; and the soft state, 217-18; and technical virtuosity, 143-44
Dictatorship, 73
Diggons, John P., 65n
Dilthey, Willhelm, 258, 282-83, 284, 286, 287-90, 293n
Disciplinary (*see also* cross-disciplinary; interdisciplinary; *and* multidisciplinary): departments, 125; frameworks, 297; game, 181; orientations, 188; paradigms, 108; Ph.D.-granting departments, 119; propriety, 159; science, 188; social science, 108, 109, 169, 258; sociology, 109, 177-91; standards, academic, 185
Disciplinary system, 29, 30, 37, 49, 96
Discipline of the machine, 102, 145
Diversity of social scientific perspectives, 209n
Division of labor: and interdisciplinary work, 186; international, 26, 32; among scientists, 128; in social science, 30; as a sociological concept, 31, 182
DNA, 171
Dobel, J. Patrick, 213, 220
Dobriansky, L.E., 65n
Dorfman, Joseph, 50, 65n
Duerr, Michael, 239
Dumont, Louis, 9
Duncan, Otis Dudley, 279, 283

Durkheim, Emile, 17, 22n, 30, 31, 34, 75, 94, 151n, 172, 173, 271; on anomie, 135; as conservative, 263; and the institutionalization of sociology, 39n; and Saint-Simon, 173; and the sociology of knowledge tradition, 76, 131
Dysfunctions, latent, 142; of softness, 212

Eakin, Marshall, 140n
Ecole Polytechnique (Paris), 16
Ecological: complex, 279n; model, 52; orientation, 53; perspective, 51
Ecology versus technology, 81n
Economic crisis, 248
Economic Development and Cultural Change, 209
Economic and Social Consequences of Automation, The (Pollack), 102
Economics, 29-30, 39n, 48, 96, 107, 121, 172, 173, 199; agricultural, 133; in the Frankfurt School's work, 107; institutional, 48; pure, 109; Small quoted on, 65n
Economic science, and institutional economics, 48; Veblen's critique of, 46, 48
Economists, First World, 218
Efficiency: and dehumanization, 34; ideology of, 139; as a sine qua non in engineering, 109; social impact, 72
Ehrlich, Paul, 246, 248, 253n
Einstein, Albert, 111, 132
Eldridge, E.T., 39n
Elite(s): academic, 175; access to, 140n; as accountable, 110; authoritarian, and security, 103; circulation of in academic sociology, 174; control of national R&D, 110; cultural, 91; definitions of progress, 103; and development, 199, 201, 298; focus in SSTS, 131, 132; in Frankfurt School's work, 102, 103, 110; and humanities, 114; interests and technology, 262; Mannheim's analysis of, 70; and mass in *Xia Fang* program, 228; in Mills' work, 82-97; and nonelites,

84; philosophers' identification with, 189n; planning, 226-28; political, 72, 202, 258, 298; political, military, and economic, 84, 102, 258; political and industrial, 138; and population aid, 251; priorities and technological innovation, 110; relationship to technology, 264-65; rule by, 69, 74; social, and development, 203, 205; as subjects of scientific study, 173; technical, 72, 144; technicians and planners, 69, 144; uses of social facts, 209n; in Veblen's work, 48; vision of technology and development, 80

Elitist theory of scientific knowledge, 188

Ellul, Jacques, xv, 24, 59, 141-52, 151n; early formulator of autonomous technology thesis, 141; *La Technique ou l'enjeu du siècle*, 141

Empiricism: abstracted (*see* abstracted empiricism); and explanations of technology, 76; as an ideology, 83; and pure sociology, 175, 273; in U.S. sociology, 42, 43, 59, 61

Enfantin, Prosper, 16

Energy: intensive technologies, 195; and society, 125

Engels, Friedrich (*see also* Marx, Karl), 17, 19, 28; *Communist Manifesto*, 18

Engineer(s): class consciousness among, 23n; humanized, 88; and in-depth knowledge of technology, 271-72; as an ineffective actor, 88; instincts of, 47; on-the-job training for, 240; their political platform, 79; and the price system, 47; of revolution, 19; social, 136; and the social system, 271; and the struggle for power, 299; as users of theory and ideology, 165; working with social scientists, 123, 125-26, 138-39, 257

Engineering, xiv-xv, 21, 119, 124, 126, 184; of death, 266, 297; design, 74, 109; and efficiency, 109; and elite rule, 298; evaluation, 43, 73; humanized, 88; and irony, 75; postacademic model of, 139n; social science as a knowledge base for, 127

England (*see also* Great Britain): birthplace of social science, 7, 9, 12, 14, 22, 170; as an early-modern world power, 11, 17, 27; Mannheim working in, 67-68; its new science and technology centers, 38n

Enlightenment, 8, 19, 22n, 108, 137, 154; concept of development, 199-202; ideals, contrasted with current societies, 72, 199-200, 206; inspired relativism, 46; intellectuals, 201, 209n; optimistic outlook of, 37; philosophers and universalism, 205; post-, 105, 199

Environmentalist, 79

Environmental Impact Assessment Review, 190n

Eros and civilization (*see also*, Marcuse, Herbert), 101

Ethics: code of, 242, 245, 248; in policy-oriented work, 280n; population, 242-53; and science, 166

"Ethics and the Implementation of Population Programs," (Warwick), 253n

Ethnomethodology, 164, 272, 282, 290

Evans, Donald, 218

Expanded Programme of Research, Development, and Development Training in Human Reproduction (annual series, World Health Organization), 252n, 253n

Fable of the Bees, The (Mandeville), 10
Fall, the, 106
Family Planning (*see also* fertility *and* population control): clinics, 250; information, 247; and the soft state, 216; in the Third World, 243
Fanon, Franz, 204
Faris, Robert L., 46
Fascism(ist): era, 111; and the humanities, 113-14; influence on Mannheim, 68; as a negative ideal (*see also* Frankfurt School), 104-6;

and the "other" society, 105, 108, 110; self-destructive nature, 111; society, 107; and Soviet-style Communism, 296; and technology, 72; totalitarianism, 297; utopian views, 69-70
Faunce, William, 141, 144, 151
Featherman, David, 269n
Federal Inventory of Population Research, 244
Ferguson, Adam, 11, 163
Fertility: control, 243, 246, 247, 252; epidemiological model, 249; in Europe, 246
Fiore, Quenton, 116
Firey, Walter, 280n
First birth: of social science and technology, 8, 12; of sociology, 6, 71
First World (*see also* Second World *and* three worlds of development): and Third World, 92; academic social scientists in, 188; development in, 196, 200, 201, 207; development concepts as ideological, 203-5; economists, 218; Horowitz's work on, 99n; multinational corporations, 222, 232; pluralism in, 260; political leaders, 222-23; population policies, 241, 246; sociology in, 180; technologies, 238; transfer of coordinated planning from, 225, 228; values in Third World planning, 217, 218-19
Fission between social and technical approaches (*see also* split), 21, 28, 75, 201, 268
Fliegel, F.C., 133
Footnotes (Newsletter of the American Sociological Association), 190n
Forces of production, 18-19, 23n, 25, 103, 108
Ford Foundation, 243, 244, 252n, 254n
Forge, John, 235
Formal rationality (*see also* rationality), 34, 40n, 140n, 216
France: birthplace of social science, 7, 13, 22n; as an early-modern world power, 27; new science and technology centers, 38n; Revolution of 1789, 13-14, 12, 22
Frank, André Gunder, 28, 209n
Frankenstein (Mary Shelley), 145-46
Frankenstein, Victor, 148, 150
Frankenstein's monster, 120, 141, 145-46, 298
Frankfurt School (Frankfurt Institute for Social Research) (*see also* Adorno, T.W.; Benjamin, Walter; Fromm, Erich; Marcuse, Herbert; *and* Pollack, Friedrich), 47, 59, 71, 79, 100-16, 123, 151n, 198, 263, 264, 267, 294, 296; as academics, 112-3; as antitechnology, 103-4; argument against the humanities, 133-34; as a counterposition to Mannheim, 70, 80, 108; critical orientation, 104-5; critique of positivism, 106-8; flight from Germany, 100; in Frankfurt and New York, 107; and interdisciplinarity, 107-8; limitations of its approach, 110-14; marginality, 111; as Marxist, 100, 112; as misanthropic, 111; moral dimensions of its work, 104-5; origins, 100; and the "other" society, 104-6; pessimism, 104, 110-12; and postacademic social science, 101, 104, 112; quoted on positivism, 105; quoted on technology, 100; return to Germany, 101; social study of technology, 102-6
Frankfurt, University of, 100-1
Franklin, Benjamin, 8
Freedman, Ronald, 249-50
Freedom, Power, and Democratic Planning (Mannheim), 68, 142
"Freedom to breed," 248
Free enterprise: capitalism, 84, 102; model of development, 200
Free-floating intellectual, 99n
Freeman, David, 128
Freeman, Orville, 241n
Free speech movement (FSM), 146-47, 150
Freischwebende intellectual (*see also* free-floating intellectual), 74, 182
Freud, Sigmund, 103, 135, 142, 163;

344 Sociology/Technology

his perspective combined with Marx's (*see also* Frankfurt School), 101, 103; as a sociologist, 116n
Freudianism, compared to Marxism, 115n-16n
Frisbee, W. Parker, 304
Fromm, Erich (*see also* Frankfurt School), 101, 103, 108, 114, 115n; on *Shabbat*, 106
Functional analysis, 163, 272
Functional rationality (*see also* rationality), 33, 40n, 140n
Funding agencies, and pure science, 172

Galbraith, John K., 142, 146
Galtung, Johann, 222
Gandhi, Indira, 253n
Gandhi, M.K., 219, 228
Gandhi, Sanjay, 253n
Garfinkel, Harold, 164, 284, 287
Gaugain, Paul, 113
Gay-Lussac, J.L., 16
Gellar, Sheldon, 202-3
Gendron, Bernard, 128, 140n
General Electric Company, 190n
Genocide, 115n, 226, 296
Geography, 29
Georgia Tech (Georgia Institute of Technology), 140n
Gere, James, 171
German Ideology, The (Marx and Engels), 23n
Germany: birthplace of the Frankfurt School, 100; birthplace of social science, 17-19, 22, 170; educational system, 38n, 273; Horkheimer's return to, 112; influence of the Frankfurt School today in, 101; Nazi, 103, 105, 296-7; Park's studies in, 51; patent holdings by, 236
Gerth, Hans, 35, 39n, 43, 68
Ghana, 216
Giddings, Frank, 45
Gies, Joseph, 33
Gillespie, D.F., 140n, 186
Ginzburg, Morris, 285
Glasser, Nathan, 70
Glenn, Norval, 283

Godwin, William, 245
Goffman, Erving, 283
Golding, Martin, 253n
Goldschmidt, Arthur, 224
Gouldner, Alvin W., 22n, 23n, 28, 35, 38n, 39n, 48, 71, 75, 104, 148, 204, 218, 274, 277; on the counterculture, 152n; as an historian of the Frankfurt School, 108; on linguistics, 78; on mock bureaucracy, 211; quoted on the control of technology, 79
Goulet, Dennis, 204, 233, 241n
Grand Theory, 95, 97, 272, 273, 277
Great Britain (*see also* England): patent holdings by, 236
"Great wall" separating sociology and technology (*see also* Ogburn, William F.), 54, 110
Greek mythological histories, 142
Greensboro, North Carolina, 162
Greep, Roy, 254n
Gresham's Law, 286
Grimshaw, Alan, 121
"Groupthink," 187
Grundrisse (Marx) 73n
Gulag Archipelago, 297
Gurr, Ted R., 48
Gurvitch, George, 64n, 174

Habermas, Jurgen, 78, 101, 115n
Hacienda system, 224
Hackel, Ernst, 51
Hagen, Everett, 199
Hall, Timothy, 125
Halty-Carrere, Maximo, 230n
Hardin, Garrett, 164, 246, 248
Harvard School of Public Health, 249
Hastings Center Report, 253n
Hastings Institute, 246, 250, 253n
Hauerwas, Stanley, 246, 253n
Hauser, Phillip, 253n
Hawley, Amos, 52
Hayek, F.A., 16
Hebrew mythological histories, 142
Hegel, G.W.F., 18-19
Heilbroner, Robert, 151n
Heitowit, Ezra, 125
Hemmer, Carl, 251
Hempel, George, 283

Hermeneutic(s), 107, 164, 259, 282, 289; as technical virtuosity, 290-91
Hetman, François, 133, 136, 140n, 233; quoted on social impact analysis, 133, 134
Hickenlooper Amendment, 222
Higgott, Richard, 230n
Hindu mythological histories, 142
Hinkle, Gisela, 45
Hinkle, Roscoe, 45
History, 127, 170; Orwell's views on, 203; theory of, 94-95
Ho Chi Minh City, 162
Hodges, H.A. 293n
Hoffstader, Richard, 45
Holborn, Hajo, 293n
"Hollow Miracle, The" (Steiner), 116n
Holocaust, 111, 296
Honolulu, Hawaii, 154
Horkheimer, Max (*see also* Frankfurt School), 100, 101, 103, 107, 114; founder of Frankfurt School, 100; return to Germany, 101, 112
Horne, Thomas A., 11
Horowitz, Irving Louis, 14, 33, 99n, 115, 126, 130, 165, 171, 172, 195, 209n, 212, 220, 222, 223, 266, 269n, 272, 274; his comprehensive work on development, 200-3; on First/Second/Third World relations, 99n, 200-1, 230n; quoted on the consensualist vision, 139; quoted on development, 8; quoted on engineering and sociology, 23n, 139n; quoted on interdisciplinarity, 207; quoted on Mills' approach, 83; quoting Mills' "Letter to a White-Collar Wife," 99n
Hough, Granville, 133
Hughes, H. Stuart, 44
Human factors: in development, 197; engineering, 39n
Human interest(s) (*see also* interest), 33, 70, 83, 115n, 138, 264, 266, 272, 273, 278, 298
Humanism(ist): commitment, 111, 273, 274, 278; ideals, 204; and partisanship, 274-77; perspective, 80; science, 125; and scientific method, 277, 278; type of inventiveness, 233; values, 275
Humanities: Frankfurt School's argument against, 113-14; prestige, 170, 172-73
Human relations: in industry, 39n; technological dimensions, 270
Humanizing dimension, 208
Human rights, 245, 251
Hume, David, 16
Hymer, Stephen, 222

IBM (International Business Machines), 232
ICP (International Communications Program, Smithsonian Institution), 254n
Idea of a Social Science, The (Winch), 282, 284
Idealism(ist): in critiques of social science, 267; development models, 203-5; moralist, 204-5; and separatism, 288
Ideological (*see also* diagnostic *and* therapy): accounts of technology's impacts, 188-89; approaches to technology, 167; character of "value free" approaches, 160; commitments, relativizing, 220; commitments of postacademic social science, 260-69, 270-80, 281, 283; conflict, 273; connotations of postacademic social science, 207, 258-59, 271, 273, 277; critics of positivism, 175; development concepts, 203-5, 297; differences and the pure/applied distinction, 170; depth in the work of social scientists, 131; dimensions of sociology, 294; evaluation of "head" versus "hand" work, 171; fascist solutions, 69-70; implications of development, 199; modes of analysis, 176; orthodoxy, 294; perspectives on technology and culture, 28, 157; perspective on technology and society, 198; purity, 265, 266; reasons and poetics, 287; traps and implications, 120; versus scientific knowledge, 28; versus technological reasons, 296; views of

346 Sociology/Technology

social change, 69; views and
  Verstehen, 286
Ideologues, The, 16
Ideologue(s), 81n, 166, 173, 175, 261;
  "ideology-free," 261
Ideology (*see also* dialectic, of
  ideology and technology): academic
  sociology as, 61; as an antonym for
  science, 28; and autonomous
  technology, 146-48; and the City
  Index approach, 157, 164-66; of
  efficiency, 139; empiricism and
  rationalism as, 83; "end of," 79,
  259; of environmental doom, 79;
  and lack of detachment, 74; in
  Mannheim's work, 70, 73-74; in
  Marx's work, 28, 44; political, 78;
  of political conflict, 291; and
  population control, 245, 252;
  positivism as, 166; postacademic
  social science, 258-59, 260-69,
  270-80; and science, 30, 38n; and
  scientism, 38n; and self-interest,
  297; sociological analysis of, 75;
  and the sociologist's role, 137; as a
  subject of study, 34; and substantive
  rationality, 73; as a symbol system,
  165; of technological impact, 120;
  and technology, 75, 77, 80;
  ultrareactionary, 254; word coined,
  28
Illich, Ivan, 233, 234
Illinois, University of, 190n
Imperial Germany (Veblen), 49
Imperialism (*see also* colonialism *and*
  neocolonialism), 196, 232
Inappropriate technology (*see also*
  appropriate technology *and*
  technology), 217-19, 226, 229, 236
India: dispute with IBM, 232; family
  planning programs, 216, 253n;
  Gandhi's movement, 219; *Jajmani*
  system, 224; Kipling's observations
  on, 113; Maharastra, 253n; as a
  recipient of population aid, 243; in
  Weber's work, 36-37; West Bengal,
  230n
Indo-China War, 90
Industrial classes, 147
Industrialism, 163

Industrialization: and distribution, 24;
  early takeoff in England and
  Scotland, 8, 9; European, 32; as an
  issue in France, 13; and
  secularization, 10; social science as
  a theory of its consequences, 20, 22;
  solution of its problems, 26; U.S.
  sociologists' views on, 46, 52
Industrial Revolution, 26, 129, 135,
  141, 196, 198, 199; and autonomous
  technology, 145; effects on its
  observers, 9, 10; in England, 8, 13;
  in Europe, 6, 36, 195; in France, 13;
  and impoverishment, 33; irony and
  technology since, 172; in Marx's
  work, 19-20
Industrial society: and autonomous
  technology, 141; bureaucratization
  of, 84; in the Chicago School's
  work, 53; development
  commitment, 28; development
  priorities, 24; human and
  technological aspects, 25; and the
  "other" society, 105; and the
  Protestant ethic, 32; as an R&D
  system, 102; in Spencer's work, 45;
  and technological applications,
  103-4
Information industry, 126
Innes, Harold, 38n, 115n, 272
Instinct(s): to emulate, 48, 56, 86; of
  the engineer, 47; of the
  entrepreneur, 47; Veblen's theory
  of, 65n; of workmanship, 48, 56,
  58, 86, 88, 92
Institutes of technology, 29, 127, 170,
  268
Intangibles, 154, 250
Intellectual(s): and bureaucracy, 77, 79;
  countercultural, 151; craftsmanship,
  83, 92, 93; and development
  models, 200; Frankfurt School as,
  112; marginal, 151; Mills'
  observations on, 82, 88, 91; Orwell
  quoted on, 203; outsiders, 150;
  political struggle among, 147
Interdisciplinarity, 110, 119, 123,
  139n, 180, 188, 195, 198; and
  Freischwebende, 182; and
  developmentalism, 207-8; growth

of, 181-83; and multiple realities, 185-86; in postacademic social science, 270, 278, 292; and problem centeredness, 183-84; science of, 188; as a topic of social research 169; technology and development studies, 207

Interdisciplinary (*see also* cross-disciplinary *and* disciplinary): and multidisciplinary activity, 277; activity, management of, 186-88; curricula, 189; degrees, 126; Durkheim as, 130; field, 12, 130; field equated with a methodology, 131; orientation, 35, 94, 123-24; orientation in Weber's work, 35, 131; pluralism, 278; postacademic social science as, 179, 257, 266, 278; professionals, 126; research, 120-21, 180-83, 186, 198, 283; science as problem-focused, 182-84; social science in Mannheim's program, 68, 70, 131, 182; social scientist, 184; sociological theory on Ogburn's work, 56, 131; sociology in Mills' work, 98; sociology in Veblen's work, 49; specialities, 128; studies in the work of the Frankfurt School, 106, 107-8, 113; study of development as, 207-8; teaching, 132, 181, 186-88; team research and teaching, 198, 263

Interest(s): of academic social science, 136-37; -bound perspectives, 73; of capitalism, 102; conservative, 74, 137, countersociological, 61; and domesticated development, 207; effects on social theory, 77; of elites and technology, 262; groups and bureaucracy, 213; groups, anti–birth control, 245; human (*see* human interests); and ideology on Mannheim's work, 73-74; influence on technology, 26; in Marx's work, 21; national, 89, 257; political, 248; and R&D priorities, 86; and research, 259; self-, and ideology, 297; serving account, 168n; serving value judgments, 160; vested, 132

International Council for Science Policy Studies (ICSPS), 128-29

International relations, 83, 89

International Union for the Protection of Scientific Property, 235-36

Intersubjectivity (*see also* objectivity *and* subjectivism), 185-86, 274, 284

Invention(s), xv, xvii, 8, 12, 83, 98; defined, xiv; evaluation of, 239; of new life forms, 171, 190n; and social and psychological processes, 270; of social science, 22

Inventor(s): early French, 16; Ogburn quoted on, 56; Marx as, 17; and the patent system, 236, 239; Smith as, 11

Inventory of Private Agency Research (HEW), 252

"Invisible hand," 12

Iowa, University of, 253n

Iron Curtain, 200

Iron law, 99n; of oligarchy, 201

Ironic focus, 71-72, 195; in art and sociology, 163; as a technique for discovery, 163

Irony (*see also* dialectic *and* unintended consequences): in application, 132-36; in Comte's view, 16; and development, 197; history of the term, 81n; as a literary device, 287; in Mannheim's work, 71, 72-75, 132-34, 136; in postacademic social science, 119, 262; in Smith's optimism, 36; and technology, 188; of technology, 134, 135, 142; as theme in early social science, 12; in *Theory of the Leisure Class*, 65n

Irrationality and corruption, 213

ITT (International Telephone and Telegraph), 232

IUD (intrauterine device), 244, 245

Jajmani system, 224

James, William, 52, 82

Janis, Irving, L., 197

Jay, Martin, 105, 108, 115n; quoted on the "critical impulse," 115

Jews: and the Holocaust, 296; concept of *Shabbat*, 106

Johns Hopkins University, 45

Johnston, Jon, 126
Journalism: in the Chicago School approach, 52, 173-74; linked with pragmatism, 53; muckraking, 53; Orwell quoted on, 203; and Robert Park, 51
Journal of Politics, 139n
Joyce, James, 114; McLuhan quoted on, 116n

Kafka, Franz, 35
Kampuchea, 297
Kant, Immanuel, 288-89
Kasarda, John D., 52, 65n
Kash, D.E., 139n
Kautsky, John H., 48
Kidd, Charles, 230n
Kilmann, Ralph, 139n
Kipling, Rudyard, 113
Klass, Morton, 224, 230n
Knight, Franklin, 43
Knowledge (journal), 139n, 190n
Koestler, Arthur, 296
Korea, 223, 243
Kornhauser, William, 70
Kranzberg, Melvin, 9, 29, 33, 125, 170; quoted on technology, 231
Kritische Theorie (critical theory) (*see also* Frankfurt School), 108
Kroeber, A.L., 210
Kuhn, Thomas, 131, 132

Labor unions, 221
LaGrange, Joseph, 16
Laissez faire, 70; development policies, 201; model, 137
Lakatos, Imre, 131
Lake, Anthony, 222
Lal, Shivwaji, 233
Lall, Sanjayya, 236, 237, 239
Landes, David, 141, 151n
Landman, David, 190n
Landon, M., 139n, 140
Lange, Oskar, 10
Lasch, Christopher, 64n, 263, 269n
La Technique ou l'enjeu du siècle (Ellul), 141
Latent dysfunctions, 142
Latifundium system (*see also* hacienda system), 224

Latin America: independence struggles, 195; population control, 252;
Law, 216
Lawless, Edward, 136; *Technology and Social Shock*, 136
Law of three stages, 15, 284
Layton, Edwin, 23n, 65n, 88, 139, 140n, 272
Lazarsfeld, Paul, 97
Leap of faith, 111
Lecky, W.E.H., 94
Legitimacy: loss of in development, 214; struggle for in social science, 59, 173; struggle for in sociology, 59-61, 109, 136, 172-74, 180-81, 188; of traditional authorities, 224
Lengermann, Patricia Madoo, 66n
Lenin, V.I., 146
Lentz, Kathleen, 253n; "China's Population Problem: A May Fourth Period Debate," 253n
LePlay, Frederic, 16
Lerner, Daniel, 10
Lerner, Max, 65n
Lévi-Strauss, Claude, 284, 287
Lewis, W. Arthur, on development planning, 215
Liberalism (*see also* conservatism; pluralism; and radicalism): and humanism, 277; as ideology, 79; and postacademic social science, 278; practicality, 98; strategies, 146; technocratic-managerial, 265; in theories of stratification, 85; "unconscious" in Chicago School approach, 51, 56, 61
Licensing of patents, 236, 237, 238, 240
Lifeboat ethic, 246
Lilley, Samuel, 128
Limits to Growth (Meadows et al.), 246
Lincoln, Nebraska, 160, 161, 162
Lindblom, Charles, 218, 225
Linguistics, 28, 78-79, 107, 183, 283
Literature, 107, 127, 170; Orwell's views on, 203
Little Science, Big Science (Price), 128
Lloyd, Iris, 218
Lowe, Adolph, 68
LRIS, 209n

Luce, Phillip, 150
Luddism, 59, 106, 148, 150, 151n; intellectual, 146, 147, 151, 235
Luddite: intellectual, 151, 164; orientation in Marcuse's work, 103
Lundberg, George, 61
Lund Conference, 197
Lund Letter, 197, 233
Lyman, Stanford M., 81n
Lynd, Robert, 81n, 83, 115n, 263

Macrae, Duncan, 280n
Madison, Winsconsin, 162
Magna Carta, 11
Maine, D., 243
Malraux, André, 276; *Man's Fate*, 276
Malthus, Thomas, 245
Malthusian catastrophe, 245, 248, 253n
Mamdani, Mahamood, 243, 248, 249
Man and Society in an Age of Reconstruction (Mannheim), 68
Man's Fate (Malraux), 276
Management, 31, 39n, 124, 149; of interdisciplinary activity, 186-88
Manager(s), 88, 123, 126, 138; and control of technology, 149
Mandeville, Bernard de, 9, 11, 13, 15, 72
Manhattan Project, 131
Mann, Thomas, 136
Mannheim, Karl, 28, 33, 44, 65n, 94, 101, 102, 103, 104, 113, 114, 123, 131, 132, 137, 139n, 144, 148, 149, 151n, 165, 182, 188, 195, 198, 199, 263, 264, 267, 269n, 273, 299; on communications, 78-79; counterposition to Frankfurt School, 70, 80, 108; critique of elite rule, 69 72, 85, 227; on democracy, 69-70, 79; on detachment, 73-75; *Diagnosis of Our Time*, 68; and educational reform, 68; and Frankfurt School, 100, 103, 112; *Freedom, Power, and Democratic Planning*, 68, 142; interest in irony, 71, 72-75, 132-34, 136; *Man and Society in an Age of Reconstruction*, 68; and Mills, 82-83, 86-87, 89; observations on development and Third World, 195-96, 200-1; on planning, 69-70; program for social reconstruction, 69-72; quoted on detachment, 74; quoted on dictatorship, 73; quoted on history of sociology, 22; quoted on postacademic social science, 257; quoted on progress, 67; quoted on social reconstruction, 297; quoted on underdeveloped countries, 196; and sociology of knowledge, 71, 75-80, 203; his "third way," 69, 70, 72, 91, 167
Manuel, Frank, 14
Mao Tse-Tung, 204, 228
Marcuse, Herbert (*see also* Frankfurt School), 100, 114, 116n, 143, 151; as antitechnology, 103-4; *One Dimensional Man*, 103, 104; pessimism, 104; quoted on technology and repression, 294; theories of technology, 103-4
Marini, Margaret, 175
Market, 25
Martin, Michael, 282
Marx, Karl, 17-21, 22n, 23n, 25, 35, 42, 44, 94, 103, 104, 136, 141, 145, 163; *Communist Manifesto*, 18; *Contribution to the Critique of the Political Economy*, 23n; delegitimation of, 137; dialogue with Weber, 85, 173; *German Ideology*, 23n; *Grundrisse*, 23n; and ideology, 28, 31, 34; influence on Veblen, 50; long-run optimism, 37; his perspective combined with Freud's, 101, 103, 107; quoted on technology, 26; and sociology of knowledge tradition, 76; and Spencer, 173; and universalism, 206; views on development, 201-2; views on proletariats compared with Ogburn's, 58; views on proletariats compared with Veblen's, 48;
Marx, Leo, 151n
"Marx for the Managers" (*see also* Burnham, James; and Mills, C. Wright), 88
Marxism, 75, 265, 293n; as a cause of cultural fission (*see also* fission *and* split), 28; compared to autonomous

technology thesis, 148; as critique of progress, 26; Mills' writings on, 83; reaction to, 32; schism over in U.S. sociology, 60; as sociology, 115n-116n

Marxist(s), 276; anti-, 83; commentary on social science, 264; critics of New Left, 150; development models, 203-5; Frankfurt School as, 112; ideal of communist society, 105; orthodoxy, 85; perspective, 222; sophisticated, plain, and vulgar, 49; theories and methods, 297; theories of alienation, 88; theories of class and class consciousness, 83; tradition, 233; world view in social science, 260

"Marxology," 265, 277

Maryland, University of, 190n

Mass(es): and elite planning, 227; in less-developed countries, 90; participation in China, 228; and public, 110

Massachusetts Institute of Technology (MIT), 190n

Mass consumption, 8, 92

"Massification," 69, 86, 87, 110

Mass production, 102

Mass society: democratization of, 87; Mannheim's influence on study of, 70; in Mills' work, 84, 87; technology in, 72

Masur, Gerhard, 293n

Matras, Judah, 243

Matthews, Fred, 51

Mattill, John, 197, 233

McLuhan, Marshall, 38n, 115n, 272; quoted on Innes, 38n; quoted on Joyce, 116n

McMath, Robert, 29

McNamara, Robert, 253n

Meadows, D.H., 163

Means of production, 23n, 79, 84, 233

Medical and behavioral approaches to population control (*see also* family planning; fertility; and population control), 246, 249-51, 252

Mencken, H.L., 97

Merton, Robert K., 22, 38n, 71, 132, 142, 212, 260; quoted on social engineers, 136

Messianic: bohemians, 16; tradition and Sabbath, 106, 115n; utopia in Frankfurt School's work, 105

Mesthene, Emmanuel, 128

Methodological: creed of abstracted empiricism, 175; issues, 174; "miasma," 167; orthodoxy, 294; separation, 290

Methodologically transfixed social thought, 287

Methodologist as virtuoso, 144

Methodology: equated with an interdisciplinary field, 130; positivist, 295; propelled orientation in U.S. sociology, 59; role of in Weber's work, 37; separatist, 295

Methodologyism, 143

Methodology of the Social Sciences (Weber), 286

Mexico City, 162

Middle East, political independence movements in, 195

Millar, John, 11

Miller, G. William, 241n

Mills, C. Wright, 23n, 35, 38n, 79, 80, 82-99, 101, 102, 104, 109, 110, 112, 114, 123, 137, 141, 167, 263, 264, 267, 269n, 271, 273, 294; as an academic, 96-98; on craftsmanship, 87-89; critique of social science and academic sociology, 91-97; on development, 99n, 198, 199, 203; influenced by Chicago School, 82; influenced by Mannheim, 68, 70, 75, 82; on international relations and the Third World, 89-91; as marginal, 91, 96-98; on stratification and elites, 83-87; on the terms *sociology, social science*, and *social studies*, 99n

Mitroff, Ian, 139n, 191n

Mock bureaucracy (*see also* bureaucracy *and* Gouldner, Alvin W.), 211-13, 229; and colonial relations, 221; in Third World public administration, 212; and traditional authority, 224

Modernization, 16, 226

Modern society: discipline of machine

in, 102; Mannheim's program for its recontruction, 68; Mills' views on, 83; and rationality, 143; social equality in, 85; technological dimensions, 38; and understanding technology, 109

Monist (journal), 253n

Montesquieu, Charles de, 16

Moore, Barrington W., 103

Moore, Wilbert E., 28, 167n

Moral, autonomy, 98; scatter, 98; weaknesses of academic sociology, 93

Moral consquences of technology, 10-12, 78

Moralist-idealist approach to development, 204-5

Moral philosophy: as early social science, 9, 11, 12; in Veblen's program, 49

Moravcsik, Michael, 125

"Muddling through," 218

Muecke, D.L., 81n, 135; quoted on technological progress, 135-36

Muller, Ronald, 222

Mullins, Nicholas, 283, 290

Multidisciplinary (*see also* cross-disciplinary; disciplinary; *and* interdisciplinary), 107; social science, 179; university programs, 122

Multinational corporation (MNC), 222-23, 231, 234, 236, 239, 240; pharmaceutical, in Sri Lanka, 238

Multiple perspectives, 134

Multiple realities, 185-86

Multiple regression techniques, 176

Mumford, Lewis, 141, 142, 143, 151

Munch, Peter (*see also* Abel/Wax debate), 243n

Murphy, G.M., 253n

Music and music criticism, 107

Myrdal, Gunnar, 199, 203, 248; on the soft state, 211, 213, 214, 219, 220

NASA (National Aeronautics and Space Administration, U.S.), 131

National Academy of Sciences (U.S.), 243

Nationalism, 103

Nation building, 209n

NATO, 200

Natural philosophy, 107

Natural science(s): dimension of sociology, 294; and historical sciences, 281-82; method, 292; models, 286; orientation, 30; and politics, 287; rational, public, and pluralist character, 292; versus subjectivism, 251, 281-93

Nazi(ism): analysis by Frankfurt School, 101, 102; Germany, 103, 105, 296-97; rise to power, 100

Nebraska, University of, 153, 160, 167

Nelkin, Dorothy, 133, 163, 187; quoted on technology assessment, 133, 134

Neocolonialism (*see also* colonialism *and* imperialism), and soft state, 219, 222-23

Nepotism, 212, 221, 223, 226

Neumann, Franz, 102, 103

Neutrality: and disciplinary identity, 61; Mills' rejection of, 93; in postacademic social science, 273-77; in Weber's work, 44

New consciousness industry, 79, 147

New Deal of Roosevelt, 64n

Newell, W.T., 140n, 186, 187

New Left, 101, 150-51, 152n

New School for Social Research: Veblen's career at, 50; Mannheim's influence, on, 68

New Science, The, (Vico), 281

Newton, Sir Isaac, 16

New York City, 161

Nigeria, 216

Nisbet, Robert, 15, 199, 269n

Noble, David, 38n

Noble unwarranted conclusion, 275-76

No-growth option, 234

Nonacademic (*see also* academic science; academic social science; adacemic sociology; postacademic social science; *and* postacademic sociology), 20; activity for sociologists and anthropologists, 169; careers of Veblen and Ogburn, 57; contrasted with postacademic, xii; as a label, 178; milieux, 138; orientations, 178; research base,

127; social science, 122-24; sociologists, 177; sociology, 268; sociotechnical work, 124

Non-social science(s): misunderstanding of, 66n; and pure/applied distinction, 171, 174, 176, 177-78; relationship to engineering, 29; shift from academic to postacademic, 122, 125-28; and social science, 98, 259; and state control of development, 203; and subjectivity, 274

Normlessness, 213

Nuclear: armaments, 103; capability, 90; destruction, 296; monopoly, 89; weaponry, 298

Oak Ridge National Laboratory, 190n

Objectivist, 259

Objectivity, 134, 137; and academic sociology, 279n; and detachment in Mannheim's work, 73-74; loss of, 137; positivist stress on, 284; in social science, 273-77; and value explicitness, 276

"Obligations to Future Generations" (Golding), 253n

O'Connor, James G., 102

Office of Technology Assessment (U.S. Congress) (*see also* technology assessment), 133

Ogburn, William F. (*see also* Chicago School of Sociology), 43, 46, 51-59, 78, 104, 110, 112, 123, 131, 142, 152n, 174, 271; as an academic, 57-59; on battle between youth and conservatism, 56, 87; comparisons with Mills, 87, 91, 97; and culture lag, 65n; his "great wall," 54, 110; as leading member of Chicago School, 53; on multiple perspective, 185; on postacademic role, 57, 65n; as president of the American Sociological Society, 51, 55; and social impact analysis, 135; as technocrat, 58-59; as technological determinist, 58-59

Ohio State University diffusion project, 210, 251, 254n

"Oil eating" bacteria (*see also* Chakrabarty, Amanda), 190n

Omaha, Nebraska, 158, 162

One Dimensional Man (Marcuse), 103, 104

"On Intellectual Craftsmanship" (Mills), 99n

Ordinary Language School (*see also* Wittgenstein, Ludwig), 284

Organization for Economic Cooperation and Development, 132

Orwell, George (Eric Blair), 88, 113, 114; quoted on British proletariat, 22; quoted on science and totalitarianism, 103, 203

"Other" society (*see also* Frankfurt School *and* fascism): and development, 209n; in Frankfurt School's work, 104-6, 296; as Messianic, 105; as opposite of fascism, 105, 108, 110; perspective, 111; and postacademic social science, 299; as Sabbath, 115n; Talmudic sense of, 106; and understanding technology, 106

Overdeveloped (*see also* development *and* underdevelopment), 89; and Grand Theory, 95; societies, 91, 98

Paradigm, 108, 131, 144, 265

Paraiso, Emile, 190n

Pareto, Vilfredo, 30, 293n

Paris Convention of 1883 (on patents), 236

Park, Peter, 274

Park, Robert Ezra (*see also* Chicago School of Sociology), 51-53; as journalist, 51

Parsons, Talcott, 43, 95, 128, 199

Partisanship in social science, 273-77, 280n

Passman, Otto (U.S. congressman), 253n

Patel, Surendra, 235

Patent(s) (*see also* invention *and* technology, proprietary), 232; convention, 236, 240; law, 235, 239; pharmaceutical, 273-78; protection, 236; system, 232, 235-37, 239, 241n; and technological monopolies, 238-39;

and technology transfer, 237
Paternalism, 221
Patronage, 223, 224, 227
Peer(s): and academic rewards, 265, 268; evaluation, 176; of Mills, 97-98
Pegasus International (corporation), 236
Pelto, Pertti, 143
Penalty of leadership, 13, 49, 50
Penrose, Edith, 236, 238
Peters, J., 212
Petersen, John H., 133
Pettit, Joseph, 171
Ph.D. degree, 170, 265, 292; disciplinary, 119, 121
Phenomenology: models of, 30; and separation, 290; and sociology, 290
Philistine, 113; versus bohemian, 147
Philosophes, 16
Philosophical Investigations (Wittgenstein), 284
Philosophy (*see also* moral philosophy), 107, 127, 170; academic, 284; and population ethics, 245; of science, 283-84; social, 199
Physician(s), 123, 125; as model for social theorist, 76
"Picture theory" of language (*see also* Wittgenstein), 284
Pierce, C.S., 82
Piotrow, P.T., 243
Planner(s), 71; and control of technology's impacts, 74, 144-45, 149; and development models, 200; education of, 73; elite, 69; political platform, 79; Third World, 218-19, 229; users of theory and ideology, 165; working with social scientists, 127, 138-39
Planning, 21, 119, 124, 149; appropriate, 228; center, 226; development, 209n, 214-19, 230n; elite, 226-27; elite, in Mills' work, 85; in Mannheim's work, 69-70; mock, 228; official, 103; organizations in Third World, 228; and population control, 246, 251; process, 215, 227; sector, 225;

social, 219; social science as knowledge base for, 127; "soft," 214, 218-19, 221
Plato, 81n, 171
Pluralism (*see also* conservative; liberal; *and* radical): critiques of subjectivism, 291; in interdisciplinary research, 187, 188; and political ends, 277; in postacademic social science, 260-62, 263, 265, 266, 269, 272, 273, 278, 283, 296; social scientists' commitment to, 188, 278; in the study of development, 208
Podhoretz, Norman, 81n
Poetics: approach, 289; and scientific method, 290; sociology as, 282, 287-89
Policy: advice, 161; analysis, 198; application, 119; development, 224, 232; humanization of, 257; military, 129; on population control, 242, 244; -oriented work, 280n; process, delineated, 201; public, 122; -relevant, 130; -saturated social change, 225; science, 173; and soft state, 229; state, and development, 200-3; in technology transfer, 210
Policymakers, 124, 139, 161, 257
Political: commitments and research, 259; by "conduction," 278; development, 209n, 213, 222; dimension in theories about technology, 77; economy (*see also* moral philosophy), 11, 12; parties, 220; philosophies, 226; power struggle in postacademic social science, 269, 299; role in postacademic social science, 260-69
Political science, 29, 96, 107, 108, 172; academic, 230n; Small quoted on, 65n
Pollack, Friedrich (*see also* Frankfurt School), 100, 102, 103, 108; *Economic and Social Consequences of Automation*, 102
Polynormativism, 214, 221
Popper, Karl, 285
Population control (*see also* family planning *and* fertility), 242-54;

agencies, 250; demographic, economic and medical aspects, 245; ethical aspects, 241, 244-54; role of cultural values in, 246-47; specialists, 243; USAID-funded, 216, 252n
Population crisis, 246, 248-49, 252
Population growth rates, 243-46, 248
Population Index, 243
Population Reference Bureau, 243
Porter, Alan, 139n, 140n, 181, 186
Positive Philosophy (Comte), 14
Positive Polity (Comte), 14
Positivism (*see also* empiricism *and* subjectivism), 85; classic, 284, 291; critics of, 290; equated with sociology, 64n; Frankfurt School quoted on, 105; Frankfurt School's critique of, 104, 105, 106-8; and hermeneutics, 289; ideological and subjectivist critics of, 175; as ideology, 166; and moral purpose, 105; its "picture theory" of language, 284; and quality of life, 163-66; Saint-Simon's doctrine of, 14; in sociology, 108; sociology of knowledge critique of, 165-66; subjectivism as critique of, 283-85
Positivist(s): approach, 164; defined, 167n; exaggerations of scientific method, 290; French, 71; methodology, 295; -operationalist models, 283, 291; philosophers, 285, 288; philosophies of science, 283; science, 287; social science, 165, 291, 296; sociology, 176, 284; versus subjectivist claims, 178; theories and methods, 297; versus *Verstehen* approach, 259
Positivist-technocrat, 120, 157, 160-62, 164, 167n-68n
Postacademic calling, 188-89
Postacademic science, 70; era of, 3, 82; transition to, 124, 189
Postacademic social science (*see also* academic social science; preacademic social science; *and* social science): alternative to development models, 206; and applied sociology, 179; approaches to technology, 4; and conservatism, 262, 264; contrasted with academic, 178; contrasted with nonacademic, xii; defined, 123-24; and development, 196, 198-208, 209n; and developmentalism, 184; effectiveness, 123; Frankfurt School's role in relation to, 101, 104, 110; and humanities, 113, 114; ideological connotations, 207, 260-69, 270-80; and ideology, 77, 260-80; institutionalization, 125, 257-59, 269; and interdisciplinarity, 169, 180-88; lack of depth, 132; Mills' delineation of field, 93, 93; Ogburn's delineation of role, 57; and pluralism, 262; and pure sociology, 177; and radicalism, 265-66; role, 123; role of scientific method in, 292; roots in Mannheim's work, 68-69, 72, 80; and sociology's heterogeneity, 294; and subjectivism/natural science debate, 281, 283, 285, 290; transition to, 72, 119-40, 169, 188, 198; Veblen's vision of, 50
Postacademic social scientist, 75, 80, 89, 101, 105, 112, 113, 137, 141, 257, 258; commitment to development, 199; and development = accumulation, 139; and dialectic of ideology and technology, 75; moral test of, 138; and orthodoxy, 294-95; and purposefulness, 278
Postacademic sociology: as applied, 111; Habermas on, 115n; and Marxism an Freudianism, 116n; relationship to academic sociology, 123; and other social science disciplines, 298
Postacademic style, 123-24, 167
Pound, Ezra, 113
Power Elite, The (Mills), 85
Power elite, 85-86; Mills' theory of, 86-87
Pragmatism: in Chicago School approach, 51, 288; linked with journalism, 53; Mills' writings on, 82-83; in Ogburn's work, 57; perspective on stratification, 85;

philosophers, 178, 288; and practical knowledge, 189n-90n; theories, 268; in U.S. sociology, 44

Preacademic social science (*see also* amateur social science *and* three births): compared to Frankfurt School, 107; continuities with Mannheim's work, 71; and development, 201-2, 204-7, 266; stress on criticism and application, 100, 172; technology studies, 53; and Weberian sociology, 35

"Predisciplinary," 180, 190n, 208

Price, Derek S., 128, 129, 139n, 140n, 241n

Price, Don K., 142

Price theory, 273

Private foundations, 127

Private sector organizations, 127

Problem centeredness, 52, 107, 108, 182, 183-84, 187, 188

Profession versus discipline in sociology, 63-64, 96, 291

Professional journals, 39n, 170

Professional: curricula, 189; schools, 127; students, 179; teaching programs, 132

Professionalization, 25, 30; of social science, 39n, 296; in U.S. sociology, 63

Progress (*see also* development): concern of French sociologists, 16; critical appraisal of, 264; and development, 198, 268; elite definitions of, 103; and growth of business, 47; haphazard, 298; industrial, 32; and irony, 72; Mannheim quoted on, 67; Marxist critique of, 26; Muecke quoted on, 136; and "one dimension," 103; and reversals, 9; scientific, and totalitarianism, 103; skeptical views of, 15; in Spencer's work, 45; and stability, 33; subjective side of, 166; in Sumner's work, 43; technocratic views of, 79; technological, 47, 88, 164; and technological advance, 33; in technological society, 85; through application of science and technology, 9, 21

Progressive, evolutionary concepts, 263; technology, 148, 149

Proletariat, 27, 48, 58, 147; dictators of, 199

Prometheus theme, 141, 145-46, 298

Promise, 93, 267

Protestant ethnic, 32, 286

Protestantism, 163

Prudhon, Pierre, 19

Pseudoscience, 94

Psychoanalysis, 107, 282

Psychology, 29-30, 39n, 96, 107, 108, 172; Marcuse's essays on, 103; pure, 109; of science, 129

Psychotherapy, 34

Public(s), 39n; absence of, 86; cooperation with social scientists and engineers, 257; and cooptation, 288; democratic, 85; education of, 73; interest, 297; issues, 41; and mass, 110; Mills on, 85-87; opinion, 42, 69; opinion surveys, 157; organizations, 127; participation, 69; sector, 130; social scientists' responsibility to, 110; and technicians, 87; technocracy's, 149

Public administration, 211-12

Public policy (*see also* policy), 122, 130; centers, 127; and population control, 252

Pugwash conference, 131

Pure/applied distinction: as invidious, 170-72; and non-social scientists, 189n; as pragmatic, 171-72; in social science, 169-70, 172-79

Pure physics model, 31, 33-34, 42, 43, 83, 109

Pure science: approach, 100; criticism by Frankfurt School, 106; model, 38, 93, 96, 188; orientation, 106, 174

Quality of life, 120, 141, 153-68; and development, 154, 209; meaning of phrase, 154-55; Mills quoted on, 88, 99; and population control, 245; quantification, 164

Quereshi, M.A., 140n

R&D (Research and Development),

appropriate technology, 231;
approach to development, 25, 39n,
196, 207; causes and effects, 5;
character and history of, 270;
cooperative programs, 240; control
of, 102, 202; and elite rule, 298;
industrial orientation, 103;
laboratories, 121, 173; military,
102; national priorities in, 86, 110;
order, 44; responsibility for, 43;
sector, 36; societal system, 34, 49,
50, 102, 199; in technologies of
death, 296-97
Radical (*see also* conservative; liberal;
marxist; *and* pluralist): approach to
innovation, 109; caucuses, 176;
critics of technology transfer, 235;
edge of development, 99n; and
humanism, 277; politics, 26; and
postacademic social science, 260,
264-65, 266, 268, 269, 272, 273,
283, 296; scholars and Frankfurt
School, 100, 108, 296; strategies,
146; theories of stratification, 85
Raffin, Jacques, 237
Ragin, Charles, 216
Rational altruism, 227
Rationality (*see also Wert*): and
bureaucracy, 84; formal (*Wert*), 34,
40n, 73, 140n; functional (*Zweck*),
33, 40n, 73, 107, 140n, 143; and
ideology, 78; Mannheim's analysis
of, 70; Mills quoted on, 88; and
ideology, 78; and postacademic
social science, 258; substantive, 73,
107, 109, 110, 258; technical, 74,
110, 143, 279n; tendencies of
technology, 34, 201; two types of,
73
Rationalism, 29, 83
Rationalization, 84; of intellectual
functions, 92
Rau, William, 279n
Raushenbush, Winifred, 39n
Ravenholt, R.T., 251, 253n
Ravetz, Jerome, 143
Recession in academic institutions, 127
Redfield, Robert, 53
Reflexivity: defined, 26; in European
sociology, 65n; in social science,
38n, 277; theory, 267
Reich, Charles, 152n
Reich, Warren T., 245, 249
Reich, Willhelm, 116n
Religion: compared to technology, xiii;
early sociology as, 17, 172; as focus
in social science, 271; in
Mannheim's later work, 70
"Reliving" (*see also Verstehen*), 282,
289
*Report of the President's Commission
on Population Growth and the
American Future*, 244
Republic, The (Plato), 81n, 171
Research: applied, 120, 183;
bureaucratic ethos, 174;
communities, 129; on
interdisciplinarity, 169;
interdisciplinary, 120, 181-83;
method, subjectivist, 283;
nonacademic, 127; and partisanship,
274; on population ethics, 242,
244-45; "real" scientific, 291; on
research, 186; social, 259; social
impact (*see also* social impact
analysis), 179; social science, 126,
177; sociotechnical, 119, 187;
specialized, disciplinary, 126;
strategies, 182; support for, 127,
188; team, 121, 134, 181, 189
Research on Research (journal), 190n
Revelle, Roger, 250, 254n
Reverse adaptation, 143-45
Revolt of the Engineers (Layton), 88
Revolution (*see also* Industrial
Revolution): in industry, polity, and
society, 201; of rising expectations,
223; scientific, 131; as the setting
for the three births, 21-22, 172;
Veblen's views on, 48;
working-class, 85
Revolutionary America, 90
Revolutionary Russia, 90
Reynolds, Janice, 294
Reynolds, Larry, 294
Rhoades, Lawrence, 46, 64n, 139n
Richards, Tudor, 217
Richardson, Jacques, 196
Rickert, Heinrich, 286
Rickman, H.P., 282, 293n

Riesman, David, 70
Riggs, Fred, 212, 213, 214, 218, 220
Riley, Gresham, 35, 274
Ritzer, George, 259
Robinson, Austin, 218
Rodgers, Willard, 153, 157
Rodney, Joseph, 27, 209n
Role model, 123
Romaticist versus Classicist, 147
Rose, Gillian, 107, 108, 115
Rose, Hilary, 9, 13
Rose, Stephen, 9, 13
Rosenberg, Bernard, 65n
Ross, E.A., 45, 94
Rosser-Owen, D.G., 241n
Rossini, Frederick, 127, 133, 139n, 140n, 182, 186
Roszak, Theodore, 147-8, 152n
Rousseau, Jean-Jacques, 189
Routinization (see also bureaucratization), 88
Rubel, Maximillian, 23n
Rudner, Richard, 81n, 274, 283, 286
Rule of technique (see also technical virtuosity), 141, 143
Rules of Sociological Method (Durkheim), 39n
Rural sociology, 133

Sabbath, 106, 115
Sai, Fred, 253n
Saint-Simon, Henri, 6, 14, 15, 16, 17, 21, 22, 41, 136, 151n, 172, 201, 298, 297; differences between him and successors, 22n; and Durkheim, 173; as ideologue, 28; influence on Marx, 18; originator of technological perspective, 35; on prescientific systems, 75; as technocrat, 47; and universalism, 206; as utopian scientist, 18; views on development, 201-2; Saint-Simonians, 16, 22n
Salamon, Albert, 16
Sampson, Anthony, 222
Sardar, Z., 241n
Sartre, Jean-Paul, 151n
Savigny, Friedrich Karl Von, 10
Savio, Mario, 147-48, 150
Saxburg, B.O., 140n

Scenarios in technology assessment, 166
Scheler, Max, 28, 73
Scheuer, James (U.S. congressman), 253n
Schiller, Friedrich Von, 205
Schneider, Louis, 10, 11, 71, 135, 138, 140n, 148, 152n; critique of Ellul, 152n; critique of Ogburn, 58; and ironic focus, 73; quoted on ironies of technology, 134-35; quoted on objectivity, 137; quoted on second-order consequences, 134; on Sumner, 45
Schumpeter, Joseph, 94, 199
Schutz, Alfred, 164, 284
Science (periodical), 253n
Science (see also natural science *and* non-social science): and aesthetics, 288; criticism of, 129; empirical, 285; exact, 203; fission with ideology (see also fission *and* split), 28, 29; as free commodity, 233; of freedom, 272; historical, 281; organizational, 67; philosophies and historians of, 283; politics of, 172; as process, 287; as product, 287; pure, 170; pure and applied, 169-72; rule by (see also technocracy), 15-16; "of sciences," 14; as social activity, 287; sociological analysis of, 75; sociology of, 172, 290; as symbol system, 165; and technology in developing countries, 129, 196-97, 232; transition from academic, 67; vulnerability to traditional authority, 201
Science advocacy, 187
Scientific: approaches to technological innovation, 167; credibility, 160; division of labor, 109; growth of knowledge, 131; perspective on technology and culture, 75; progress and totalitarianism, 103; views of social change, 69
Scientific American (periodical), 195
Scientific method, 42, 66n, 96, 166, 170, 175, 179, 180, 185, 276, 278, 287, 290, 291, 292
Scientism, 38n, 59, 166, 266-67, 287

358 Sociology/Technology

Scientist(s): decisions regarding purity and application, 172; and "despecialization," 130; experimental, 171; lack of autonomy, 94; morality of, 223-34; pure, 175; versus elites, 110
Scientometrics, 130, 132
Scotland, 7, 9, 11, 12, 14
Scott, James, 212, 220
Scott, Marvin B., 81n
Scottish moralists (the Scots) (*see also* Ferguson, Adam; Millar, John; *and* Smith, Adam), 9, 11, 13, 15, 17; influence on Marx, 18-19
SDI (Social Development Issues—journal), 154
Seckler, David, 48
Second birth (*see also* first birth; third birth; *and* three births): of sociology, 13, 17; and technology, 15
Second-order consequences, 134, 145, 262
Second World (*see also* First world; Third World; three worlds; *and* U.S.S.R.), 92; development in, 196, 200, 201, 207; development concepts, 203; leaders, 222, 223; population policies, 246; state bureaucracies and enterprises, 222, 232; technologies, 238; transfer of coordinated planning from, 225; values in Third World planning, 217, 218-19
Seidman, Robert, 212, 216
Selznick, Phillip, 228
Separate method debate (separatism) (*see also* methodology), 62, 290; Dilthey's views on, 282; and idealism, 288-89; and methodology, 295; Winch's views on, 282
Serendipity, 178
Service role for social science, 121-22
Shabbat, 106, 115
Sharma, K.D., 140n
Sharma, N.C., 133
Sheiber, Jane, 125
Sheldon, Eleanor, 167n
Shelley, Mary, 145-46
Shills, Edward, 43, 53

Shipman, John R., 241n
SIA (*see also* social impact analysis), 133, 188; and irony, 133-36; Third World, 133-34
Simmel, Georg, 51
Simonds, A.P., 68
Singer, Hans, 133; quoted on technology transfer, 231
Skolimowski, Henry, 140n
Skolnikoff, Eugene, 241n
Slater, Phil, 115n, 116n
Small, Albion, 45, 46, 61; quoted on applied sociology, 81n; quoted on sociology's image, 64n
Small, H.C., 140n
Smith, Adam, 6, 21, 22, 31, 42, 136, 163, 172, 201; influence on Marx, 18; influenced by Mandeville, 10; as major Scottish moralist, 11; originator of technological perspective, 35; and universalism, 206; views on development, 201-2
Smith, Dusky Lee, 109
Smith, G.L., 139n, 140n
SMSA (Standard Metropolitan Statistical Area), 162
Snow, C.P., 128
Snowmobile, 143
Social control movement, 233, 234-35, 240
Social Darwinism, 45
Social evolution: concepts in academic social science, 263; doctrines of Spencer and Sumner, 46; Durkheim and Sumner on, 39n; Mills' views on, 82; in Ogburn's work (*see also* culture lag), 57, 65n; Veblen's revision, 47
Social feasibility studies, 198
Social impact analysis (assessment) (*see also* SIA), 83, 131, 133, 142, 149, 179, 184, 231, 232; in Frankfurt School's work, 102; in government agencies, 126; irony in, 75, 71; Ogburn's innovations in, 56; by Sofranko et al., 241n
Social indicators, 153, 158-62, 167n; movement, 167n; school, 141
Socialism (*see also* communism *and* Marxism), 34; replaced by planned

technological innovation, 69; Saint-Simon's doctrine, 14; scientific, 201; scientific versus utopian, 19; utopian, 17
Socialist society, 202
Social phenomena as recalcitrant, 274
Social physics, 14, 30, 294
Social problems: in Chicago School approach, 53, 173-74; orientation in U.S., 42; "specialist," 178
Social psychology of knowledge, 108
Social Research (journal), 115n
Social responsibility movement, 233-34, 240
Social science (*see also* postacademic social science): academic (*see* academic science), 188; its acceptance, 127; applied, 35, 70, 83, 125, 188, 209n, 266; applied versus theoretical, 209n; classic European, 207; conservatism in, 262-64; constraints, 289; and decontextualized language, 285; and detachment, 73; and development ideals, 199; developmentalist, in Mannheim's program, 68; disciplinary, 108, 109; disciplines and technology, 109; dominance of system concept in U.S., 52; in France, 14; homocentric, 290; and humanization of technology, 257; and ideology, 166; as ideology, 77-78; and innovation, 7; interdisciplinary, 70, 169, 179-91, 207; Mills' critique of, 91-96; modern origins, 12, 22, 129; natural laboratories of, 173; *The New Science* as a foundation document, 281; nonacademic, 122-24; and objectivity, 274; Orwell's views on, 203; pluralism in, 261-62; positivist, 106, 165; as product, 287; pure, 109, 123, 188; radicalism in, 264-65; role in social change, 70; and sociology and social studies, terms compared, 99n; and technology, origins, 14, 22, 44
Social Science Research Council (U.S.), 189n
Social scientist(s) (*see also* scientist):

academic, 203; academic, First World, 188; and academic problems, 178; as affected by interests, 77; applied, 190n; and ethical issues, 242; ideological commitments of, 263; interests of, 257; interdisciplinary, 184; and knowledge in the human interest, 70; Mills' writings on, 92; in nonacademic activities, 121-22, 125-28; political role, 35; positivist, 291; postacademic (*see* postacademic social scientist); postwar, 90; and purposefulness, 277; responsibility to the public, 98; role in technological innovation, 270; secular commitments, 276; transcendence of disciplines, 189; as users of theory and ideology, 165; versus elites, 110
Social solidarity, 216
Social studies: and sociology and social science, terms compared, 99n
Social Studies of Science, 139n, 190n-91n
Social theory: as a basis of technology, 32; technological revolution's influence on, 75
Sociological careers, 128
Sociological Practice, 123, 190n
Sociological Imagination, The (Mills), 85
Sociological imagination, 99, 267
Sociological variables, 181, 183
Sociological way of looking at the world, 138-39, 140n, 180, 267
Sociologie, 107, 194
Sociologist(s): academic, 35, 179, 175, 268; academic, early U.S., 45; amateur, 179; applied, 105, 108, 111, 137, 196; bureaucratization of, 94; certification and evaluation of, 176, 181; communication gaps with technicians and publics, 120; early French, 14-47, 18, 19; as "freaks and faddists," 64n; and interdisciplinarity, 139n; as ironist, 71; and linguistics, 78; nonacademic activity for, 169, 177; and population ethics, 245;

postacademic (*see also* postacademic social scientist), 111; in practice, 122; and research, 259; and scientific method, 289-90; "strange" language, 61; teaching, 127; of technology, 35, 71, 82, 98, 105, 108, 111, 175, 179, 268, 269

Sociology: academic (*see* academic sociology); academic ideals, 278; applied, 38, 57, 60, 98, 104, 122, 169-77, 189, 190n; authentic, 292; classic development orientation, 64; craft versus profession, 294; defined by Mills, 85; departments established in U.S., 41-44; developmental, 83, 119; disciplinary, 37, 108, 109, 180-81; Frankfurt School, 107; equated with positivism, 64; founders of, 80; golden age of (in U.S.), 44; growth, 126; heritage, 267; heterogeneity, 281, 294; in human interest, 83; institutionalization, 39n, 44; institutionalization in U.S., 45, 60; isolation from other fields, 61; journals, 39n; of knowledge (*see* sociology of knowledge) as knowledge for development, 93, 137; Mannheim's influence on, 70; Marcuse's essays on, 103; new, 267; Orwell's views on, 203; as poetics, 282, 287-90; positivist, 66n, 108, 284; pragmatist, 82-83; problem selection in, 39n; professional activity in, 66n; pure, 94, 109, 110, 123, 169-79, 257; quest for academic legitimacy, 59-61, 109, 136, 172-74, 180-81, 188; relationship to science and industry, 61; role of Chicago School in U.S., 51; roles of service and mastery, 137; rural, 133; of science, 172, 290; as scientific, 53; social status, 62; sociological critique of, 260; of sociology, 283; struggle to appear scientific, 95; teaching and research, 121; technological applications, 39n; and technology, 22, 24, 28, 62, 110, 111, 112, 207; of technology, 35, 37, 40n, 43, 59, 55, 57, 59, 102, 119, 123, 125, 144, 189, 269n; in U.S., 41-44, 61-64; versus sociology, 294

Sociology of knowledge, 15, 68, 141, 152n, 267; critique of positivism and technocracy, 165-66; and interdisciplinarity, 182; in Mannheim's work, 71, 75-80; in Mills' work, 82, 91-96; perspective, 195; in postacademic social science, 119

"Sociology of the streets," 173

Sociology/technology, 76-77, 119

Sociotechnical: fields and technology transfer, 210; issues, 125; orientation and development, 197; problems and interdisciplinarity, 184; professionals, 169; research, 187, 189; research and education, 119; work, 124, 188; work, nonacademic, 124

Sofranko, Andrew, 133, 241n

Soft state, 200-1, 210-30; and colonialism, 219-22; and corruption, 211-14; and development planning, 216, 225, 230n; and mock bureaucracy, 211; and neocolonialism, 222-23; and policy, 225-29; social benefits, 226; traditional and local sources of, 223-25

Solomon, Jean-Jacques, 126, 186, 278

Sorokin, Pitirim A., 64, 130, 132, 154, 167, 295; quoted on academic sociology, 140n

South America, 224

Soviet Union: conditions for democracy, 90; as "core" country, 230; technology, 89; as totalitarian state, 103, 296-97; transfer of bureaucracy and planning from, 228

Specialization: and abstracted empiricism, 96; as aspect of industrial society, 25; and growth in science, 128; and pure sociology, 176; in social science, 29, 31, 61, 258; in U.S. sociology, 43

Spencer, Herbert, 94; evolution views, 43, 45, 46; influence on U.S. sociology, 42; and

institutionalization of sociology, 39n; and Marx, 173
Spiegel-Rösing, Ina, 128-29, 130, 140n, 241n; quoted on scientometrics, 131
Spinoza, Baruch, 210
Split (ting) (*see also* fission), 30, 75, 172, 198, 199
Stanley, Manfred, 73
State: autonomy, 210; authorities, 202; and authorization of technology, 207; control, 103; as institution, 262; officials, 223; policy and development, 200-3, 205; power, 132, 202, 209n, 264, 296; relationship to private sector, 102; soft (*see* soft state); and technical rationality, 279; totalitarian, 103
Steady state, 234
Steck, R., 139n
Steiner, George, 114, 116n; quoted on criticism, 116n; quoted on the humanities, 113
Sterilization, 250, 251; compulsory in Maharastra, India, 253n
Steroids, 243, 244
Stolnitz, George, 253n
Stratification: centrality in social science, 23n; concept, 174, 259, 271; effects on technology, 26; Marx's focus, 20, 23n; in Mills' work, 82, 83-87; and partisanship in social science, 274; and sociology of knowledge, 79; traditional systems, 227, 229; Veblen's theory, 50; in Weber's work, 35-36
Straube, Win, 236
Studies in Comparative International Development, 139n, 209n
Study of Science, Technology, and Society (SSTS), 120, 124, 128-32, 136, 179, 188
Subjectivism (*see also* separate method debate *and Verstehen*), 281-93; in academic social science, 282-83; as critique of positivism, 283-85, 290; false, 280
Subjectivist(s), 164-66, 259, 296; critiques of positivism, 164, 175; dimension of sociology, 294; model of research method, 283; program, 289; versus positivist claims, 178
Sumner, William Graham, 43, 45; evolutionary views, 46; as *homme de lettres*, 50; influence on Veblen, 46
"Supply side" approaches to population aid, 251
Sweezy, Paul M., 222
Suppe, Frederick, 131
Switzerland, patent holdings by, 236
Sypher, Wylie, 151n
Systems theory, 183

TA (*see also* technology assessment), 133, 188; and irony, 133-36; literature, 140n; research in Third World, 134
Taiwan, 243
Talmudic sense of the "other" society, 106
Tapinos, G., 243
Tarr, Joel, 140n, 152n
Taylor, F.W., 65
Taylorism, 116n
Taylor, J.B., 139n, 187
Teachman, Jay, 243
Team research (*see also* research), 121, 122, 134, 181, 189; Frankfurt School's commitment to, 107
Technical: classes versus business classes, 47; classes and industrial classes, 147; community, 198; credibility, 160; fields, 258
Technical virtuoso(ity) (*see also* rule of technique), 92, 96, 130, 138, 143-44, 151, 176, 265, 266, 277, 278; and hermeneuticism, 290, 292
Technicians(s): appeal to, 233; awareness of technology's social impacts, 207; certification of, 270; class consciousness among, 23; communication gaps with sociologists, 120, 122; and control over technology's impacts, 74, 145; elite, 69, 71; and in-depth knowledge of technology, 271-72; perceptions of innovation, 132; and postacademic social science, 125, 257-58; power struggle, 150; power

struggle with bureaucrats, 79; power struggle with entrepreneurs, 81, 88; power struggle with political and economic elites, 87; and progressivism, 263; and pure/applied distinction, 177-78, 273; and academy, 170-71; and reflexivity, 26; separation from sociologists, 94; and social scientists, 262; and state control of technology, 203; transcendence of disciplines, 189; and the unplanned, 72; versus elites, 110; white-collar, 81

Technics-out-of-control (*see also* autonomous technology), 142, 151; as ideology, 146

Technocracy, 131; its children, 148; Marx's views, 20; publics, 149; and quality of life, 163-66; Saint-Simon's vision, 15, 25; sociology of knowledge critique, 165

Technocracy, Inc., 48

Technocrat(s): children of, 148; defined, 167n, 168n; and ideology, 79; Ogburn as, 58-59; Veblen and Saint-Simon as, 47

Technocratic managerial liberalism, 265; solutions, 262; utopias, 264

Technological change, 17, 35, 47

Technological delivery system, 69

Technological determinism, in Ogburn's work, 58-59

Technological fix, 149; orientation, 233, 240; syndrome defined, 164-65

Technological imperative, 143-45

Technological innovation(s), 3, 36, 39n; and City Index, 153; conservative approaches, 262; control-extending, 86; darker side of, 164; design and evaluation, 4, 31, 43, 149; in development, 50, 154, 196, 198, 203; in developed countries, 133; dimensions, 208; domesticated process, 203; and elite priorities, 110; and evolution in Frankfurt School's work, 102; and human values, 232; and ideology, 261; industrial and-, 161;

labor-saving, 102; liberating potential, 135; in Marx's work, 20; monopolies, 238-39; and multiple realities, 185; and New Left, 150; Ogburn's views, 56; perception of, 132; planned, 69; policies, 238; political component, 144; and postacademic social science, 124-25, 268, 270; reorientation, 298; research, 177; and second-order consequences, 145; and social science as a knowledge base in, 83, 184, 266; social theories about, 77; sociology's role, 41, 70; as source of distortion, 74; statistical indicators, 120; Adam Smith's studies of, 12; as technically rational process, 73; and technical virtuosity, 143; and technocracy, 167n-68n; "third way" to, 69; as threat to life and liberty, 272; the unplanned in, 17-72

Technological monopolies, 238-39

Technological revolution (*see also* Industrial Revolution), 75-76

Technological society, 24, 41, 111, 189, 257, 264, 299; control in, 151; critics of 157, 267; development in, 199; evolution toward massification, 110; in Frankfurt School's work, 104; humanized, 264; in Mannheim's work, 69-70; in Mills' work, 80, 84, 85, 89, 90, 92; and postacademic social science, 124

Technologue, 79

Technology(ies): and accumulation, 48; agricultural, 218; -aided planned social change, 68; alternate, 231, 234, 235; as amoral, 205; antihuman character of, 36; applied study of, 114, 120, 177, 266; appropriate (*see* appropriate technology); appropriate planning, 228-29; autonomous (*see* autonomous technology); based on physics, etc., 25; for basic needs, 133, 232; in bureaucratic society, 36; capital-enriching, 138; capital-intensive, 36, 37, 196; comparative impact, 49; communications, 204;

control-extending, 298; "convivial," 234; of coordinated planning, 225; critical, applied approaches to, 109; critiques of effects, 34; as cultural phenomenon, xvi, 109; and culture, 75; of death, 296; defined, xiii, 21, 23n 292; dehumanization of, 103, 149; and development, mutual influence, 202; for development, 90, 195-97, 207; of development and monitoring development, 206; diffusion (*see also* diffusion), 143; disruptive effects, 231; distorted theories of, 120; domesticated role, 86; efficiency, 69; as elaborated technique, 205; elite vision of, 86; energy-intensive, 196; energy-related, 204; energy-using, 37, 263; environment-destroying, 298; ethical impact, 245, 247; excluded factors, 33-34; expanded view, 72; explanations, 76, 80; exploitative potential, 240; as extension of man, 26, 272; in fascist, communist, and mass societies, 72; favoring elite/minority rule, 69, 298; focus on, in postacademic social science, 270-73, 278; as free goods, 234; and government, 17; growth, 297; of Holocaust, 296; humanization of, 35, 138, 141, 257, 293, 295, 299; as ideology, 87; indigenous, 231, 234; industrial, 27-28, 33, 34, 48, 204, 218, 233; institutionalized, 38; of intellectual productivity, 95; intermediate, 231, 234, 239; ironies of, 134, 141; and irony, 74-75; as knowledge, xiii-xiv, 124, 202, 278; labor-intensive, 234; labor-saving, 37, 103, 196, 203; leisure-trivializing, 103; liberating possibilities, 53, 272; low-level, 234; market, 236; in Marx's theories, 19; medical, 204, 233; military, 89, 233; modern, 46; moral impact, 7, 10, 12, 78-79; mystification of, 298; as neutral, 231-32; improper sociological term, 109; as object, 36; of oppression, 296; and "other" society, 106; ownership and control of, 232-41; as part of environment, 31; passive, academic study of, 51, 180; patented, 238; pessimistic views of, 37; and policy, 70, 71; and political action, 149-51; for political and economic ends, 20; population control, 242-54; and power, 145; power to enervate, 201; privacy-invading, 298; and progress, 298; as proprietary, 232-33, 235, 240; as purposive, 72; rationalizing tendencies, 34; relationship to other idea systems, 75; regulation, 279; research, 35; role in war, 68; role in development programs, 211; runaway (*see also* autonomous technology), 146; secondary and objective focus on, 38; side effects, 298; social, 205, 226; social antecedents and consequences, 83, 86; and social change, research, 100; social context, 79; social dimensions, 279; as social force, 127; social impact (*see also* social impact analysis), 28, 31, 141, 244; and social science, 14, 29; and social science reconciled, 299; and society, 146; and society in Mannheim's work, 67; and society in Marx's work, 18-19, 23n, 37; and sociology, 3, 22, 24, 59; sociology of (*see* sociology of technology); as source of concern, 31; as source of freedom, 106, 109, 270-71, 279n, 296, 298, 299; as state prerogative, 203, 207, 297; substantive rationality, 110; stratified effects, 20; as theory and praxis, 19; in Third World (*see also* Third World), 210; and totalitarianism, 296-99; trivialization, 103; as a variable, 271; "victim" of, 136; vulnerability to traditional authority, 201
Technology and Culture, 209n
Technology and Social Shock (Lawless), 136

364 Sociology/Technology

"Technology assassination," 135
Technology assessment (*see also* TA), 18, 26, 131, 132, 142, 149, 165, 184, 231; and development, 198; irony in, 71, 75, 133-36; methodological literature, 280n
Technology policy (*see also* policy), 124, 132, 189, 207, 278
Technology transfer (*see also* diffusion), appropriate, 240; for basic needs, 232; control of, 234; of coordinated planning, 225; inappropriate, 218; and lack of development, 240; and MNCs, 222; among MNCs and developed countries, 240; normative aspects, 233; and patenting, 237; population control, 242-54; radical critics, 235; social and human consequences, 231; and soft state, 217-19, 231; in Third World countries, 240; to Third World, 210-11; in Veblen's work, 49
Telegraphy, 171
Television, 171
Telos (journal), 115n
Territorial imperative in science, 180
Texas A&M, 99n
Textbook structure, 125
Textron Corporation, 241n
Theology, 14, 15, 107; and population ethics, 245
Theory and Society, 115n
Theory of the Leisure Class (Veblen), 48; as ironic, 65n
Theory of Moral Sentiments (Smith), 11
Theory of history, 94-95
Therapy (*see also* diagnosis), as dimension of ideology, 61, 76-77, 146, 147, 173, 204
Third birth (*see also* first birth; second birth; *and* three births): as Marxist, 20, 24; of sociology, 17
Third Reich, 296-97
"Third Way": to development, 72, 204; Mills as proponent, 91; to technological innovation, 69; in technology studies, 70
Third World (*see also* First World; Second World; *and* three worlds): acceptance of social science in, 127; and appropriate technology (*see also* appropriate technology *and* technology transfer), 36; bureaucracy (*see* bureaucracy, Third World); capitals, 217; challenges to patent system, 238-40; and colonialism, 218-21; and development, 195-97, 207; development, compared to Europe's, 13; development models, 200, 201, 204; and development of industrial societies, 109, 202; and development planning, 214-19, 226, 228; governments and "compassionate behavior," 200; Horowitz on, 99n; Mannheim's awareness of, 71, 99n; in Mills' work, 89-91, 99n; and modern technology, 210; nationals, 217; and neocolonialism, 222-23; patents in (*see* patents); patron/client system, 224; planners, 218-19; as polynormative, 221; population control, 242-43; population growth, 243, 248; population policies, 246; and postacademic social science, 125, 266; public bureaucracy, 211-13; rise of, 195; and soft state, 211-30; TA and SIA research, 134-34; and technological independence, 239; in technology studies, 166-67; technology transfer to, 210-11, 231, 233-35, 237; and traditional sources of corruption, 223-25; as "underdeveloping" countries, 27-28; values, 247; Veblen's views on, 50
Three births: and development, 21, 199, 205; of sociology, 7-19
Three worlds of development, 200-5, 298; domestication of development in, 202; operationalization of development in, 201
Thompson, A.R., 136
Thompson, Victor, 218, 220
Timasheff, Nicholas, 39n
Tiryakian, Edward, 39n
Todd, Ralph, 153, 157, 161
Toffler, Alvin, 164

Tolstoy, Leo, 135
Tönnies, Ferdinand, 34
Torah, 81n
Totalitarianism: in China, 230n; and elite planning, 227; fascist, nonfascist, and antifascist, 297; in Frankfurt School's work, 103; and the human interest, 115; in Mannheim's work, 68, 70; in Nazi Germany and Soviet Union, 103; Orwell quoted on, 103, 203; and the "other" society, 299; and technology, 296-99
Transdisciplinary (*see also* interdisciplinary *and* multidisciplinary), 107, 138
Trend, The (journal), 140n
Truzzi, Marcell, 286, 293n
Tubal ligation, 244, 245
Turgot, Anne Robert, 16
Turner, Teresa, 238
TVA (Tennessee Valley Authority), 228

Uncertainty, 131; in development planning, 216; of modernization, 226
Underdeveloped, 91; countries (*see also* Third World), 196; societies, 98
Underdevelopment, 36, 145
Uniform Crime Report (U.S. Federal Bureau of Investigation), 162
Unintended consequences (*see also* irony): in academic social science, 78; and autonomous technology, 142, 146, 149; in work of Mandeville, 10; in work of Mannheim, 71; in postacademic social science, 119, 298; and technology's effects, 163, 165, 262
United Nations (UN), 133; Conference on Science and Technology for Development (UNCSTAD), 196-97, 241n; Development Program (UNDP), 196; Educational, Scientific, and Cultural Organization (UNESCO), 134, 196; International Children's Emergency Fund (UNICEF), 231; Office of Population, 243; Office of Science and Technology, 197n; Population Program, 252n; Statistical Office, 243
United States: acceptance of social science, 127; academic recession, 127; Census, 162; cities, 153-55 (Table 1); cold-war mentality, 89; conditions for democracy, 90; Congress, 133; corporations in support of science, 67; depression, 64n, 174; domestic population policy, 244, 253n; Ellul's disciple in, 142; focus on the unplanned, 71; Frankfurt School's theories applied to, 103; higher education, 48; influence of Frankfurt School in, 101; involvement in Indo-China War, 90; patent holding by, 236; plantation system, 224; pluralism, 260; and Third World population control, 242-53; population specialists, 243; postacademic era, 266; pure science, 171; sociology, 41, 44-66, 96, 110, 121, 125; study of mass society and pluralism, 70; technology, 89
United States agencies and organizations: Agency for International Development (USAID), 126, 216, 231, 243, 252n, 253n; Department of Energy (DOE), 126; Department of Health, Education, and Welfare (HEW), 243, 252n; Department of Housing and Urban Development (HUD), 126; Department of State, 233, 240n; House of Representatives Appropriations Committee, 253; House of Representatives Committee on International Relations, 244; House of Representatives Select Committee on Population, 253n; National Science Foundation, 189n; Patent Office, 190n; Senate Committee on Commerce, Science, and Transportation, 197n; Space program (*see also* NASA), 191n; Supreme Court, 190n

Universalism, 205-7
Utopia(n), 19; accounts of technology's impact, 188-89; communist, 70; concepts, 105; in Frankfurt School's work, 105; interests, 74; replaced by social science, 70; socialist, 19-20; socialism, 17; technocratic, 264; views of social change, 69

Vaikos, Constantine, 236
Value-explicitness, 180, 208, 270, 273-77, 278, 279, 280n, 292; and objectivity, 276
Value-free, 134, 157, 173; as "theory-free," 160
Value judgments: interest-serving, 160; and development, 208; in science, 274, 275; in social science, 105, 269
van Dam, André, 239
Vasco da Gama era, 8
Vasectomy, 244, 245
Veatch, Robert, 244, 253n
Veblen, Thorstein, 13, 40n, 43, 45, 46, 78, 94, 109, 110, 112, 114, 131, 137, 263, 264, 266, 267, 268, 271, 294; as academic, 50, 64n; antiprogressivism, 65n; approach to technology, 49-51; compared to Mills, 82, 86-88, 91, 92, 97, 99n; compared to Ogburn, 53-58; contrasts with Marx, 48, 50; critique of academic social science, 49-51; as a "failure," 65n; instinct theory, 65n; as ironic, 65n; as marginal, 65n; Marxist orientation, 46, 65n; as technocrat, 47; theory of revolutionary intellectuals, 48
Verstehen (*see also* Dilthey, Willhelm; separate method debate; *and* subjectivism): and Abel/Wax debate, 293n; in Dilthey's philosophy, 282; school, 166; and scientific method, 290; versus positivism, 259; in Weber's work, 285-86
Vico, Giovanni Battista, 259, 281
Vidich, Arthur, 70
VITA, 241n
Voluntary organizations, 221

Wallerstein, Immanuel, 8, 27, 209n
Wall Street crisis, 64n
Walsh, W.B., 139n, 140n, 186
Walton, J.D., 36, 190n
Ward, Lester Frank, 45
Warsaw Pact, 200
Warwick, Donald, 247, 251, 252n, 253n
Washington University (St. Louis), 140n
Washington, University of, 140n
Wax, Murray (*see also* Abel/Wax debate), 26, 38n, 103, 138, 282, 293n, 295
Wealth of Nations (Smith), 11
Weber, Max, 8, 23n, 30, 35, 39n, 44, 75, 94, 163, 173, 174, 271; compared to Veblen, 47; debate with Dilthey, 259, 282-83; dialogue with Marx, 85, 173; focus on bureaucracy, 43, 84, 88; influence on contemporary sociology, 38, 42; influence on Park, 51; and Mills, 84; quoted on subjective explanation, 286; and sociology of knowledge tradition, 76; on subjective explanation, 284-86, 293n; his system, 39n; theory of technology, 34, 36-7
Weberian: orthodoxy, 85; sophisticated, plain, and vulgar, 99n
Weimar Republic, 100
Weinstein, Jay, 36, 154, 190n, 213, 242, 248, 253n, 254n, 269n, 277
Wert (rationality) (*see also* formal rationality *and* rationality), 34, 275
"Whiggishness," 263, 269n
White, Lynn, Jr., 128
White, M.K., 228
White Collar (Mills), 85
Whitehead, Alfred North, 128
"Whose Side Are We On?" (Becker), 274
Willhelm, Sidney, 269n, 279n
Winch, Peter, 164, 282, 284-86, 287, 293n; *The Idea of a Social Science*, 282, 284
Winner, Langdon, 15, 134, 141-51, 152n; as Ellul's disciple, 142; on

Frankenstein's monster, 145-46
Wionczek, Miguel, 197
Wirth, Louis, 53, 65n, 109, 137; quoted on U.S. sociology, 62-63
Withey, S.B., 167n
Wittfogel, Karl (*see also* Frankfurt School), 102, 213
Wittgenstein, Ludwig, 283, 284, 285, 287; *Philosophical Investigations*, 284
Wolf, Robert, 103
Wolff, C.P., 133
Wolff, Kurt, 68, 81n
Working class (*see also* proletariat), 83, 163; agitation, 27; and dialectical materialism, 19, 25; revolution, 85
World Bank, 254n
World Fertility Survey, 243
World Health Organization (UN), 252n
"World of Nations," 281
World Population Conference (Bucharest, 1974), 224-45, 251, 252, 254n
World Population Plan of Action, 245, 248
World Population Tribune (Bucharest, 1974), 253n

Xia-Fang (downward transfer), 228

Yin and Yang, 78
You and Machines (Ogburn), 56

Zeitlin, Irving, 22n, 78, 39n
Zelditch, Morris, 127
Zohar (cabalistic writings), 282
Zucker, E., 139n-40n